I0031435

AT THE
LEADING
EDGE

The ATLAS and CMS LHC Experiments

AT THE

LEADING

EDGE

The ATLAS and CMS LHC Experiments

Editor

DAN GREEN

Fermi National Accelerator Laboratory, USA

World Scientific

NEW JERSEY · LONDON · SINGAPORE · BEIJING · SHANGHAI · HONG KONG · TAIPEI · CHENNAI

Published by

World Scientific Publishing Co. Pte. Ltd.

5 Toh Tuck Link, Singapore 596224

USA office: 27 Warren Street, Suite 401-402, Hackensack, NJ 07601

UK office: 57 Shelton Street, Covent Garden, London WC2H 9HE

Library of Congress Cataloging-in-Publication Data
At the leading edge : the ATLAS and CMS LHC experiments / edited by Dan Green.
    p. cm.
    Includes bibliographical references.
    ISBN 978-9814277617 (alk. paper) -- ISBN 978-9814304672 (pbk : alk. paper)
  1. Large Hadron Collider (France and Switzerland) 2. Nuclear counters. 3. Symmetry (physics)
4. Particles (Nuclear physics) I. Green, Dan.

    QC787.P73A8 2010
    539.7'7--dc22

                                                                    2009045137

British Library Cataloguing-in-Publication Data
A catalogue record for this book is available from the British Library.

Copyright © 2010 by World Scientific Publishing Co. Pte. Ltd.

All rights reserved. This book, or parts thereof, may not be reproduced in any form or by any means, electronic or mechanical, including photocopying, recording or any information storage and retrieval system now known or to be invented, without written permission from the Publisher.

For photocopying of material in this volume, please pay a copying fee through the Copyright Clearance Center, Inc., 222 Rosewood Drive, Danvers, MA 01923, USA. In this case permission to photocopy is not required from the publisher.

It is an essential characteristic of experimentation that it is carried out with limited resources, and an essential part of the subject of experimental design to ascertain how these should best be applied.

— *Sir Ronald A. Fisher*

Our observations of Nature must be diligent, and our experiments exact.

— *Denis Diderot*

CONTENTS

Chapter 1

INTRODUCTION: HOW PHYSICS DEFINES THE LHC ENVIRONMENT AND DETECTORS

D. Green

CMS Department, Fermilab
Batavia, IL 60510, USA
dgreen@fnal.gov

1. Introduction

Much has been written[1] about the general purpose experiments ATLAS and CMS, which have recently been completed and are installed in the Large Hadron Collider (LHC) at the CERN laboratory. The accelerator itself has a URL[2] and a comprehensive Technical Design Report (TDR).[3]

The experiments themselves have home pages,[4,5] which are sources of considerable public information. In particular, the physics capabilities of the detectors are made available there. In addition, the technical details of the experiments have recently been written up as a reference source for specifying the detectors.[6] The ATLAS and CMS experiments have also been described and compared in a review article,[7] which gives many detailed descriptions of the detector performance.

Given this wealth of information, there yet remains a need to explain why the search for new physics beyond the Standard Model (SM) defines the requirements for detectors at a proton–proton collider and why specific choices were made by the ATLAS and CMS experimenters in order to meet those requirements. Indeed, the "why" of these detectors is the main subject of this volume. This introductory chapter explores the environment in which the detectors must operate and allow the extraction of the "new physics" that the LHC will make accessible.

2. Electroweak Symmetry Breaking (EWSB)

It has been known for more than 20 years that the electroweak bosons exist and are massive, with the Z having 91.2 GeV of mass while the W has 80.4 GeV. Since these masses imply a breaking of the electroweak gauge symmetry, an explanation which extends beyond the SM is needed. The necessary energy scale to be explored is set by looking at a process such as $e^+ + e^- \rightarrow W^+ + W^-$. In the SM this electroweak process has an S wave amplitude, A_0, which violates perturbative unitarity at high center of momentum, (C.M.), energies $\sqrt{s}$:

$$A_0 = \alpha_W s / 16 M_W^2 < 1,$$
$$\alpha_W = \alpha / \sin^2 \theta_W \sim 1/31.6. \tag{1}$$

Therefore, new physics must intervene at or before the "terascale," where $\sqrt{s} < 4 M_W / \sqrt{\alpha_W} = 1.8\,\text{TeV}$. The angle θ_W is the Weinberg angle and α_W is the electroweak coupling constant, while α is the electromagnetic fine structure constant.[8] Units with $\hbar = c = 1$ are used in all the equations.

Another indication of the appropriate energy scale for new physics is set by considering a weakly interacting, neutral, stable particle which would make up the "dark matter" which has been inferred in cosmological and astrophysical studies. If such a particle had a mass of order $1\,\text{TeV}$, it would have the correct relic density to be a candidate particle for dark matter. It is to be hoped that this fact is not simply coincidental.[9]

The historical approach to this high energy scale is shown in Fig. 1, where the energy available for making new particles is plotted as a function of the year when a new accelerator facility came on-line. The exponential growth of available C.M. energy with time has clearly slowed lately. It is also clear that the new LHC accelerator, with a C.M. energy of $14\,\text{TeV}$, is the only facility now available which can decisively probe the necessary terascale. The C.M. energy of the scattering of the fundamental constituents is less that the proton–proton C.M. energy because the protons are composites composed of quarks and gluons.

It should be noted that the simplest extension of the SM is to posit a scalar particle which couples to mass, the Higgs boson, which then has a vacuum expectation value which breaks the electroweak symmetry, couples to the gauge bosons and gives them a prescribed mass.[10] Further postulating a Higgs Yukawa coupling to fermions, albeit with undetermined coupling strength, allows for an SM extension which compactly gives mass to all SM particles. Nevertheless, this is only the simplest SM extension and remains a hypothesis until the terascale is fully explored at the LHC and the "Higgs boson" is either discovered or not.

3. LHC Luminosity and Energy

Given the importance of the terascale, and the breakdown of electroweak vector boson scattering at that scale, a requirement of the LHC is to enable

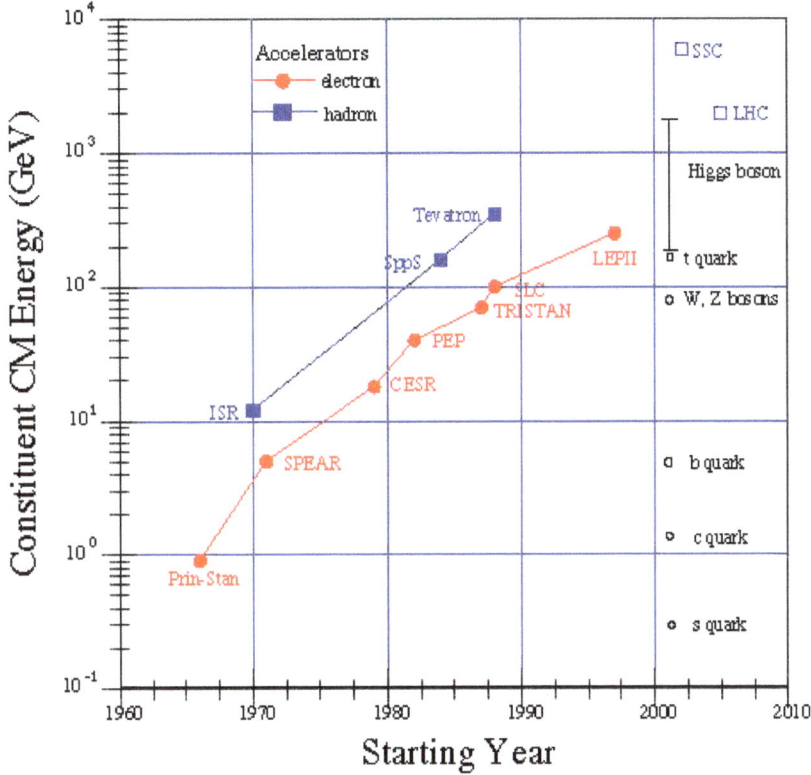

Fig. 1. The available C.M. energy at recent accelerator facilities as a function of the starting year of the facility. The first data-taking run at the LHC will be in 2009. The masses of the quarks and bosons are indicated, as well as the possible range of the Higgs mass.

the study of vector boson scattering at a C.M. energy $\sim 1\,\text{TeV}$. Exploring the dynamics of pairs of vector bosons with a pair mass $\sim 1\,\text{TeV}$ would also probe the triple and quartic interactions of the gauge bosons at the desired mass scale.

The vector bosons decay to either quark or lepton pairs. However, the enormous backgrounds which exist at the LHC due to strongly produced QCD processes make the detection of leptonic decays experimentally favored. This fact explains why LHC detectors tend to focus on lepton detection. The branching ratio for a W to decay to a muon plus neutrino is $B_\mu \sim 1/9$. A crude estimate of the cross section for electroweak $W + W$ production at the LHC with subsequent W decay to muons is

$$\sigma(p + p \rightarrow W^+ + W^- \rightarrow \mu^+ + \nu_\mu + \mu^- + \bar{\nu}_\mu) \sim (\alpha_W^2/\hat{s})B_\mu^2. \qquad (2)$$

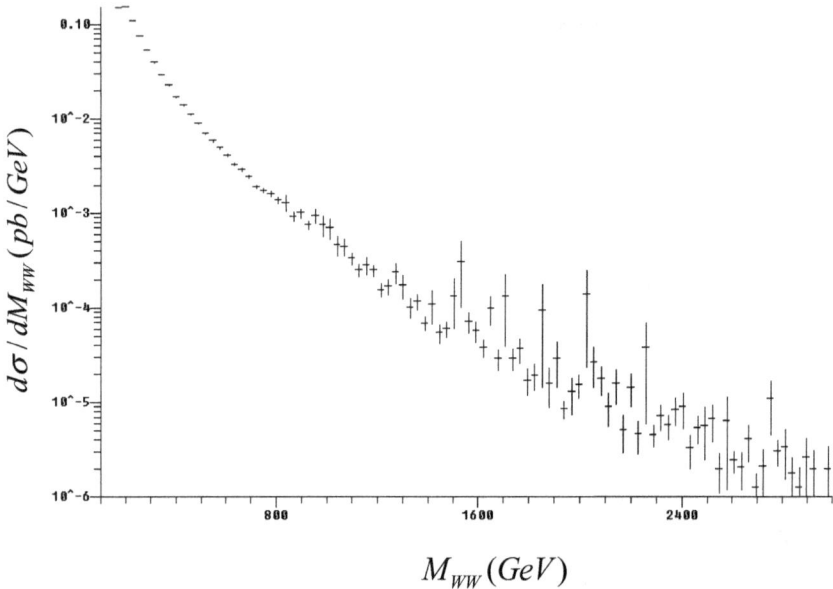

Fig. 2. COMPHEP[11] calculation for the cross section for the production of W pairs as a function of the pair mass at a C.M. energy of 14 TeV in $p + p$ collisions.

This estimate gives a 5 fb cross section times branching ratio squared for a W pair mass of $\sqrt{\hat{s}} = 1$ TeV.

A detailed Monte Carlo model[11] result for $W + W$ pair production is shown in Fig. 2. The cross section for $W + W$ mass above 1 TeV is 640 fb. Requiring two muons in the final state yields a cross section of 7.9 fb, in reasonable agreement with the previous estimate. In order to have sufficient statistical power in studying this process, the LHC should provide 100 fb^{-1}/year. Taking a running time, T, of 10^7 s/yr ($\sim 30\%$ of the calendar year), there will be ~ 790 $W + W$ events produced per year with a mass above 1 TeV which decay into the experimentally favorable final state containing two muons. A similar event sample will be available in the two-electron final state and twice that in the muon-plus-electron final state.

Therefore, the LHC was required both to have a C.M. energy of 14 TeV to reach and explore well the terascale where a mechanism for EWSB must be found, and to have a luminosity, L, of

$$L = 10^{34}/\text{cm}^2\,\text{s}, \qquad T = 10^7\,\text{s},$$
$$LT = 10^{41}/\text{cm}^2\,\text{yr} = 100\ \text{fb}^{-1}/\text{yr}. \tag{3}$$

Note that the LHC energy is seven times that of the existing Fermilab Tevatron and the full design luminosity is about 30 times larger. Thus, the LHC represents a major improvement in discovery potential in the study of EWSB over the presently operating hadron collider facility.

4. Global Detector Properties

The high luminosity which is required of the LHC by our need to explore terascale physics means that the detectors will be exposed to high particle rates. Therefore, LHC experiments will require fast, radiation-hard and finely segmented detectors. The speed of the detectors sets a scale for the accelerator radiofrequency (r.f.) bunch structure. It is assumed in what follows that all the detectors can be operated at a speed which can resolve the time between two successive r.f. bunches, which is 25 nsec at the LHC.

Consider, for example, a silicon solid state detector operated with a bias voltage which creates an electric field, E. The charge collection time is $\tau = d/\mu E$, where d is the thickness of the detector and μ is the hole or electron mobility. Numerically, for $d = 300\,\mu$m, the electron drift velocity at a typical depletion voltage of $\sim 50\ V = dE$ is $\mu E \sim 42\,\mu$m/ns, leading to a charge collection time of ~ 7 ns. The holes are ~ 3 times slower, so a charge collection time of 21 ns is well matched to the LHC bunch crossing spacing. Shorter bunch crossing times are not useful, because the detectors would simply be forced to integrate over the multiple bunch crossings occurring within their resolving time.

Having progressed from the need to study EWSB decisively to the C.M. energy and luminosity of the LHC, a "generic" detector will now be considered. This detector is defined simply to serve as a mechanism whereby general features of LHC detectors can be explored, while detailed detector concepts and technological choices are deferred to later chapters of the text.

Some of the relevant cross sections as a function of C.M. energy are shown in Fig. 3. The scale shown covers many orders of magnitude, which reinforces the fact that the LHC experimenters must explore rare processes in their study of the terascale. Indeed, the clean and background-free detection of the rarely produced leptons implies that redundant measurements of the muons and electrons are required. For example, W leptonic decays are indeed rare: $\sigma(W \to \mu + \nu)/\sigma_{\text{tot}} \sim 10^{-7}$.

Note that the total inelastic cross section is ~ 100 mb. With a luminosity as defined above the total inelastic reaction rate R is, $R = \sigma L = 10^9/\text{s} = 1$ GHz. Having chosen the r.f. bunch spacing to be 25 ns, each crossing contains ~ 25 inelastic events at full luminosity. This leads to experimental

D. Green

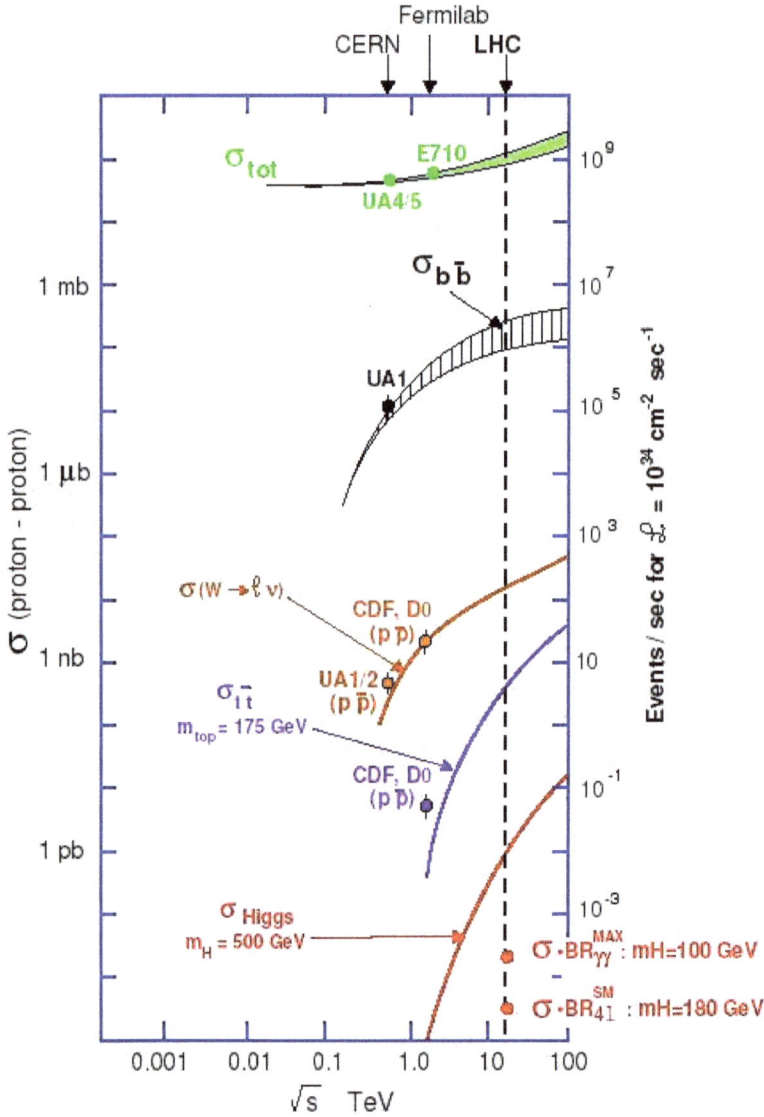

Fig. 3.   Cross sections as a function of C.M. energy for hadron colliders. Note that the cross section scales shown cover more than 13 orders of magnitude.

issues but the luminosity that is required at the LHC and the size of the inelastic cross section make a "pileup" of $N_I = 25$ inelastic events in each bunch crossing unavoidable.

In designing the generic detector it is crucial to know what part of phase space the produced particles occupy. Note that single particle relativistically invariant phase space for a particle of mass M, momentum $\mathbf{P}$ and energy E is

$$d^4 P \delta(P^2 - M^2) = d\mathbf{P}/E = \pi dy dP_T^2,$$
$$E = M_T \cosh y, \qquad M_T^2 = M^2 + P_T^2, \tag{4}$$
$$y \to \eta = -\log(\tan \theta/2), \qquad P \gg M.$$

Therefore, if simple one-particle phase space is followed by the produced particles in a typical inelastic reaction, they can be expected to be found uniformly distributed in rapidity, y. The momentum transverse to the proton beam directions is denoted by P_T. For light particles, $M/P \ll 1$, the rapidity y can be approximated by the pseudorapidity η.

Note that most of the produced particles are pions which have a mass of 0.14 GeV, while their mean transverse momentum at the LHC is expected to be ~ 0.8 GeV. This estimate comes from extrapolation of data at lower energies such as the data shown in Fig. 6. Therefore, the angular variable, the pseudorapidity η, is a good approximation to y in most cases.

Data from hadron colliders on the pseudorapidity distribution of produced particles are shown in Fig. 4. Indeed, production data at $\eta \sim 0$ (i.e. near 90° in the $p + p$ C.M. system) display a roughly uniform distribution. Note also that the density, $D = 1/\sigma(d\sigma/d\eta)$, of particles in the constant region and the width of the constant region rise slowly with increasing C.M. energy.

A COMPHEP[11] calculation of the rapidity distribution of gluon jets is shown in Fig. 5. The kinematic limit for the 7 TeV incident protons is $y_p =$

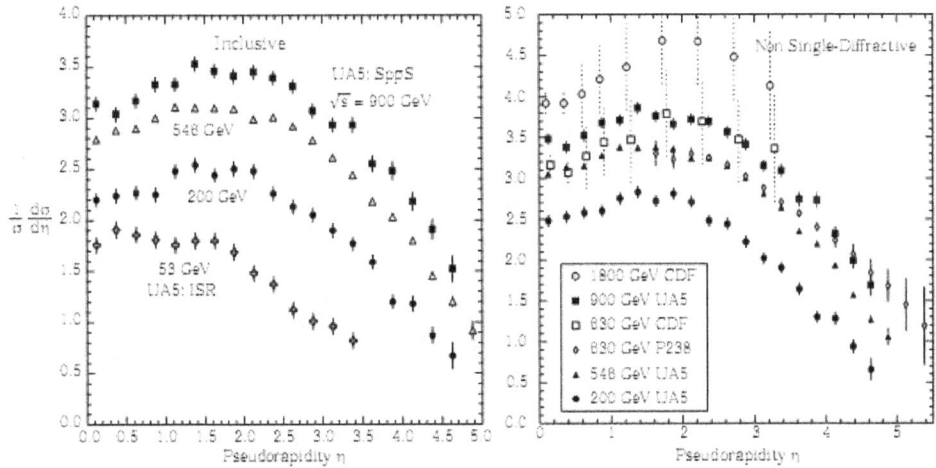

Fig. 4.   Distribution in pseudorapidity for produced charged particles in inelastic collisions at hadron colliders.[8]

D. Green

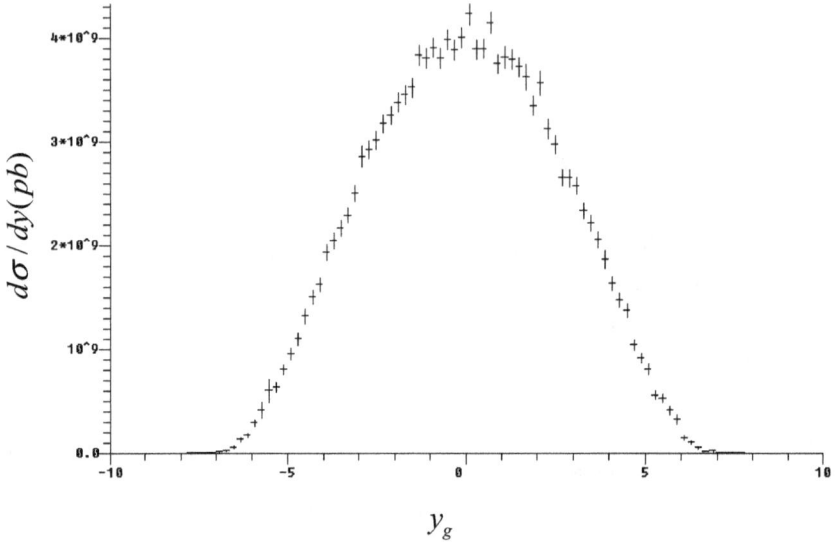

Fig. 5.   Rapidity distribution of low transverse momentum gluon jets at the LHC. Generic detector coverage and the incoming proton rapidities are indicated by the arrows.

±9.6. Clearly, requiring that the LHC detectors detect most of the produced particles means that a rapidity coverage of $2y_{max} \sim 10$ is needed.

One also needs to know the energy distribution of the produced pions. The distribution of the transverse momentum of the inclusively produced particles is shown in Fig. 6. The scale for transverse momentum is low and also slowly increasing with C.M. energy. With this coverage the transverse energy in a collision which is lost in the region $|y| > 5$ does not cause a significant loss of physics capability.

Extrapolating these increases in particle rapidity density and transverse momentum to the LHC, a generic detector should cover a full rapidity range of 10. The particle density per unit of rapidity will be about nine, six charged pions and three neutral pions, under the assumption that all produced particles are pions and all charge states of the pions are equally probable. The mean transverse momentum of the pions will be $\sim 0.8\,\text{GeV}$. Therefore, in each time-resolved r.f. bunch crossing there are 2250 particles in the complete detector coverage:

$$D \sim 9\,\pi/\text{unit of } y, 6\pi^{\pm} + 3\pi^{\circ},$$
$$\langle P_T \rangle \sim 0.8\,\text{GeV}, \tag{5}$$
$$N_I(2y_{max})D = 2250\,\text{particles}, 1.8\,\text{TeV total } \sum P_T.$$

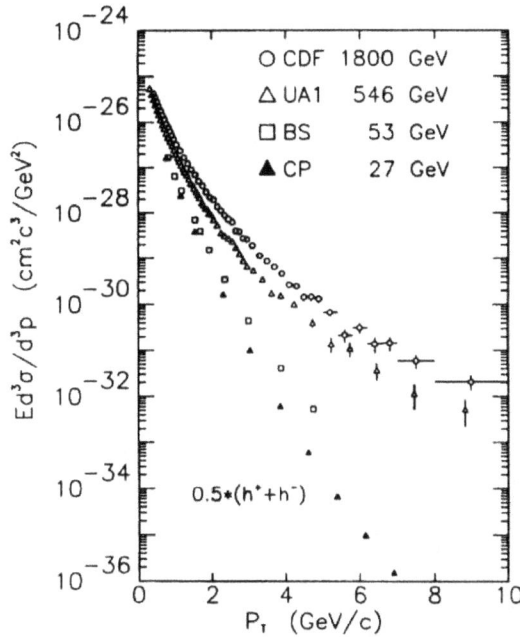

Fig. 6. Transverse momentum distribution of inclusively produced charged secondary particles at hadron colliders.[8]

The generic detector will see 2250 pions in each bunch crossing carrying a total of 1.8 TeV of transverse momentum. These particles are backgrounds and contribute to the "pileup" of energy which occurs within the resolving time of the detector, in addition to those particles from processes which are of interest to the experimenters.

The high mass scale processes of interest populate phase space somewhat differently. For example, a hypothetical 2 TeV mass recurrence of the Z boson decaying into electrons preferentially appears at small rapidities, as shown in Fig. 7. Therefore, the generic detector will put more stress on the low $|y|$ regions of phase space and deploy precision detectors to cover rapidities $|y| < 2.5$. In fact, the more precise generic detectors are limited to the region $|y| < 2.5$, due to their rate limitations and their ability to withstand radiation damage both to the detectors and to the front end electronics. The larger $|y|$ region is covered only by radiation-hardened calorimetry.

The kinematic maximum value of rapidity for a given mass M and C.M. energy can be estimated using energy and momentum conservation in the production of that mass by the incident quarks or gluons contained in the proton and moving with momentum fraction x_1 and x_2 of the proton beam.

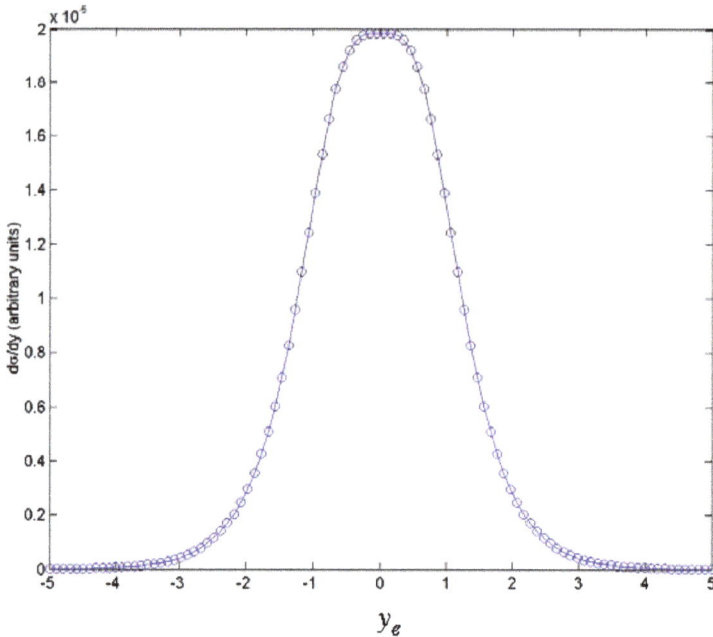

Fig. 7. Rapidity distribution of the electrons resulting from the two-body decay of a hypothetical 2 TeV mass recurrence of the Z boson. The limit for precision detection systems at $|y| < 2.5$ is indicated by the arrows.

These two particles fuse together to produce mass M moving with momentum fraction x and rapidity y:

$$x_1 x_2 = M^2/s, \qquad x_1 - x_2 = x,$$
$$x_1 = [M/\sqrt{s}]e^{-y},$$
$$x_2 = [M/\sqrt{s}]e^{y},$$
$$y_{\max}(M) \sim \ln(\sqrt{s}/M).$$

$$(6)$$

For the LHC, operating at a C.M. energy, $\sqrt{s}$, of 14 TeV, the maximum rapidity of $y_{\max}(2\,\text{TeV}) = 1.9$ for a mass of 2 TeV occurs when $x_2 \to 1$, which is in rough agreement with the y, distribution of the leptons from the assumed two-body decay of that massive particle, as shown in Fig. 7. Clearly, the generic detector covers this reaction with good detection efficiency. The logarithmic dependence of the maximum rapidity on C.M. energy was already seen in the pion y behavior in Fig. 4.

There are also specific processes which relate to vector boson reactions, which are a primary focus of LHC physics studies. Specifically, the virtual emission of a W by an initial state quark in both of the protons will lead to the scattered $W + W$ plus the two quarks recoiling after W emission. This

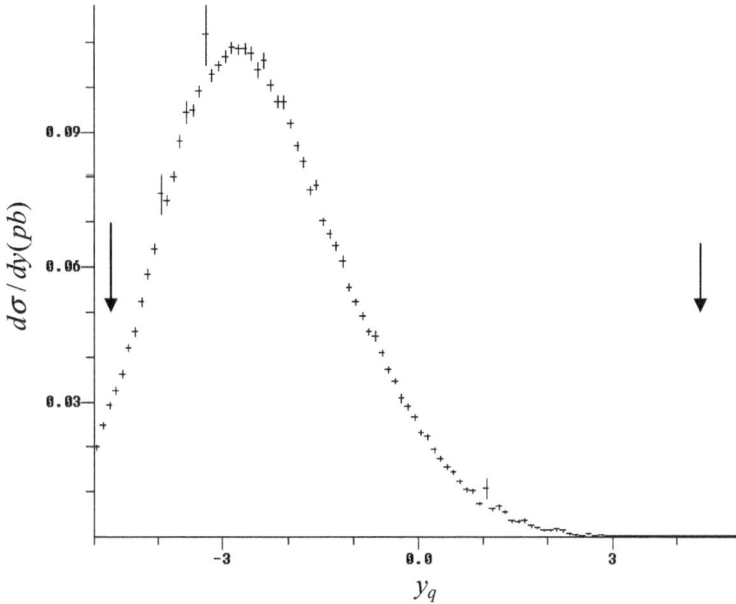

Fig. 8. Rapidity distribution of the final state quark after virtual emission of a W boson in the formation of a $W + W$ resonance of 0.2 TeV mass via the vector boson fusion mechanism. Coverage of ± 5 units of rapidity is needed for efficient detection of this process.

process is called "vector boson fusion" and is a critical component of the LHC physics program because it has a signature which allows the experimenters to tag and isolate vector boson scattering. The rapidity distribution of the final state quarks is shown in Fig. 8. Clearly, this is a second reason to have generic detector coverage out to ± 5 units in rapidity, albeit with somewhat less precise measurement capability and no tracking.

5. The "Generic" Detector

The needed full detector coverage in angle followed from the properties of inelastic events and from the vector boson fusion process. The generic detector consists of several subsystems with specific roles in particle detection and identification, and is described below. The idea here is simply to observe from the physics requirements what a typical general purpose detector might look like. It is generic in the sense that specific design choices are not made but rather the general needs of the several detector subsystems are examined.

The overall coverage in rapidity of ± 5 units follows from the distribution of inclusive inelastic secondary particle production and the desire to detect the forward-going quarks from the vector boson fusion mechanism.

This coverage should be "hermetic," which means that all produced particles should be detected and their positions and momenta should be well measured. If this is achieved, then the production of neutrinos may be inferred by vectorially adding all the observed particle transverse momenta in the final state because the initial state contains almost no transverse momentum. If an imbalance remains the existence of "missing transverse momentum" or MET implies that an undetected, noninteracting, neutral stable particle (or particles) has been produced which carried off the MET.

The generic detector has detection elements which are roughly of uniform extent in pseudorapidity and azimuthal angle since then they will have roughly equal probability to be struck by a secondary particle. If a detector "tower" size of

$$\delta\eta = \delta\phi/2\pi = 0.1 \tag{7}$$

is chosen, then there will be 6300 independently read-out "towers" reporting a measurement of particle passage over the full coverage of the detector, $2y_{max} = 10$. Since there are 2250 produced pions at full LHC luminosity in the detector acceptance, the mean probability for a tower to be occupied is $\sim 36\%$ (24% charged, 12% neutral).

The connection between the interval in polar angle and the pseudorapidity interval follows from Eq. (4). The interval at small $|\eta|$ is approximately the same for rapidity and angle.

$$\begin{aligned}
d\eta &= [e^{\eta}d\theta]/(1 + \cos\theta) \\
&\rightarrow d\theta, \qquad \eta \sim 0 \\
&\rightarrow e^{\eta}d\theta/2 = d\theta/\theta, \quad \eta \gg 1.
\end{aligned} \tag{8}$$

The generic detector is shown in Fig. 9. There are several specific subsystems which are indicated. A general purpose detector, such as those installed at the LHC, aims to measure all the particles of the SM from each interaction as well as possible. This follows because, whatever form the physics beyond the SM takes, the final state particles will ultimately decay to the bosons, quarks, gluons, charged leptons and neutrinos of the SM. The role of the subsystems in detection and identification of the SM particles is indicated in Table 1.

The generic detector focuses on the "barrel" or wide angle region, $|\eta| < 1.5$, because heavy new particles are kinematically expected to be produced there. Coverage by the vertex, tracking and electromagnetic calorimetry is limited to $|\eta| < 2.5$ because of the fierce radiation field which exists at smaller angles. That field must, however, be confronted because of the

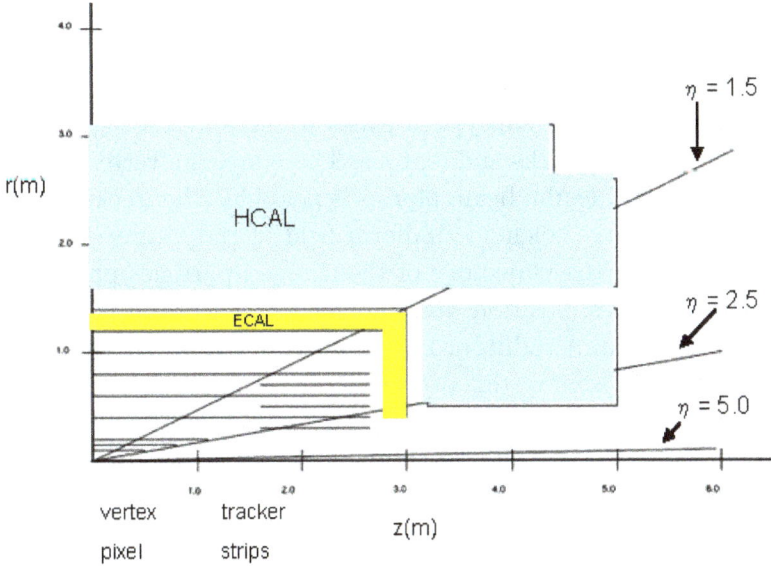

Fig. 9.   A "generic" LHC detector which covers ±5 units of pseudorapidity. Only the central ±2.5 units of coverage are shown here. The remaining small angle calorimetry is at $z = 10\,\mathrm{m}$. The muon detection system is also not shown.

Table 1.   Particles of the SM and detection and identification in detector subsystems.

Particle		Signature	Generic subsystem
$u, c, t \rightarrow W + b$	Quarks	Jet of hadrons (λ_0)	Calorimeter
d, s, b			ECAL + HCAL
g			
e, γ		Electromagnetic shower (X_0)	Calorimeter ECAL tracker
ν_e, ν_μ, ν_τ $W \rightarrow \mu + \nu_\mu$		Missing transverse energy (MET)	Calorimeter ECAL + HCAL
$\mu, \tau \rightarrow \mu + \nu_\tau + \bar{\nu}_\mu$ $Z \rightarrow \mu + \mu$		Only ionization dE/dx	Muon absorber and detectors tracker
c, b, τ		Secondary decay vertices	Vertex + tracker

high luminosity needed to explore EWSB at the terascale and the need for hermetic coverage.

The size and location of the detector subsystems follow from the physics requirements. The vertex subsystem exists to measure the secondary decay vertices of the heavy quarks and leptons. Therefore, the vertex subsystem is as near to the LHC vacuum beam pipe as is possible. The tracking subsystem is immersed in a large solenoid (uniform field in the z direction) magnetic field. By measuring the trajectory of the charged particles produced in the collisions, the charges, positions and momenta of all produced charged particles are determined. A radius of 1 m is allocated to the vertex and tracking systems in order to achieve this objective with sufficient accuracy.

The size of the calorimetry is dictated by the characteristic distance which is needed to initiate an interaction, either electromagnetic (ECAL) or hadronic (HCAL). These sizes are

$$(X_0)_{\text{Pb}} = 0.56\,\text{cm},$$
$$(\lambda_0)_{\text{Fe}} = 16.8\,\text{cm}. \tag{9}$$

The showers in ECAL are fully developed and contained in about 20 radiation lengths, X_0. Allowing space for shower sampling and readout, 20 cm in depth is used for ECAL. The HCAL follows in depth with 1.5 m of steel or 8.9 nuclear absorption lengths, λ_0. The focus in angular coverage of all the subsystems is in the "barrel" region, $|\eta| < 1.5$. Coverage by the vertex, tracking and ECAL subsystems goes only to $|\eta| < 2.5$.

In the region $2.5 < |\eta| < 5.0$ the radiation field at the LHC precludes all but very radiation-resistant calorimetry. This forward calorimeter region is not shown explicitly, but is thought to reside at $z = 10$ m, a large distance which reduces the radiation dose. If it were stationed at $z \sim 3.2$ m, the dose would be ~ 9.8 times larger with a substantial added radiation resistance required of the specific technology chosen for the calorimetry.

The high rates of background processes, as shown in Fig. 3, imply that the leptons must be measured in robust and redundant systems. For example, muons are measured in the tracking systems and then again in redundant specialized muon systems which have lower rates of fairly pure muons because almost all other SM particles have been absorbed in the thick calorimeters. Electrons are also measured redundantly, in this case using the tracking systems and the electromagnetic calorimeters. In this way, the generic detector allows for clean lepton identification and measurement even though the cross sections of most interest, such as Higgs decays, are $\sim$ fb or smaller, while the inelastic backgrounds are $\sim$ mb, a factor of at least 10^{12} larger.

6. The Generic Vertex Subsystem

The subsystem at the smallest transverse radius, r, is the vertex subsystem. It consists of pixels of silicon deployed from radius 10 cm to 20 cm in three layers in the generic model. The vertex detector is used to efficiently find and identify the secondary decay vertices of the heavy quarks and leptons, as shown in Table 1. The lifetimes of these particles set the scale for the pixel size.

$$(c\tau)_\tau = 87\,\mu\text{m},$$
$$(c\tau)_b \sim 475\,\mu\text{m}, \tag{10}$$
$$(c\tau)_c \sim (123, 312)\,\mu\text{m} \quad (D^0, D^\pm).$$

Assuming a pixel size of $\delta z = \delta s = 200\,\mu\text{m}$, where s is the distance in the azimuthal direction (on a cylindrical vertex pixel layer) and z is along the beam direction, the vertex detector spatial resolution is sufficient to identify and measure the decay vertices. Note that an analog readout of the energy deposited in neighboring pixels yields better spatial resolution than a simple digital readout result, which is the pixel width divided by $\sqrt{12} = 3.46$. That factor is used to convert from the full extent of a uniform distribution to the r.m.s. of that distribution.

As an example of the utility of b vertex identification, the strong production of top pairs is a large background to the measurement of the rarer electroweakly produced W pairs, which we wish to measure at 1 TeV W pair mass. In this case, tagging of b quarks is useful since it serves to indicate that the W pairs arise from the copious top background in the process $p + p \rightarrow t + \bar{t} \rightarrow W^+ + b + W^- + \bar{b}$.

An event at the Fermilab Tevatron is shown in Fig. 10. The Lorentz boost, $\gamma_b = E_b/M_b$, means that, for example, b decays are separated from the primary vertex by distances measured in millimeters, $\gamma_b(c\tau)_b$.

The probability for a pixel to be struck by a charged pion in a single bunch crossing is

$$P_{\text{pix}} \sim D_c N_I (\delta z/r)(\delta s/2\pi r). \tag{11}$$

With a density, D_c, of six charged pions per unit of y, and $N_I = 25$ inelastic events per bunch crossing at full luminosity, the pixel occupation probability at the innermost layer at radius $r = 10$ cm is ~ 0.000095. The low occupation probability is achieved at a cost of the independent readout of many silicon pixels. Roughly, for three layers extending over $|z| < 80$ cm there are 8000 pixels per layer in z and 4700 pixels in azimuth for a total of 110 million pixels covering $|\eta| < 2.5$. The sparse occupation of the pixel system makes it an

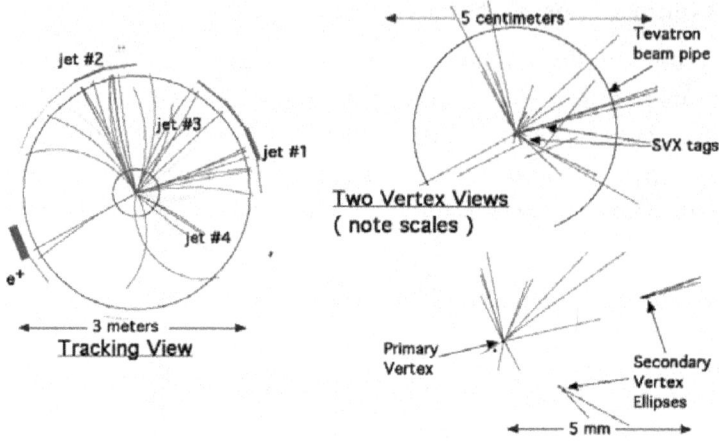

Fig. 10. An event in the CDF detector with two secondary vertices coming from decays of B mesons, which are seen to be separated from the primary production vertex by transverse distances of order mm.

ideal location to start finding track patterns. Indeed, the required high rates at the LHC were what caused ATLAS and CMS to adopt pixel detectors, which was the first time pixel detectors were deployed at hadron colliders.

Clearly, there is a tradeoff in radius between the number of pixels and the radiation field. If the vertex system could start at a radius of 5 cm, then the number of pixels would be reduced to 28 million and the radiation would be increased fourfold due to the r^2 behavior of the number of pixels and the $1/r^2$ behavior of the dose. A smaller radius also reduces the position error made in extrapolating from the inner pixel layer to the actual decay vertex because of multiple scattering in the beam vacuum pipe and the detectors, which causes angular track errors.

The radiation field can be approximately evaluated using the number of charged tracks leaving ionization in the silicon detectors, ignoring increases due either to secondary interactions of those tracks, photon conversion, or their multiple passage through the detectors due to particles being curled up in the magnetic field.

$$(\text{Dose})_{\text{pix}} \sim \sigma_I L T D_c (1/2\pi r^2) dE_{\text{ion}}/d(\rho_{\text{Si}} \delta r). \tag{12}$$

The dose is the energy deposited per weight and 1 mrad = 6.2×10^{10} GeV/gm. The number of inelastic events per second is $\sigma_I L = 10^9$ Hz and the exposure time is taken to be a "year," $T = 1$ yr, defined to be 10^7 s. With a density, D_c, of six charged particles per interaction and vertex detectors at $r = 10$ cm, the charged particle fluence is $9.5 \times 10^{14} \pi^{\pm}/\text{cm}^2$ yr. Each minimum ionizing

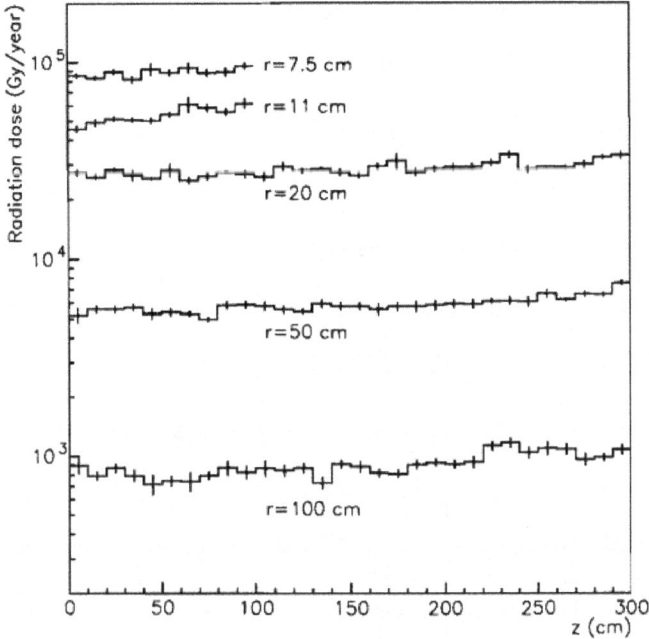

Fig. 11. Radiation dose in a tracking system as a function of z at various radii.[4]

particle at normal incidence deposits $dE_{\text{ion}}/d(\rho_{\text{Si}}\delta r) = 1.66\,\text{MeV}/(\text{gm}/\text{cm}^2)$ or $0.12\,\text{MeV}$ in a $\delta r = 300\,\mu\text{m}$ thick detector. The estimated dose is then $2.5\,\text{mrad/yr}$ for the inner vertex detectors.

The results of a detailed Monte Carlo model are shown in Fig. 11. A Grey is equal to $100\,\text{rad}$. Note that the dose is about $5\,\text{mrad/yr}$ at a radius of $10\,\text{cm}$, is roughly flat in z and falls roughly as the square of the radius, as expected from the simple estimate made above.

The pixel vertex detectors could be deployed as a barrel covering $|y| < 1.5$ and an endcap covering the region $1.5 < |y| < 2.5$. In that case the barrel inner radius of $10\,\text{cm}$ would extend to $\pm 21\,\text{cm}$ in z and the endcap at $\pm z = 21\,\text{cm}$ would have an inner radius of only $3.4\,\text{cm}$. Thus, the radiation dose would be 8.6 times higher in the inner end-cap in this configuration.

However, the long barrel pixel system has a problem at small angles too. The $\delta r = 300\,\mu\text{m}$ thick detector is traversed $(1/\sin\theta \sim 6.1, |\eta| = 2.5)$ with a length of $1800\,\mu\text{m}$ so that the deposited charge is shared over 9 pixels, which degrades the position resolution. Analog readout of the charge sharing can recover some of this loss, but the occupation probability increases and the signal to noise ratio of each pixel decreases. In fact, both ATLAS and CMS chose the barrel/endcap geometry for these reasons.

7. The Generic Solenoid Magnet

The vertex and tracking subsystems are assumed to be immersed in a strong solenoid field. Roughly speaking, the current I flowing through n turns/length of conductor leads to a field; $B = \mu_0 n I$. Suppose that there were conductors of size 2 cm in z all stacked by 4 in r for a total of $n = 200$ turns/m. Then, to achieve a field of 5 T, 20.8 kA of current is required. A plot of the stored energy per mass versus the stored energy of some of the solenoids used in high energy physics is shown in Fig. 12. Clearly, LHC magnets are a step above what has been previously achieved in the Tevatron (CDF and D0) as regards the solenoids used in the detectors.

Note that in the generic detector it is just assumed that the vertex and tracking systems are immersed in a 5 T magnetic field. The choice of coil location is finessed; it could be just outside the tracker or just outside the calorimetry. No consensus on the location exists and therefore it is ignored here while exploring how a generic detector relates physics requirements to detector design. ATLAS and CMS have made different choices, as will be explained later.

8. The Generic Tracker Subsystem

The requirement that the LHC experiments decisively confront EWSB has been seen to imply a complete study of $W + W$ scattering at a mass of ~ 1.0 TeV. In that case the transverse momentum of the W is ~ 0.5 TeV and

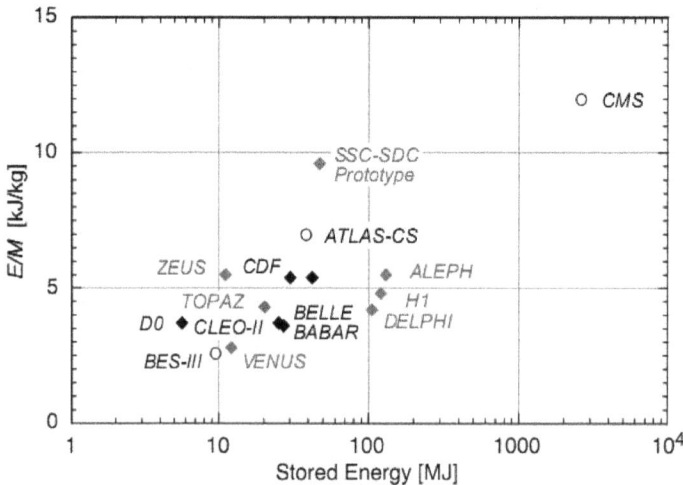

Fig. 12.   Energy per unit mass vs. total stored energy for a variety of solenoid magnets used in high energy physics experiments.[8]

that of the lepton from the two-body W decay is $\sim 0.25\,\text{TeV}$. The radial size of the tracker is then set by the need to measure the lepton momentum well at this mass scale.

The approximate transverse momentum impulse imparted to a particle of charge e in a magnetic field B in traversing a distance r is

$$(\Delta P_T)_B = er B = 0.3 r B, \tag{13}$$

where the units are GeV, Tesla and m, and an electronic charge is assumed. The bend is in the azimuthal direction and the angle through which the particle is bent, $\Delta\phi_B$, is approximately the impulse divided by the transverse momentum, $(\Delta P_T)_B / P_T$. The error on the inverse transverse momentum is therefore

$$d(1/P_T) = dP_T/P_T^2 = d(\Delta\phi_B)/(\Delta P_T)_B \sim ds/er^2 B, \tag{14}$$

where ds is the spatial resolution in the azimuthal direction. Note that the quality factor for the momentum resolution scales as $1/r^2 B$, which argues for a strong magnetic field and a large radius for the cylindrical tracking subsystem.

Assuming a $5\,\text{T}$ magnetic field extending over $r = 1\,\text{m}$, the magnetic impulse is $1.5\,\text{GeV}$. The bend angle is then $6\,\text{mrad}$ for the $0.25\,\text{TeV}$ lepton from the W decay. If the tracker spatial resolution is taken by assuming the digital readout of strips of $400\,\mu\text{m}$ width, the resolution r.m.s. is $ds \sim 400/\sqrt{12}\,\mu\text{m} = 115\,\mu\text{m}$. The momentum will then be measured to 1.9%. Later in this chapter, Eq. (16), it is shown that a 1.2% accuracy is needed in order not to degrade the Z peak natural-width-to-mass ratio, which implies analog readout of the tracker detector.

The momentum is measured to 10% accuracy at $1.30\,\text{TeV}$, which makes a sufficiently good measurement at the terascale. A smaller radius or lower field tracking system would require a more accurate measurement than the ds value quoted above if the momentum accuracy were to be maintained. Smaller strip size, more tracking layers, or analog readout could be used to improve the spatial resolution.

At a lower momentum scale there is another limitation on the measurement accuracy besides position resolution. The multiple scattering transverse impulse, $(\Delta P_T)_{\text{MS}}$, due to scattering in the tracker material compared to the magnetic field impulse sets a limit of

$$dP_T/P_T \sim (\Delta P_T)_{\text{MS}}/(\Delta P_T)_B,$$
$$(\Delta P_T)_{\text{MS}} = E_s \sqrt{\sum L_i/2X_0}, \tag{15}$$

where the scattering energy, E_s, is 21 MeV, X_0 is the radiation length, and the sum is over all material in the tracker. For a tracker plus vertex system having silicon detectors of $400\,\mu$m thickness in 11 layers — 8 layers in 4 stations with small angle stereo in the tracker and 3 vertex layers — the scattering impulse is 3.2 MeV.

There are two terms in the momentum resolution, $dP_T/P_T = cP_T \oplus d$, which are folded in quadrature. There is a term due to measurement error, $c = 0.000078\,\mathrm{GeV}^{-1}$ [Eq. (14)], and a term due to multiple scattering, $d = 0.0021$ [Eq. (15)], which limits the low momentum measurements. The crossover transverse momentum where the errors are equal is 27 GeV in this example and the total resolution at this momentum is $\sim 0.3\%$. Clearly, keeping the contribution due to multiple scattering low allows one to improve the momentum measurement accuracy at low momentum. By the way, only counting the detectors themselves and ignoring power leads, cooling leads and readout electronics makes this a rather optimistic estimate.

A complete Monte Carlo model prediction for the CMS transverse momentum resolution is shown in Fig. 13. The resolution increases roughly as P_T for momenta above 10 GeV because of the position measurement resolution, while there is no improvement at momenta below 10 GeV due to the multiple scattering limit caused by the tracker material itself.

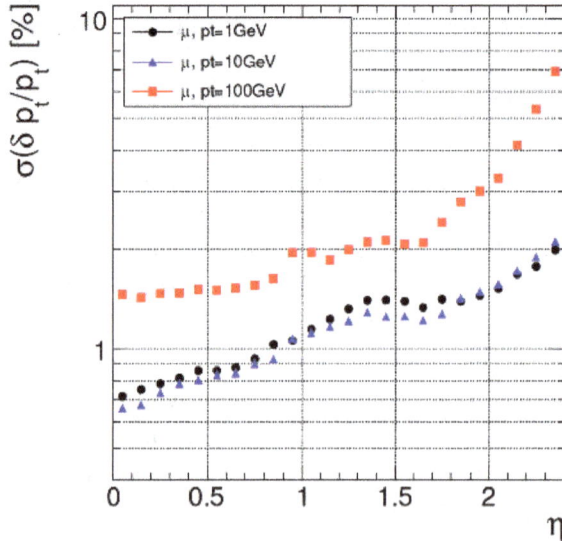

Fig. 13.   Transverse momentum resolution in the Monte Carlo model of the CMS detector.[4]

The solenoid field chosen for the generic detector provides a fractional transverse momentum error which is independent of the polar angle for $|\eta| <$ 1. Note that the force on a charged particle is $\mathbf{F} = e(\mathbf{v} \times \mathbf{B})$, so that more forward-going particles see a reduced force and a resulting deterioration in the momentum resolution.

There is additional information and improved spatial resolution available if analog information is recorded from the tracker. For example, a measurement of the deposited energy in the silicon and the momentum as inferred from the helical trajectory results in a mass determination of the secondary particles. A plot of deposited energy, E_{ion}, versus particle momentum is shown in Fig. 14. Since $dE_{\text{ion}}/dx \sim 1/\beta^2 = (E/P)^2 \rightarrow (M/P)^2$, the method works well at low momenta where the particles are nonrelativistic. For slow particles the deposited energy goes as M^2 at fixed P. At high momenta all particle velocities approach c and particle mass discrimination is lost.

Most of the information provided by the tracker is contained in the measurement of the azimuthal coordinates (the "bend plane"), while the helical paths are less demanding in the z direction. Therefore, in order to reduce

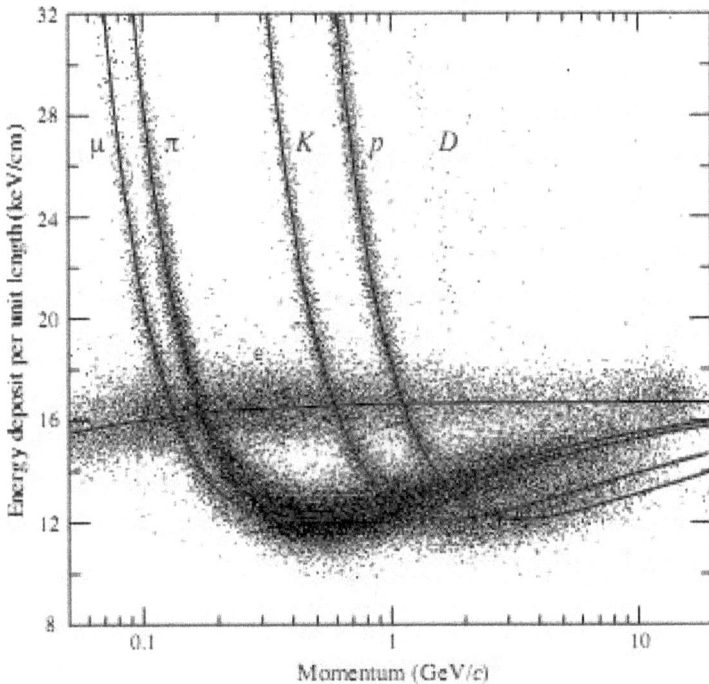

Fig. 14.   Deposited energy in a tracking system as a function of particle momentum for several types of particles.[8]

costs, the generic tracking detector is fashioned from silicon strips, 10 cm long in z and 400 μm wide in s. Each of the four stations has a strip layer oriented along z as the long axis and a second at a small angle to the first. Using Eq. (11), the probability for a strip to be occupied ranges from 0.006 to 0.00096 as the strip radius goes from 40 to 100 cm. This level of occupation is sufficiently sparse to allow for robust track pattern recognition and trajectory fitting. The total number of strips in one layer at an average radius of 70 cm is ~ 52 in z, $|z| < 2.6$ m, times $\sim 11{,}000$ in r. Therefore, in eight layers there are roughly 4.6 million independent strips to record.

Note that gaseous tracking detectors can also be used. A typical drift velocity is $v_d \sim 5$ cm/μs, or a drift distance of 0.12 cm per 25 ns bunch crossing. Therefore, a gaseous tracking system will likely integrate over several bunch crossings in time, thus increasing the occupancy and making the pattern recognition more difficult.

The radiation dose scales from the pixel dose as $1/r^2$, which means that the tracker silicon strip detectors suffer a far smaller dose than the pixel detectors. In estimating the dose it was assumed that neutral photons are not converted to electron–positron pairs in the tracker material and that the charged pions do not interact to produce multiple charged pions. It was also assumed that particles are not bent and pass only once through a given radius.

None of these assumptions is correct. In particular, if the transverse momentum impulse imparted to a particle is twice the transverse momentum of the particle, that means it has been curled up in the field. In that case, $(\Delta P_T)_B = 2(P_T)_{\text{loop}} = erB/2$. Particles with transverse momentum of 0.75 GeV never get to radii greater than the tracker radius. They are "loopers" which curl up in the tracker volume and make multiple hits in the detectors, thus increasing the dose and the occupation probability. Since the mean transverse momentum of produced pions is ~ 0.8 GeV (Fig. 6), about one half of all charged particles produced in an inelastic collision never travel beyond the generic tracking system.

As will be mentioned below, this has negative implications for measuring the momentum carried by low energy particles in a "jet" of hadrons because they will not even reach the calorimetry. A more sophisticated use of calorimetry plus tracking, termed "particle flow," is called for where the tracker is used to recover the "loopers" which are lost to the calorimetry and to potentially improve the momentum of charged tracks which are well measured in the tracker and not as well in the calorimetry.

9. The Generic ECAL

The generic electromagnetic calorimeter has the role of measuring the energy and position of the photons and electrons created in the LHC collisions. It is the first of the calorimeters to be struck by outgoing particles in the generic detector. Calorimeters measure energy by initiating interactions of the incident particles and completely absorbing the resulting energy, which appears as a geometrically growing "shower" of particles. As compared to the tracking and vertex systems, which absorb only the tiny ionization energy deposited in them, the calorimeter readout is "destructive" in that the showering particle is totally absorbed.

The ECAL supplies a redundant measurement of the electrons. The tracking system first measures the electron momentum, charge, position and direction. The ECAL redundantly measures the electron position and energy. The electrons are important in the study of vector boson interactions because they are decay products of the W and Z. Redundancy is needed in order to be able to cleanly study the rare interactions of vector bosons using the final state electrons in a background of strongly produced neutral pions and the photons resulting from their decay.

The electrons appear as decay products of the electroweakly produced vector bosons, $Z \rightarrow e^+ + e^-$, $W^- \rightarrow e^- + \bar{\nu}_e$. The ECAL energy resolution requirement is set by the natural width, Γ_Z, of the Z resonance. The Breit–Wigner full width of the Z resonance is irreducible and the resolution of the ECAL energy measurement should not increase the width of the dielectron mass peak beyond the irreducible scale set by the natural width. The factor 2.36 is used to convert full width at half maximum to r.m.s.

$$\Gamma_Z = 2.5\,\text{GeV}, \quad M_Z = 91.2\,\text{GeV},$$
$$(dE/E)_{\text{ECAL}} < \Gamma_Z/(2.36 M_Z) = 1.2\%. \tag{16}$$

It is customary to decompose the calorimetric energy resolution into a stochastic coefficient a and a constant term b and fold the two contributions in quadrature, $dE/E = a/\sqrt{E} \oplus b$. The constant term arises from the nonuniformity of the medium and the requirement is then clearly to have $b \ll 1\%$. This means that the manufacturing process control of the ECAL must be very stringent.

For the stochastic coefficient, the energy of the Z decay electrons is $E \sim 0.25\,\text{TeV}$ for $1\,\text{TeV}$ mass $Z + Z$ scattering, so that the coefficient a should be $< 19\%$. To achieve that it is important to first understand the electromagnetic showering process in better detail. The critical energy for an electron is that energy where radiative processes are comparable in probability to nonradiative processes such as ionization. In Pb the critical energy, E_c, is 7.3 MeV. At that energy the electron is quite relativistic, having a Lorentz boost factor of $\gamma = 14.3$. The radiation length, which sets the length scale for radiative processes, is $X_0 = 0.56$ cm in Pb.

The development of a shower is parametrized[8] using the depth, L, in radiation length units, $t = L/X_0$ and energy in critical energy units, $y = E/E_c$. The incident energy is E_0 and $a \sim 1 + (\ln y)/2$. The shower development in depth is then

$$dE/dt = E_0 b(bt)^{a-1} e^{-bt}/\Gamma(a),$$
$$t_{max} = (a-1)/b, \qquad b \sim 0.5.$$

(17)

A typical shower profile for 30 GeV electrons is shown in Fig. 15. There is a rapid rise due to the geometric shower growth. Because the energy of the electrons and photons in the shower is shared over more particles as the shower grows with depth, the average particle energy falls with depth until a point comes where radiative shower growth stops, at the "shower maximum" at a depth t_{max} where the average particle energy is near the critical energy. The shower then dies out at greater depths by losing energy due to ionization and photoelectric absorption. This behavior is due to the geometric behavior of the shower, with N_s total particles in the shower.

$$N_s \sim (E/E_c) \sim 2^{t_{max}},$$
$$t_{max} \sim \ln(E/E_c).$$

(18)

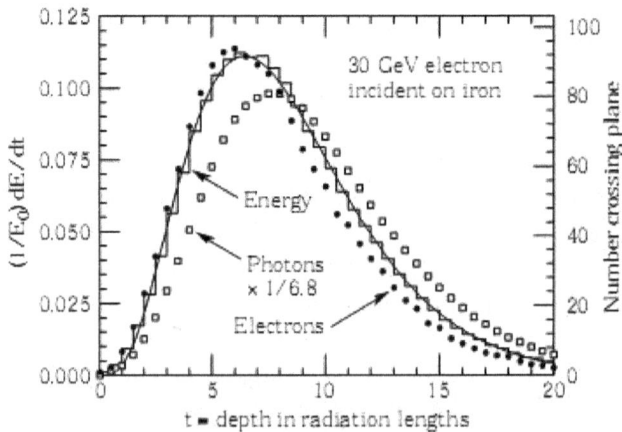

Fig. 15.   Longitudinal electromagnetic shower development for 30 GeV electrons incident on iron and showering into photons and electrons.[8]

For the tracker [Eq. (14)], the fractional momentum resolution scales as $1/r^2$ and increases linearly with the momentum. For calorimetry the needed depth scales logarithmically with the energy. The fractional energy error also improves or stays constant with energy. Therefore, calorimetry tends to do the best job at high momenta while tracking excels at lower momenta.

As seen in the figure, the shower is fully contained over a depth $t \sim 20$. It develops rapidly, so fine sampling of the shower, $\delta t \sim 0.5$, or a total of 40 depth samples is called for to achieve good energy resolution. The shower maximum occurs at $t_{\max} \sim 7$ and the full width of the shower development is $\Delta t \sim 7$.

The number of showering particles N_s is large, peaking in Fig. 15 at ~ 100 per sample. With fine sampling and a very uniform medium the stochastic fluctuation of the number of shower particles, N_s, can largely determine the energy resolution, $dE/E \sim 1/\sqrt{N_s} \sim \sqrt{E_c/E}$. For $E = 0.25\,\text{TeV}$, the number for Pb is 34,245, with a 0.54% fluctuation, which meets the energy resolution requirement. The stochastic coefficient is estimated to be $a = \sqrt{E_c} = 8.5\%$, where GeV energy units are used. This is near the ultimately achievable resolution, and most sampling calorimeters do not have as good performance. However, only 19% is required. At lower mass scales such as $0.2\,\text{TeV}$ a better stochastic coefficient, $\sim 8.5\%$, would be needed.

The ECAL provides particle identification for photons and electrons in addition to energy and position measurements, in the sense shown in Table 1. An incident hadron begins to interact over a characteristic distance, λ_0, which is 10.2 cm in Pb. Thus, the fully contained ECAL shower in $t = 20$ presents only a depth of 1.1 λ_0 to an incident hadron. This difference in the length scale of hadronic and electromagnetic interactions provides the ability to distinguish between electrons and photons and the hadrons. Independent readout of several depth segments of each ECAL tower improves the particle identification performance.

The transverse shower development is important too, as it defines the requirement for the ECAL tower size and also provides additional particle identification capability. The Moliere radius, r_M, is the radius of a cone within which 90% of the shower energy is deposited. In Pb the radius is $r_M = 1.6\,\text{cm}$ and that sets the tower size. Finer tower segmentation is not called for since no new information in the dense core of the shower can reasonably be extracted. Particle identification follows from the fact that hadron showers are much wider transversely with a radius $\sim \lambda_0$.

In the generic detector the barrel ECAL exists at $r_E = 1.2\,\text{m}$ and $|z_E| < 2.8\,\text{m}$, covering the range $|\eta| < 1.5$. The ECAL towers are defined by the

Moliere radius in lead, $\delta\eta = \delta\phi \sim (2r_M)/r_E = 0.027$. The barrel has 175 towers in z and 236 towers in azimuth. If there are three depth segments read out independently in order to measure the shower development in depth t, there are a total of 124,000 ECAL readout channels in the barrel alone.

It is desirable to distinguish between the electromagnetically produced photons and the strongly produced, much more copious, neutral pions. The pions decay into two photons, with a typical opening angle of $\theta \sim m_\pi/P_\pi$. With the ECAL transverse segmentation, pion momenta of $< 4.2\,\mathrm{GeV}$ give resolvable showers. At higher momenta the showers are not resolved and therefore appear to be isolated photons. Better rejection requires the use of a detector with finer segmentation located earlier in the shower before the shower spreads to fill the Moliere radius. Such "preshower" detectors are not explored further in the generic detector example.

The probability of having an overlapping neutral pion in an ECAL tower due to event pileup is $(D_0 = 3)(N_I = 25)(\delta\eta)^2/2\pi = 0.0087$, assuming a fast readout with time resolution of only one bunch crossing. Since each pion carries only $0.8\,\mathrm{GeV}$ of transverse momentum, pileup on top of a $0.25\,\mathrm{TeV}$ shower is not severe and does not degrade the energy resolution significantly. However, the desire to simultaneously have high speed and low noise electronics is somewhat mutually exclusive. The devil is in the details and they will be provided in subsequent chapters.

The radiation field is severe since, in contrast to the tracking, the entire particle energy is now absorbed by the medium. The radiation dose in the barrel due solely to neutral pion absorption by the ECAL medium is

$$(\text{Dose})_{\text{ECAL}} \sim \sigma_I L T D_0 (1/2\pi r_E^2)\langle P_T\rangle/(\rho_E \Delta t X_0). \tag{19}$$

With three neutral pions per unit of rapidity, D_0, depositing $0.8\,\mathrm{GeV}$ into ECAL with density ρ_E of Pb over a region $\Delta t X_0$ in depth, $\Delta t \sim 7$, there is a dose of $\sim 0.097\,\mathrm{mrad/yr}$ in the ECAL barrel. The dose in the endcap is, however, much higher. Fundamentally that increase is due to the higher energy pions in the endcap (at fixed transverse momentum, $E_\gamma = \langle P_T\rangle/\sin\theta$) and the smaller radii covered by the endcap which is located at z_E. The dose ratio between the barrel and the endcap for ECAL is approximately

$$(\text{Dose})_{\text{endcap}}/(\text{Dose})_{\text{barrel}} \sim (r_E/z_E)^2/\theta^3. \tag{20}$$

For the endcap the minimum angle is at $|\eta_E| = 2.5$, where the dose ratio is about 42, which means that the endcap ECAL has a yearly dose of about $4\,\mathrm{mrad}$. This rapid dose increase with rapidity sets the limit in the generic ECAL on the angular coverage outfitted with precision electromagnetic calorimetry.

The ECAL shower development itself is very rapid, so that the speed of data collection is defined by the readout. It has been assumed to have a resolving time of less than one bunch crossing, similar to the speed already quoted for the silicon tracking and vertex devices. For $\delta t = 0.5$ sampling and a depth of $20\,t$, there would be 40 samples ganged together. With a tower having an area $3.2 \times 3.2\,\mathrm{cm}$, each sample would have about $9\,\mathrm{pF}$ of source capacity if the sampling gap is $1\,\mathrm{mm}$, assuming a unity gain detector. If the entire tower is summed, $360\,\mathrm{pF}$, then with a $50\,\Omega$ connecting readout cable the rise time of the pulse would be $18\,\mathrm{ns}$. If three longitudinal segments are read out, this reduces the signal rise time to $6\,\mathrm{ns}$. Therefore, some care is needed in ensuring that the calorimeter readout is sufficiently fast.

10. The Generic HCAL

The hadron calorimeter, HCAL, measures the energy of the strongly interacting quarks and gluons by absorbing the jets of particles that these fundamental particles hadronize into (Table 1). In addition, the "hermetic" calorimetry measures the energy of all the secondary particles within a range of $|y| < 5$, which is sufficiently complete coverage (Fig. 5) that a large missing transverse energy, MET, indicates neutrinos in the final state and thus provides both particle identification and a measurement of the neutrino transverse energy.

In principle the requirements on the HCAL are very stringent, again being set by the natural width of the W boson, where a dijet resonance is now reconstructed from the quark rather than the leptonic decay of the W boson.

$$W^+ \rightarrow u + \bar{d}, c + \bar{s},$$
$$\Gamma_W/M_W = 2.6\%, \tag{21}$$
$$(dE/E)_{\mathrm{HCAL}} \sim 1.1\%.$$

In practice, such precise energy performance, with an implied stochastic coefficient of $\sim 17.4\%$, at an energy of $0.25\,\mathrm{TeV}$, is not attainable. One reason is the small number of particles in a hadronic shower with the resulting large stochastic fluctuations in that number. The analog of the critical energy for ECAL is the threshold energy to make more secondary pions, $E_{\mathrm{th}} \sim 2m_\pi = 0.28\,\mathrm{GeV}$. That energy scale leads to an estimate for the number of particles in the hadronic shower of $\sim E/E_{\mathrm{th}}$, which implies a stochastic coefficient of 53%, or 3.35% fractional energy resolution, dE/E for $E = 0.25\,\mathrm{TeV}$, far from the requirement set by the natural resonance width of 1.1%.

One attainable goal for HCAL is, rather, to do a good job of isolating the strongly produced top pair reducible background to the electroweakly produced W pair signal. The vertex system provides help in removing this background by making "b tags" of events with secondary vertices for b jets, which can then be removed. A clean sample of top pairs with hadronic W decay also allows for a measurement of the dijet mass resolution at the W mass scale and an *in situ* energy scale calibration point for the calorimetry, $t \rightarrow W^+ + b \rightarrow (u + \bar{d})_W + b$.

The total depth of the HCAL need not be as deep as that needed to fully contain the ECAL shower, because the hadron calorimetry does not achieve the desired 1.1% precision energy measurement. A plot of the depth needed for an average 95% and 99% shower containment as a function of energy is shown in Fig. 16.

Since the physics requirement is to fully explore $W + W$ scattering at 1 TeV mass, the quarks from the W decay have a transverse energy of about 0.25 TeV. Asking for 99% shower containment requires about 10 λ_0 total calorimeter depth, or about 9 λ_0 in the HCAL. The generic HCAL has an absorber depth of >1.5 m of steel or >8.9 λ_0, which meets this specification. In fact the "jet" of hadrons from the quark hadronization means that the quark energy is shared over several final state hadrons, which makes the depth requirement somewhat less demanding.

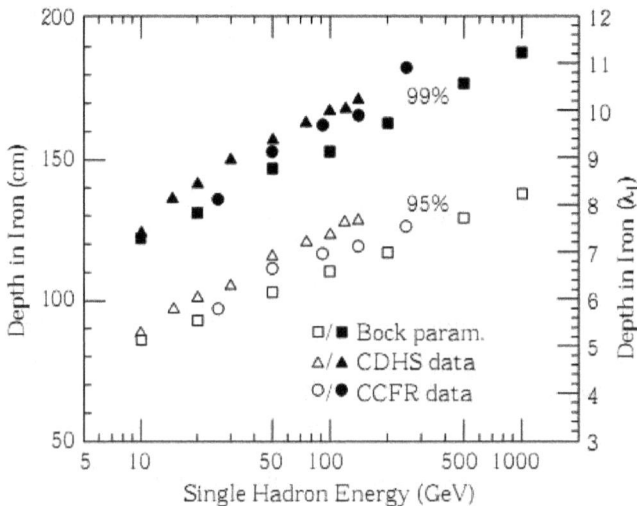

Fig. 16.   Depth needed for a shower energy containment of 95% and 99% as a function of hadron energy. Note the logarithmic dependence of depth on incident energy.[8]

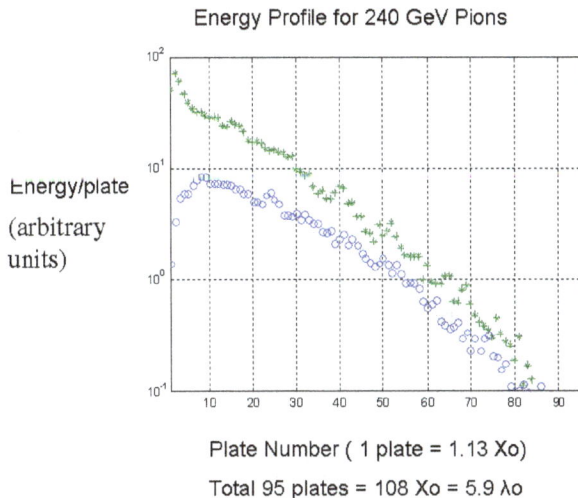

Energy Profile for 240 GeV Pions

Plate Number (1 plate = 1.13 Xo)

Total 95 plates = 108 Xo = 5.9 Λo

Fig. 17.   Data on the energy profile for charged pions with 240 GeV energy, without subtracting the initial interaction point, o, and after subtraction,*.

Data on the longitudinal shower development for 240 GeV pions in lead are illustrated in Fig. 17. The existence of two distance scales is clear. At each interaction in the developing shower neutral pions are produced, which are then absorbed over a distance scale X_0. The produced charged pions move the energy in depth to the next interaction point with a distance scale λ_0, which is about 20 times larger than the radiation length for lead. To study the hadronic shower development in detail as a function of depth would require fine sampling on the scale $\delta t \sim 1$, which is roughly the scale of the sampling frequency of the data points in Fig. 17.

The hadronic shower is contained transversely in a cone radius roughly of size $\sim \lambda_0$. Assuming that overlapping hadronic showers cannot be resolved, a size $\delta z = \delta s = 15$ cm is chosen in the generic HCAL. The generic HCAL begins at a radius $r_H = 1.6$ m. The HCAL towers then subtend a rapidity and azimuthal interval:

$$\delta\eta = \delta\phi = 0.094 \sim \lambda_0/r_H. \tag{22}$$

The larger transverse tower size, quoted earlier in Eq. (7), leads to a larger probability for an HCAL tower to be occupied by a charged pion, $(D_c = 6)(N_I = 25)(\delta\eta)^2/2\pi = 0.21$. The mean transverse energy in an HCAL tower is then 0.16 GeV.

The number of barrel HCAL towers in z is $10\,\mathrm{m}/0.15\,\mathrm{m} = 67$ and in azimuth is 67 or 4490 HCAL readout towers in the barrel. With three depth segments read out independently, there are 13,470 towers in total. Note that the showers in both the ECAL and HCAL calorimeters develop longitudinally on the scale of the radiation length so that a complete knowledge of the hadronic shower should logically have that depth segmentation. However, that level of detail has not been attempted in the LHC calorimeters.

As regards radiation, the HCAL with respect to the ECAL is at larger barrel radius but there are twice as many charged as neutral pions. The hadronic shower energy is more spatially spread out than the electromagnetic shower, Fig. 17, where the initial interaction point has a mean of one and an r.m.s. of one λ_0. Therefore, the full width of the deposited energy is roughly $\Delta\lambda_0 \sim 2\lambda_0$. Taking the factors together, the dose in HCAL is about $1/3$ that of ECAL at the same rapidity.

However, the need to cover angles as small as $|y| = 5$ arose from the requirement that the generic detector measure most of the inclusive inelastic pions and measure the forward jets from the vector boson fusion process. This requirement places a large radiation burden on the forward calorimetry. It covers the range in $|y|$ from 2.5 to 5.0. At $|y| = 4$ the average pion energy is $22\,\text{GeV}$. Approximately $77\,\text{W}$ of power is deposited by secondary particles that heat the forward calorimetry. The luminosity can be measured in the forward calorimeter with a thermometer!

At $|\eta| = 5$, the polar angle is $0.75°$. At a location of $z = 10\,\text{m}$ for the forward calorimetry the radius in the (x, y) plane is only $13\,\text{cm}$ and a "tower" with the same $\delta\eta$ as the barrel HCAL would have an extent of only $1.2\,\text{cm}$ in r and $7.7\,\text{cm}$ in azimuth, s. Clearly, such fine segmentation, given the physical size of a hadronic shower, is not very useful. Indeed, the pileup probability for a tower of transverse size $\sim \lambda_0$ is quite a bit higher than that for the barrel HCAL.

The radiation dose in the forward calorimetry may be roughly estimated to be [Eqs. (19) and (20)]

$$(\text{Dose})_{\text{forward}} \sim \sigma_I L T D_0 (1/2\pi z_F^2 \theta^3)\langle P_T\rangle/(\rho_{\text{forward}}\Delta t X_0). \qquad (23)$$

Taking only the most densely deposited neutral energy to be the full radiation dose, the z for the calorimetry, z_F, to be $10\,\text{m}$ and assuming that it is made of steel, the dose is approximately $280\,\text{mrad/yr}$ at $|y| = 5$. The size of the dose sets the limit on the calorimetric coverage; $|y| < 5$ is the limit for long term survivability of the forward calorimetry. As can be seen in Fig. 5, there is a falloff of inclusive particle production at large $|y|$, which somewhat reduces this estimated dose.

The primary HCAL function is to measure the jets of hadrons from the "hadronization" of colored quarks and gluons. Data on the fraction of the

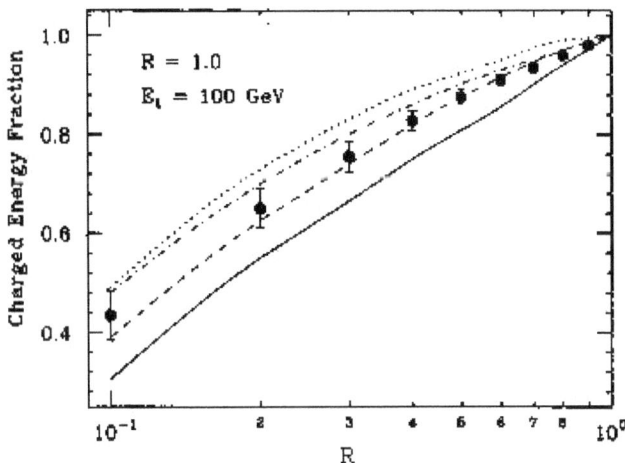

Fig. 18.   Fraction of the total charged particle energy for a jet which is found within a cone of radius R for 100 GeV transverse energy jets.

jet energy carried by those hadrons and captured in a cone of radius R ($R = \sqrt{\delta\eta^2 + \delta\phi^2}$) are shown in Fig. 18. In order to capture more than 90% of the jet energy for a 100 GeV jet, a cone of $R > 0.5$ is required. Indeed, the colored quarks and gluons can always radiate a gluon by the process of "final state radiation" (FSR). The fraction of jets which have radiated a fraction ε of their energy outside a cone of radius R is approximately[12]

$$\text{Fract} \sim (\alpha_s/\pi)[3\log(R) + 4\log(R)\log(2\varepsilon) + \pi^2/3 - 7/4], \qquad (24)$$

where α_s is the QCD fine structure constant, numerically ~ 0.1. For example, a 10% radiation of the total jet energy outside a cone of $R = 0.5$ occurs $\sim 12.5\%$ of the time. Clearly, a limited cone size, if used to estimate the true jet energy, has a limited energy resolution due to fluctuations in the energy radiated outside the cone.

A Monte Carlo model of the W hadronic decay into quark pairs which then hadronizes into dijets is given in Fig. 19. Clearly, the FSR of gluons is an irreducible limitation on doing precision dijet spectroscopy in the environment of a hadron collider.

In order to reduce the problem of FSR the cone size could be increased to capture the radiated energy. However, the pileup energy for inelastic interactions increases with the cone area. For example, a cone of radius $R = 0.5$ contains ~ 1 unit of rapidity and azimuthal angle or $N_I D \langle P_T \rangle / 2\pi = 28.6$ GeV of pileup pions. The pileup energy will fluctuate event by event. On average there are 36 pions in the cone and if their number is stochastic the

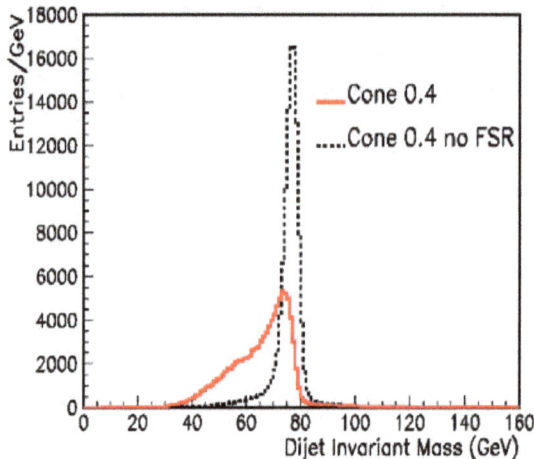

Fig. 19.   Monte Carlo study of W decay into dijets and the resulting mass resolution both with and without FSR.

fluctuations are 6 pions or 4.8 GeV in transverse energy. That energy fluctuation implies a limitation on the achievable dijet mass resolution and argues against increasing the cone size much beyond $R \sim 0.5$ at full LHC luminosity.

It might be imagined that setting a low threshold on the tower energies would remove the pileup issue and allow an increase in R so as to recover the FSR energy radiated at large R. However, low momentum or "soft" particles are crucial for the correct measurement of jet energies. A jet hadronizing into hadrons which carry a fraction z of the jet momentum can be described by a distribution $D(z)$ parametrized as $D(z) \sim (1 - z)^a/z$, where $a \sim 5$.[13] The fraction, F, of the momentum of the jet contained in particles with momentum fraction $< z_{min}$ is

$$F \sim 1 - (1 - z_{min})^{a+1}, \qquad z_{min} = (p_{had})_{min}/P_{jet}. \tag{25}$$

As an example, a 50 GeV jet has $\sim 45\%$ of its energy carried by hadrons with momenta less than 5 GeV and $\sim 12\%$ carried by hadrons with momenta less than 1 GeV. Thus, the soft hadrons from the jet are easily confused with the soft pions from the pileup, which then limits the achievable jet energy resolution.

An additional issue is that the low transverse momentum charged hadrons do not even reach the calorimetry and register their energy. These "loopers" must be efficiently detected in the tracking system and the measured jet energy incremented to properly account for them.

These considerations imply that precision multijet spectroscopy is difficult at the LHC. In a more benign environment such as the proposed ILC,[14] with much less pileup and no "underlying event," improved calorimetry with greatly expanded numbers of shower samples has been proposed,[15] which aims to improve the calorimetric energy resolution by a factor of roughly 2 with respect to the LHC detectors. Using the more precise tracking measurement of low charged particle energy ("particle flow") can also improve the overall energy resolution of the detector. Another potential path to improved performance is "dual readout calorimetry" where energy measurements of the charged and neutral components of a hadronic shower are measured independently, thus allowing the different calorimetric response of these two components to be compensated for.[16]

The calorimetric performance limitations mentioned above have implications for the reducible top pair background to electroweakly produced W pairs. At the LHC, assuming an HCAL stochastic coefficient of 60% and a constant term of 5%, a top pair event where one W decays leptonically has a mean MET of 52 GeV with a MET error due to the calorimeter resolution of ~ 17 GeV, while the W mass distribution from quark dijets has a standard deviation of ~ 5 GeV even in the absence of the pileup.

11. The Neutron Field

One of the characteristics of a hadron collider detector is the existence of a large neutron field. Both the detectors and electronics must be designed to be cognizant of this field. The inclusive particle flux in the small angle region is the major source of the neutrons. The results of a detailed Monte Carlo model are shown in Fig. 20.

The charged particle flux is estimated to be $\sigma_I LT D_c(1/2\pi r_F^2) = 9.5 \times 10^{11} \pi^{\pm}/\text{cm}^2\,\text{yr}$ at $r = 1\,\text{m}$, in rough agreement with the results of Fig. 20. The expected steep falloff with increasing r is also observed. Note that at an angle of $\theta \sim 0.1$ ($r = 1\,\text{m}$ at the forward calorimeter starting location of $z = 10\,\text{m}$), the pions have a mean energy of 8 GeV.

At a hadron collider there are many neutrons produced because the hadrons during the strong interaction showering process excite the calorimetric medium. That leads to neutrons when the nuclei de-excite. As a crude rule of thumb there are about five neutrons with a-few-MeV kinetic energy produced per GeV of absorbed hadrons. This then leads us to estimate a neutron flux of $3.82 \times 10^{13}\,\text{n/cm}^2\,\text{yr}$ at a 1 m radius, which is again in reasonable agreement with the precise results shown in Fig. 20.

The neutrons will elastically scatter, slow down and thermalize. In an elastic scatter off a nucleus with mass number A the mean fractional energy loss for a neutron is $\sim 1/(A + 1)$. Clearly, for heavy nuclei, the neutrons simply scatter and do not lose energy. On light nuclei, the neutrons transfer

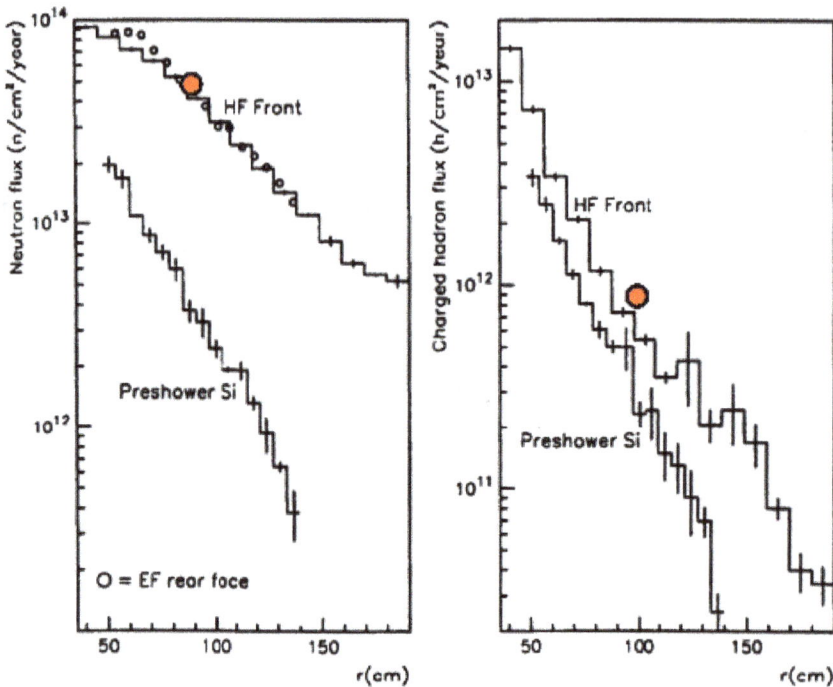

Fig. 20. Charged particle flux (right) and neutron flux (left) as a function of the radius for calorimetry at $z = 10$ m.[4] The dots are the approximate predictions.

a significant energy fraction to the nucleus. The neutrons can be described as a fluid and this fluid can "flow" to all parts of the detector.

Detector design must take the neutrons into account as they can be exothermically captured and are a source of an additional dose for both detectors and electronics. In fact, the design of shielding is quite important at the LHC. Complex shielding using polyethylene, borated concrete and lead is a feature of the experiments. The polyethylene is light and slows down the neutrons, the boron captures them and the lead absorbs the emitted capture photons.

If the forward calorimetry were brought closer to the interaction point in z to 3.2 m to be an extension of the endcap geometry (Fig. 9) the dose would then be higher by a factor ~ 9.8. Moreover, a tower of size $\delta r \sim \lambda_0$, which was chosen on the basis of hadronic shower extent would subtend an angle of 0.052 radians, whereas a tower of size $\delta \eta = 0.094$ subtends an angle $\delta \theta = 0.0013$ at $\eta = 5$. Clearly, it becomes more difficult to contain the hadronic shower in a tower with a small pileup occupancy as the rapidity increases or as the tower distance from the origin decreases.

In addition, the calorimeter is an immense neutron source and the neutrons can flow into other systems. Specifically, they can make an impact on the tracking and other precision systems unless special precautions are taken.

12. The Generic Muon System

A major physics goal at the LHC is to explore vector boson interactions. Leptonic decays of the bosons are the favored mode, since the lepton backgrounds are low and the leptons can be cleanly and redundantly measured. The muon vector momentum and position are measured first in the tracking system. The muons then deposit only ionization energy in the ECAL and HCAL, while the other final state particles are almost completely absorbed. In fact, the calorimetry can perform a useful muon identification role if the muons are isolated from other final state particles. Ideally, in the muon system which follows the calorimetry in depth, only muons exist along with noninteracting neutrinos.

Muons at low transverse momentum arise from the decays of inclusively produced pions and kaons. These muons can largely be removed as a background by requiring a good tracker fit to the hypothesis of no decay "kink" in the found track. For example, in the decay $\pi \rightarrow \mu + \nu$ the muon transverse momentum relative to the initial pion direction is ~ 0.03 GeV. Therefore, a 10 GeV pion which decays in the tracker will have a 3 mrad "kink" in the full track. In discussing the generic tracker an angular resolution of 0.12 mrad was quoted. Therefore, decays can be removed using the tracker goodness of trajectory fit up to a few hundred GeV. Since the probability of decaying in the tracker falls with increasing momentum, pion and kaon decays are not a major problem at the high momentum scales which are of major interest to the LHC experimenters.

A second source of muons is heavy flavor decay, such as $b \rightarrow c + \mu + \nu$, which occurs much more promptly and within or prior to passage through the vertex detectors (Fig. 10). The calculated LHC cross section in leading order for production of a b quark pair where one b decays into a muon with transverse momentum greater than 10 GeV is $\sigma_\mu \sim 60 \, \mu b$. At the LHC design luminosity that cross section corresponds to a rate of $R_\mu = \sigma_\mu L \sim 0.6$ MHz, which is small with respect to the total inelastic LHC rate of 1 GHz but too large with respect to an acceptable trigger rate.

The requirements for the muon system are similar to those for the tracking system. There should be a good momentum measurement up to ~ 0.25 TeV for the muons. In addition, the system must produce a trigger so as to reduce the rate of background muons from the heavy flavor decays.

The ionization energy loss in the calorimetry provides a lower momentum cutoff for the muons. The generic HCAL contains 1.5 m of steel, which stops all muons with a momentum less than ~ 1.7 GeV. However, that cutoff is not sufficient to reduce the muon rate in the muon system to an acceptable level. Therefore, the muon system must measure the muon transverse momentum accurately, set a threshold value — say, 10 GeV — and report it to the trigger system.

In the muon system nearly all the particles are muons, but they are of low transverse momentum and arise from heavy quark flavor decays. Since the ionization range cutoff is in energy, while the intrinsic muon rate is controlled by the transverse momentum, the rates in the forward muon systems are much larger than the rates at wide angles. The beam halo from upstream interactions in the LHC accelerator also makes the forward muon region more difficult. For these reasons, the generic muon system is thought to cover the same limited rapidity range as the tracking system, $|y| < 2.5$.

The steeply falling muon spectrum from heavy flavor decays means that the muon trigger system momentum resolution is very important. Poor resolution lets in a large number of triggers from lower momentum muons measured to have a higher momentum because of finite resolution, thus increasing the trigger rate. For example, if the real muon rate is characterized by an exponential falloff, $d\sigma/dP_{T\mu} = ae^{-P_{T\mu}/P_0}$, the observed rate is increased by a factor due to feeddown of $e^{(\Delta P/P_0)/2}$, where ΔP is the muon trigger system resolution. Clearly, the trigger resolution should be rather less than the characteristic muon falloff momentum scale, P_0.

The physics process of interest is $W + W$ electroweak pair production. This process has a cross section evaluated at leading order of ~ 80 pb and top pairs provide a strongly produced reducible background with a cross section of about 680 pb in leading order perturbation theory. The muon spectrum from the decay of the W is shown in Fig. 21. The muon spectrum is strongly peaked at a transverse momentum of ~ 40 GeV, a "Jacobean peak" arising from the kinematic behavior of the two-body decay of W bosons which themselves have a small transverse momentum, $P_{T\mu} \sim M_W/2$.

There is a complication for very high momentum muons. At the LHC their momenta are sufficiently large that radiation is an issue. The muon is a heavy electron, so that it experiences the same forces as the electron, but it has a much-reduced acceleration. Radiation scales as the square of the acceleration, so that the critical energy for a muon is much larger, but still finite, compared to the critical energy for an electron.

$$(E_c)_\mu \sim (E_c)_e (m_\mu/m_e)^2. \tag{26}$$

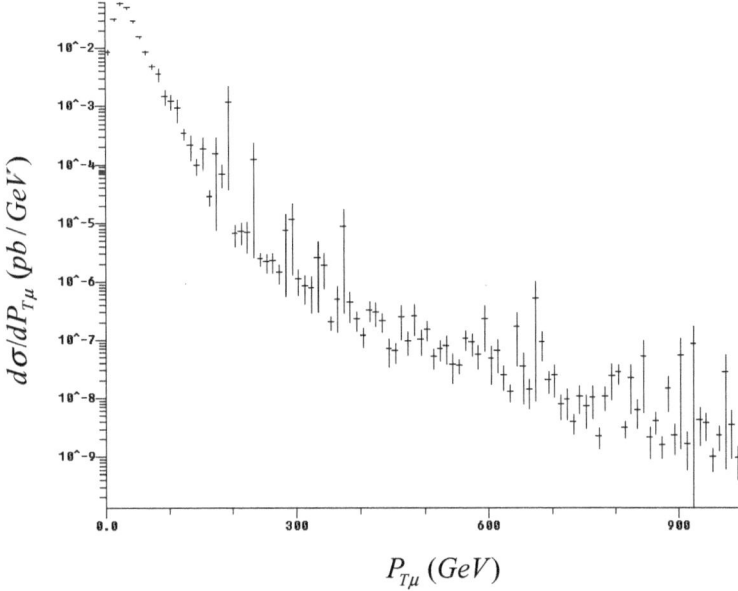

$P_{T\mu}$ (GeV)

Fig. 21.   Muon transverse momentum spectrum in W pair electroweakly produced events with no cuts on the W pair mass.

For iron, the electron critical energy, 22.4 MeV, scales approximately to 957 GeV for muons. The exact energy loss for muons in iron is shown in Fig. 22. Clearly, for muons with ~ 300 GeV momentum the radiative processes of bremsstrahlung emission and pair production are as important as ionization. This makes good tracking more difficult for the muons if the muon system tracking stations are in the solenoid flux return yoke. Note that muons in the Jacobean peak of Fig. 21 with $P_{T\mu} \sim 40$ GeV have a momentum of ~ 240 GeV at $|y| = 2.5$, at the limit of muon angular coverage. Hence, in the endcap region of the muon system many of the muons in the desired signal will be above the critical energy.

In general, those muons will radiate, which then requires multiple measurements of their trajectories since the associated shower of collimated particles will spoil some of the measurements. A robust set of measurements can be achieved by isolating each set of measurements by interposing material of depth $\Delta t \sim 10$, which serves to decouple each set from the others. An alternative is to operate in a large magnetic field in air which acts to sweep away the low energy showering particles.

The particle rates outside the calorimetry are sufficiently low that drift chambers with large drift distances can be used, as opposed to the other subsystems in the generic detector. If the chambers are operated in "air,"

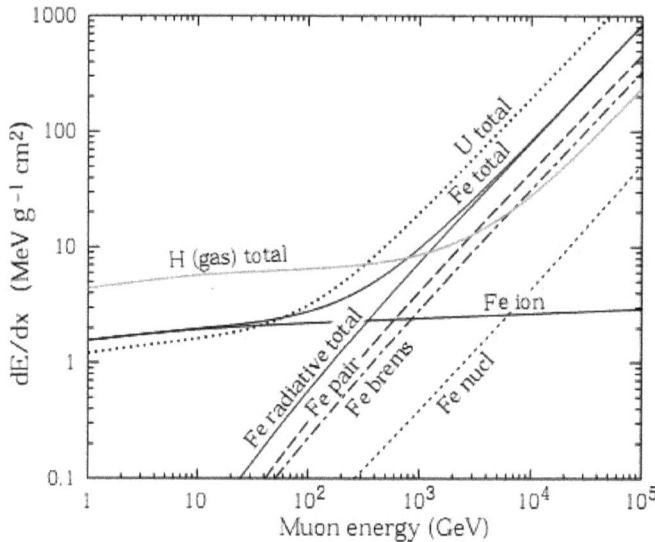

Fig. 22. Energy loss of muons in iron showing ionization, pair production and bremsstrahlung contributions to the total rate of energy loss as a function of muon energy.[8]

then the accuracy required is similar to that for the tracker and position resolution and alignment specifications are stringent. If the chambers are operated in the flux return yoke of the solenoid, then the redundant muon momentum measurement is multiple-scattering-limited [Eq. (15)]. In that case the fractional momentum error is constant for momenta below which the alignment errors are a negligible contribution

$$dP/P = (E_s/\sqrt{2})[1/eB\sqrt{LX_0}]. \eqno(27)$$

For example, with $B = 2\,\mathrm{T}$ in steel with an $L = 2\,\mathrm{m}$ return yoke, the muon momentum is measured to 13%. The tracking provides [Eq. (14)] a redundant momentum measurement with $< 10\%$ momentum error for muons with momenta $< 1.3\,\mathrm{TeV}$ in comparison. At still higher momenta the tracker resolution increases as P, so the muon system will improve the tracker muon momentum resolution, if good alignment of the muon chambers can be maintained so that the measurement remains multiple-scattering-dominated.

For the generic detector no attempt is made to specify the muon system in detail as to detectors, the magnetic field type, or the medium in which the detectors are immersed. Many different choices are possible and defensible, and there is no consensus on the design choices. Specific choices for ATLAS and CMS are discussed in subsequent chapters.

13. Up the Food Chain

The discussion so far has been about the connection between physics needs and the resultant generic detector requirements. Many of the detailed choices made by ATLAS and CMS differ. In many cases those choices have broken new ground in detector development, and these aspects of the detectors and the associated design decisions will be the focus of later chapters.

The front end electronics at the LHC experiments is clocked at the r.f. bunch crossing frequency of 40 MHz. As seen in the discussion above, the radiation field is typically 1 mrad/yr or larger, which means that the electronics on the detector must be quite radiation hard. The exact front end electronics choices are very detector-dependent, so that a "generic" discussion is not very illuminating and has not been attempted here. The electronics must also be resistant to the large neutron background, which is a characteristic of hadron colliders.

The inclusive rate at LHC design luminosity is 1 GHz. The number of interactions capable of being stored in permanent media is ~ 100 Hz. Therefore, a reduction in the rate of a factor of 10 million is needed. This is accomplished in steps. First, the rate is reduced by "triggering" on leptons and jets above some transverse momentum thresholds. Imposing these thresholds on the events reduces the rate to ~ 100 kHz. In order not to incur dead time a front end "pipeline," which stores all the data until the first stage trigger decision is made, is provided for each channel of data which is read out. The pipeline length is, say, 128 LHC clock cycles or 3.2 μs. There are ~ 100 million independent channels of information, mostly analog, in the generic detector. Typically, after this initial decision the full event is sent from the detector front ends off the detector by means of digital optical fiber data transmission to a set of digital electronics accessible to the experimenters.

More incisive trigger decisions are then made which reduce the event rate to ~ 100 Hz. Even after suppression of detector elements with no hits or with signals below some low threshold, the large number of particles per event and the large number of events piling up lead to a typical event size of 1 Mb. Assuming a data-taking run of ~ 4 months per year, 1000 Petabytes/yr, or ~ 1 million CDs, are stored for offline analysis. The trigger and data acquisition techniques for both ATLAS and CMS are more or less specific to the experiment and thus are not susceptible to a "generic" analysis. Again, discussions specific to the experiments appear in the later chapters of this volume. In fact, after the first level of triggering, commercial off-the-shelf modules are mostly deployed. Examples are telecommunications switching networks employed to assemble the full event using input from the different subsystems, and "farms" of commodity PCs which are used for online high level triggering and for offline analysis. Thus, this is a good point to turn the exposition to the detailed choices made by ATLAS and CMS in reaching for their goal of decisively confronting the new physics which will be opened up at the LHC.

Acknowledgments

The subsequent chapters in this volume refer to the design decisions made by the ATLAS and CMS collaborations. A detailed description of the detectors and their respective full authorship lists are given in Ref. 6. All the ATLAS and CMS authors are here thanked for their contributions to the design, construction and installation of these two state-of-the-art detectors.

References

1. http://en.wikipedia.org/wiki/ATLAS_experiment and links therein; http://en.wikipedia.org/wiki/Compact_Muon_Solenoid.
2. http://lhc.web.cern.ch/lhc.
3. http://ab-div.web.cern.ch/ab-div/Publications/LHC-DesignReport.html.
4. http://public.web.cern.ch/PUBLIC/en/LHC/CMS-en.html.
5. http://atlas.web.cern.ch/Atlas/index.html.
6. The CERN. Large Hadron Collider: accelerator and experiments, *J. Instrum.* **3** (2008) SO8001–SO8007.
7. D. Froidevaux and P. Sphicas. General-purpose detectors for the Large Hadron Collider, *Annu. Rev. Nucl. Part. Sci.* **56** (2006) 375.
8. C. Amsler *et al.* Review of particle physics, *Phys. Lett. B* **667** (2008) 1.
9. http://en.wikipedia.org/wiki/Dark_matter and links and references therein.
10. J. Gunion, H. Haber, S. Dawson and G. Kane. *The Higgs Hunter's Guide,* (Westview, 2000).
11. http://en.wikipedia.org/wiki/CompHEP and links and references therein.
12. G. Sterman and S. Weinberg. Jets from quantum chromodynamics, *Phys. Rev. Lett.* **39** (1977) 1436.
13. Fragmentation. Underlying event and jet shapes at the Tevatron CDF), in *Hadron Collider Physics 2005* (Springer, 2006).
14. http://www.linearcollider.org/cms/?pid=1000000 — home page for the ILC.
15. http://physics.uoregon.edu/~lc/wwstudy/concepts/ — links to four design concepts.
16. N. Akchurin *et al.*, *Nucl. Instr. Meths. A* **537** (2005) 537.

Chapter 2

THE CMS PIXEL DETECTOR

W. Erdmann

Paul Scherrer Institute, Laboratory for Particle Physics
5232 Villigen PSI, Switzerland
wolfram.erdmann@psi.ch

The inner tracking detectors at the LHC operate in a region of unprecedented particle rates. Thousands of hits must be detected, time-stamped and stored in every 25 ns bunch crossing interval. The silicon pixel detector for the CMS experiment has been designed to meet the requirements of position resolution, rate capability and radiation tolerance with a minimal amount of material. Its unique ability to provide three-dimensional high precision space points plays an important role in the tracking system of the CMS detector.

1. Introduction

The identification of long-lived particles through the detection of secondary vertices and impact parameters is an important ingredient of a high energy physics experiment. Vertex detection of particles with submillimeter decay length requires precise track measurements close to the production point. The feasibility of a tracking detector in the hostile environment near the beam of the LHC was unclear during the initial stages of planning of the CMS experiment.[1]

Energy and luminosity targets for the LHC lead to particle fluxes far beyond those encountered at other colliders. Every cm^2 of a detector close to the beam pipe is expected to be hit by 100 million particles per second when the LHC runs at the full design luminosity of $10^{34}\,cm^{-2}s^{-1}$. The smallest radius at which a traditional silicon microstrip detector can function in such an environment is limited by occupancy and radiation damage. Reducing the size of the sensor elements brings occupancy to a manageable level. It also mitigates the problems of radiation damage of the sensors. Smaller detector capacitance permits high signal-to-noise ratios even when charge collection degrades. Leakage currents per channel remain small enough to be absorbed in the front-ends. Extremely short strips or pixels are therefore an inevitable choice for a vertex detector at the LHC.

Pixels with comparable dimensions in the transverse plane and the z direction become two-dimensional arrays and introduce a new quality to tracking and technology. Hits detected by the pixel detector are true space points with considerable benefits for pattern recognition. The absence of ambiguities connected with projections makes the pixel detector the preferred starting point for track finding in CMS.[2] The relative simplicity and robustness of reconstruction have also led to applications in the High Level Triggers.[3] The CCD detectors of SLD and NA32 had proven the power of pixelated tracking detectors, but this technology cannot be pushed to cope with LHC rates. New territory for electronics, sensor and interconnect technology was entered for building the CMS pixel detector.

The CMS pixel project started in 1994 as a collaboration of PSI with Swiss and Austrian universities.[a] A parallel effort existed at UC Davis. US groups[b] took responsibility for the forward endcaps of the pixel detector.

The work of the first years focused on very basic R&D on sensors and radiation-hard readout chip concepts. Detailed studies of charge collection and Lorentz drift in heavily irradiated silicon sensors were made using the grazing angle method.[4,5] Power consumption, speed, and noise of analog stages realized in radiation-hard CMOS processes were evaluated.[6,7] Building blocks of the readout and small prototype chips followed. A fine pitch bump-bonding process suitable for the target pixel size was developed at PSI.[8]

The mechanical detector layout and a detailed module design were developed for the Technical Design Report[9] (TDR) in 1998. A first realization of the readout architecture in a full readout chip was a milestone in 2002. The PSI43 chip,[10] which uses the radiation-hard $0.8\,\mu m$ DMILL[11] process, came close to meeting the requirements but would have needed at least one more iteration. By then it had become clear that deep submicron processes can provide the required radiation tolerance[12] while offering a number of advantages.[13] The PSI43 architecture was improved and translated to a commercial $0.25\,\mu m$ process. After test submissions in 2003 and 2004 the main production of the PSI46 chip[14] started in 2006.

Detector modules were produced between mid-2006 and the end of 2007 at PSI for the barrel, and in the US for the endcaps. The forward detector was integrated at Fermilab and shipped to CERN, where it was fully

---

[a]ETH Zürich, University of Zürich, University of Basel, HEPHY Vienna.
[b]Northwestern, Rutgers, FNAL, Purdue, John Hopkins, Mississippi, Davis. Later joined by Nebraska, Iowa, Kansas State, Cornell, Milan.

commissioned and tested in 2008. The barrel detector was assembled and tested at PSI and shipped to CERN in June 2008. Both detector parts were installed and commissioned[15-17] successfully in July 2008.

The following section gives an overview of the pixel system and defining requirements. More detailed discussion of individual topics will follow in subsequent sections.

2. Overview

One main motivation for a tracking detector close to the beam-line is the identification of secondary vertices of b quarks and possibly τ leptons. This requires an impact parameter resolution on the order of $100\,\mu m$ or better. The determining factors for the impact parameter measurement are the position resolution of the innermost detector and the extrapolation uncertainty. The latter is largely due to multiple scattering in the beam pipe and detector, and it grows with the distance between the interaction point and the first measurement. The obvious primary goal is therefore to provide a position measurement with a precision much better than $100\,\mu m$ at the smallest possible radius with the smallest possible amount of material. The power dissipation of the electronics must be minimized, because cooling and cabling make a significant contribution to the material budget. Material in outer layers is less important for the impact parameter resolution, but must be kept minimal, too, in order not to degrade the overall performance of the experiment.

More specific for a CMS pixel detector are the requirements of radiation hardness to at least $6 \times 10^{14}\,n_{eq}/cm^2$ and (Subsec. 2.3) a readout architecture that handles 40 MHz bunch crossing frequency with 20 simultaneous pp collisions. Only a small fraction of the bunch crossings will be readout, but all hits must be stored during the $3\,\mu s$ latency of the CMS trigger. Trigger rates up to 100 kHz are foreseen.

2.1. *The pixel shape*

A central decision in the development of the CMS pixel detector was to exploit charge sharing among pixels to improve the position resolution. Significant charge sharing is a consequence of the Lorentz drift in the strong magnetic field of 4 T inside CMS. Charge carriers released by the ionizing particle in the silicon sensor do not follow the electric field lines to the collection electrodes, but are deflected by the Lorentz force (Fig. 1). The n side readout of the silicon sensors, chosen for its radiation tolerance, leads to a

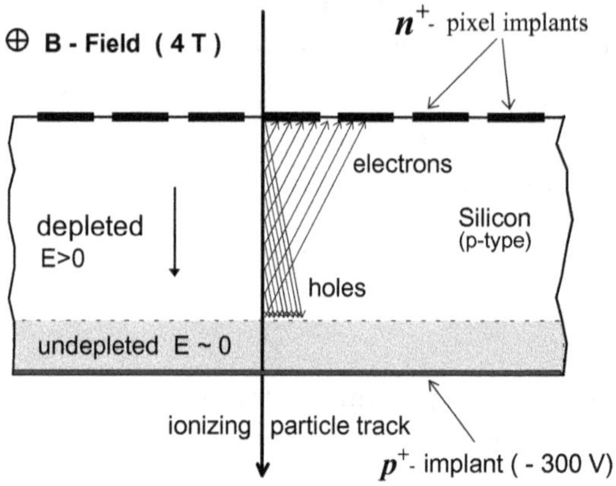

Fig. 1. Illustration of charge sharing induced by Lorentz drift in the CMS pixel barrel detector.[9] For improved position resolution the charge should be collected by at least two pixels. After irradiation, the detector stays functional beyond the point where it cannot be fully depleted, but the amount of charge and the charge sharing are reduced.

larger Lorentz angle because of the higher mobility of electrons compared to holes. Interpolating positions between pixels based on the amount of the collected charge requires the transmission of analog pulse-height information. The analog optical link developed for the CMS strip tracker naturally offers that possibility without the need for digitization in the front-end.

The position resolution of single-pixel hits is given by the pitch divided by $\sqrt{12}$. Two-pixel clusters and interpolation allow a much better resolution, limited only by fluctuations of the charge deposition. Dividing the signal charge among more than two pixels increases the data rate and reduces the signal charge per pixel but does not improve the resolution. An ideal choice of the pixel size in the direction perpendicular to the magnetic field $(r\phi)$ is therefore given by length over which charges are spread when they reach the surface of the sensor. For the usual $\sim 300\,\mu m$ sensor thickness and a Lorentz angle of $25°$ this amounts to $\sim 150\,\mu m$. A slightly smaller size of $100\,\mu m$ was chosen to maintain charge sharing, and hence resolution, after irradiation (Sec. 3).

The area of a pixel must be large enough to accommodate the readout electronics. With one dimension fixed by the Lorentz drift, this leads to a more or less quadratic shape of $100\,\mu m(r\phi) \times 150\,\mu m(z)$. Traditionally, collider detectors have much better resolution in the $r\phi$ plane, where the transverse momentum is measured from the bending of charged tracks. From the point of view of vertex reconstruction or pattern recognition there is

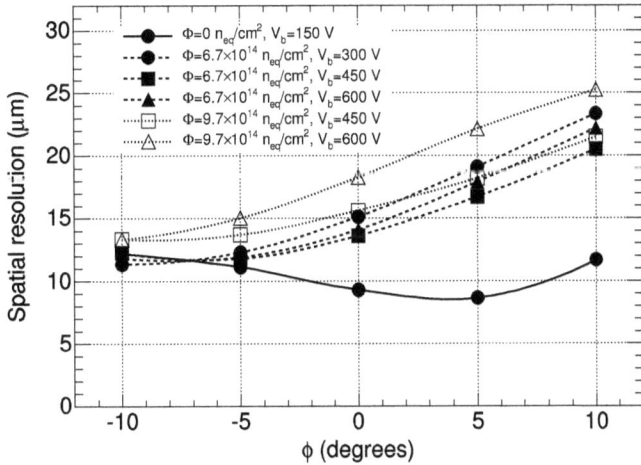

Fig. 2. Expected position resolution of the CMS pixel barrel in the r–ϕ plane as a function of the track angle of incidence.[2] The resolution degrades with increasing fluence and bias voltage because of the reduced charge sharing.

no reason to favor the transverse coordinate. The pixel shape of the CMS pixel detector results in comparable resolution in both directions. Another advantage of an (almost) square shape is the small circumference for a given area. This minimizes the pixel capacitance, which is important for noise, speed, and power dissipation.

There is no Lorentz drift in the direction parallel to the magnetic field (z direction), but sufficiently inclined tracks are detected in more than one pixel, allowing interpolation in both directions. At high rapidity, where tracks hit the barrel detector at low angles, the small z size is a disadvantage because increasing cluster size in the z direction is only beneficial for the z resolution until it exceeds two pixels. Higher multiplicities put a burden on the readout system without improving the resolution. Pixel disks therefore complement the barrel detector. Tilting the disk sensors away from the $r\phi$ plane introduces an angle between electric and magnetic field and hence Lorentz drift. Sufficient charge sharing is achieved with a tilt angle of 20° to reach resolutions of approximately $15\,\mu$m.

2.2. *Detector layout*

The inner layer of the detector should be mounted as close as possible to the interaction point. The smallest possible radius allowed by the CMS beam pipe is 4 cm. Further layers should be as far away as possible, as long as reliable matching of the outer track stub with the innermost hit is possible.

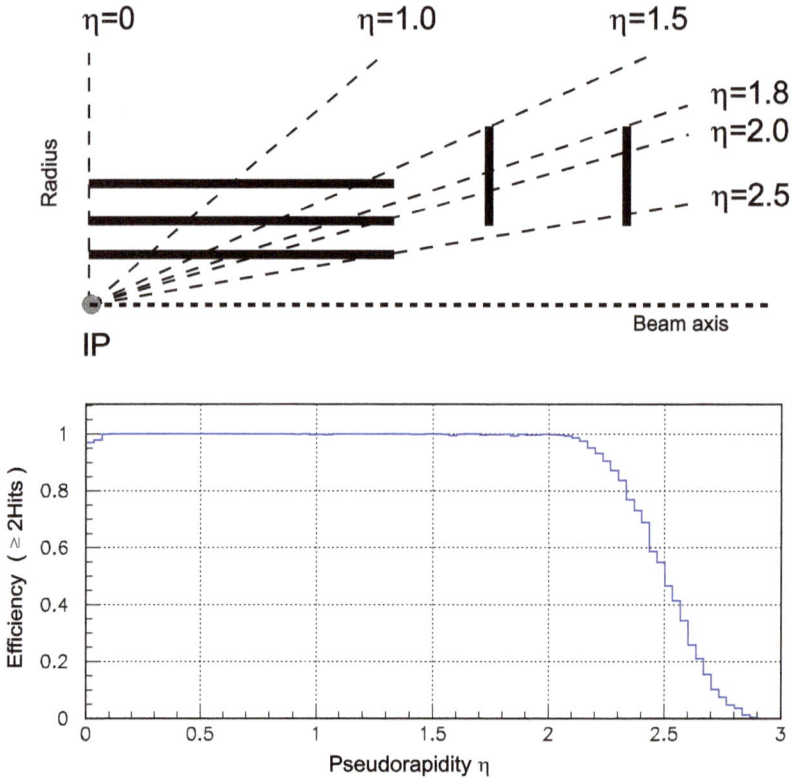

Fig. 3. Layout of the CMS pixel detector[18] (top). Three barrel layers at radii 4 cm, 7 cm, and 11 cm cover the central region. Two endcap disks at $z = 34$ cm and $z = 46$ cm extend the coverage to high rapidity. The region in which tracks are measured with at least two hits in the pixel detector (lower plot) is well matched to the acceptance of the outer tracker of CMS.

Two additional layers are placed at radii 7 cm and 11 cm. The full radius of the available volume was not used, in order to limit the total area (and hence cost) while providing three pixel hits per track.

The CMS tracker covers a pseudorapidity range of $|\eta| < 2.5$. The 4.4 cm layer, which is essential for the impact parameter resolution, nominally covers this range with its length of 56 cm. The acceptance begins to drop a little bit below $|\eta| = 2.5$ because of the length of the interaction region. A longer barrel is disfavored because the tracks crossing the detector at very shallow angles traverse more material and produce longer clusters.

The full η coverage in the outer layers would require a barrel length of up to 130 cm. The same coverage is achieved with much less detector area, and hence material, by supplementing disks in the forward region. A disadvantage of this configuration is that the bulkhead of the barrel, where a lot of material resides, comes to lie within the tracking volume. Two disks are placed on each side of the barrel at $z = \pm 34.5$ cm and $z = \pm 46.5$ cm. They have an inner radius of 6 cm and an outer radius of 15 cm. The disk positions optimize the probability of measuring two or more pixel hits on any track inside the tracker acceptance.

2.3. *Radiation hardness*

The radiation inside the CMS detector is a spectrum of charged and neutral particles coming from primary proton–proton collisions and from secondary interactions in the material of the detector.

Ionizing radiation leads to the accumulation of charges in CMOS electronics and sensor surface structures. The consequences of that must be mitigated by appropriate designs. All components of the radiation generate bulk damage in the silicon of the sensor, which determines the lifetime of the detector. The amount of damage depends on the particle type and energy. It is collectively expressed in terms of an equivalent number of neutrons with 1 MeV kinetic energy (n_{eq}).

The 4 cm layer in CMS is expected to be exposed to a fluence of $3 \times 10^{14}\, n_{eq}/\mathrm{cm}^2$ per year when the LHC operates at full luminosity. Radiation-hard electronics available at the time of the TDR was believed to be functional up to $6 \times 10^{14}\, n_{eq}/\mathrm{cm}^2$. This was taken as the target tolerance to be achieved by sensors and other components. The yearly fluences for the 7 cm and 11 cm layers are expected to be $1.2 \times 10^{14}\, n_{eq}/\mathrm{cm}^2$ and $0.6 \times 10^{14}\, n_{eq}/\mathrm{cm}^2$, respectively. While the lifetime of the 11 cm layer exceeds the projected 10 years of LHC operation, the 7 cm layer would need to be replaced at least once and the 4.4 cm layer every two years. Extracting the pixel detector in such intervals or even more frequently is also suggested by the possible need to bake out the beam pipe. Accommodating extraction and reinsertion during a few months' shutdown has an impact on the mechanical structure and the cabling of the pixel detector and CMS. Details of the installation procedure needed to be defined at an early stage of the project.

Originally, a staged installation was foreseen in which only the inner two layers were present during a low luminosity phase of the LHC. Later, when the LHC had reached full luminosity, the 11 cm layer would have been added and the damaged 4.4 cm layer removed. The staging scenario eventually became obsolete as the LHC startup slipped and the full three-layer detector was installed in 2008.

Advances in sensor material engineering and the availability of very-radiation-hard deep submicron CMOS technology have led to a detector that will be able to withstand significantly higher doses than anticipated at the time of the TDR.

3. Sensors

The sensors for the forward and barrel detectors were developed independently in cooperation with two different vendors. Both adopted the n-in-n concept, where the pixels are formed by high dose n-implants introduced into a highly resistive n-substrate. The junction is formed with a p-implant on the back-side. After space charge sign inversion, the bias voltage needed to fully deplete the sensor rises with irradiation. In contrast to a p-side read-out, the n-in-n design allows significantly underdepleted operation because the region of the high electric field has moved to the readout side after irradiation. Extremely high operating voltages can therefore be avoided. This reduces the problems of leakage currents and high voltage breakdowns in a highly miniaturized environment.

The n-in-n sensor implies the collection of electrons with their larger mobility compared to holes. This leads to the relatively large Lorentz angle that is a central part of the detector concept. It also reduces trapping, which begins to play a role at the highest targeted fluences.

The pn-junction on the back-side of the sensors must not extend into the diced edges of the sensor. Therefore, the back-side must be structured and double-sided processing of the wafers is mandatory. This leads to significantly higher costs but has another important advantage. The guard ring structure implemented on the back-side keeps the sensor edges near ground potential. This removes the risk of high voltage sparks across the 15 μm air gap between the sensor edge and the readout chip.

Additional processing is also needed on the readout side to electrically isolate the n-implants from each other. The electron accumulation layer induced by ionizing radiation otherwise tends to short-circuit the pixel implants. Two interpixel isolation techniques were evaluated, with good results. Open p-stops[19] were selected for the forward pixel detector and moderated p-spray[20] for the barrel detector.

P-stops are rings of high dose p-implants surrounding the pixels that interrupt the conducting layer. P-spray consists of a medium dose p-implantation "sprayed" without mask over the whole wafer. In the region of the lateral pn-junction between n-implant and spray, the dose is reduced

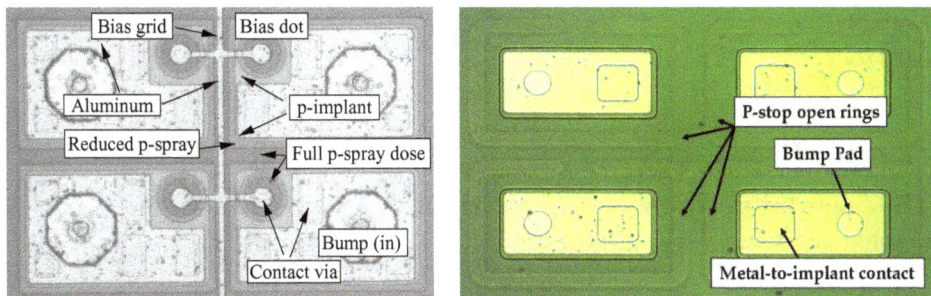

Fig. 4. Photo of four pixel cells with moderated p-spray (left) and open p-stop design (right). The pixel size is $100\,\mu m \times 150\,\mu m$.

by a factor of 2–3 ("moderated") to improve the high voltage stability of un-irradiated devices.

In order to keep unconnected pixels at a well-defined potential, the p-stop rings have openings allowing a controlled high resistive path between pixels. This path is important for testing of sensors before they are bump-bonded to readout chips. In case of individual bump-bonding failures it prevents the unconnected pixel from affecting the neighbors. The pixels in the p-spray technology are connected to a bias grid via a punch-through resistor for the same purpose. The resistor is formed by a small n-implant in the corner of each pixel that is separated from the pixel implant by a small gap.

The area between the pixels and the edge of the sensors is covered by an n-implant and connected to the chip ground via bump bonds. This ensures that the sensor edge is kept at ground potential and drains leakage currents from the guard ring region. On the back-side the high voltage is dropped over a series of guard rings between the edge and the p-implant of the active region. The distance between the outermost pixels and the edge of the silicon is 1.2 mm, which is small enough to allow wirebonding to the readout chips. The barrel sensors are arranged without overlap in the z direction, and to minimize the dead region an even smaller distance of 0.9 mm was chosen for those sides.

The high voltage stability of the sensors exceeds the maximum operating voltage of 600 V of the installed high voltage supplies. The detectors are going to be operated below room temperature to reduce the effects of reverse annealing. Oxygen diffusion as recommended by the ROSE collaboration[21] was applied to reduce the depletion voltage of highly irradiated sensors. Both types of sensors have been shown to be functional up to the original target fluence and beyond. A detection efficiency above 95% was observed after a fluence of $1.2 \times 10^{15}\, n_{\mathrm{eq}}/\mathrm{cm}^2$ with a detection threshold of 3000 electrons.[22]

4. Front-End Electronics

The front-end chip is a key component of the pixel detector, and its development accounts for a large fraction of the detector R&D. Its basic tasks are:

- Registering the signals produced by particles in the sensor;
- Storing time, position, and the amount of collected charge of all channels during the trigger latency;
- Sending out data for bunch crossings selected by the first level trigger.

A full readout of all channels, like it is used in the strip tracker, is ruled out by the large number of pixels. Zero suppression at a very early stage is mandatory, because buffering or transferring every channel for every bunch crossing is impossible. Ignoring signals of insignificant magnitude reduces local data rates by more than three orders of magnitude. However, good control of detection thresholds is needed in order to maintain efficiency and — because charge sharing is exploited — resolution of the detector. Active compensation of inevitable pixel-to-pixel and chip-to-chip variations complicates the design and operation of the detector considerably.

The size of the readout chip is limited by the manufacturer's reticle and yield considerations and is coupled to the size of the detector module. Multiple readout chips need to be grouped together to form modules of practical size. The floor plan is fixed by the need to tile chips without insensitive gaps. It is a matrix of pixel cells extending all the way to the edges of the chip on three sides. All auxiliary components, including input/output, are in the periphery on only one side (Fig. 5). High volume data flow from the pixels to the periphery suggests a column-oriented design.

The required radiation hardness originally seemed to make the use of a dedicated radiation-hard process mandatory. Early developments started with the Honeywell and DMILL process. The DMILL technology is a $0.8\,\mu\mathrm{m}$ process with two metal layers. Compared to more recent deep submicron technology those processes had large device sizes and very limited connectivity. The initial choice of the pixel size and readout architecture was made with these limitations in mind. To ensure the feasibility of the full readout chip, all layouts of test devices and prototypes were done from the beginning with radiation-hard processes.

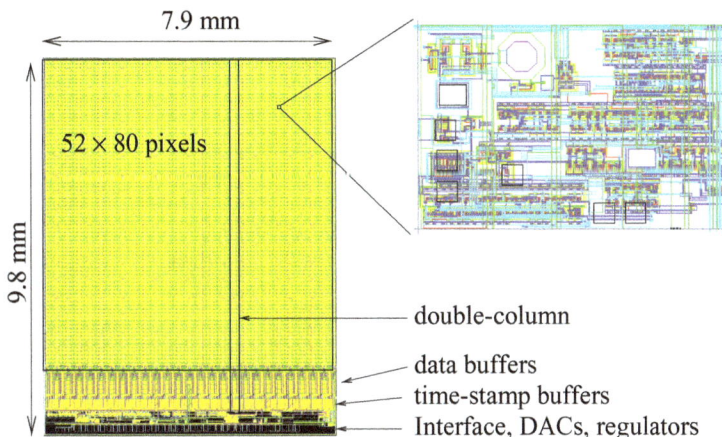

Fig. 5. Floor plan of the PSI46 readout chip. The active matrix of 80 × 52 pixels has a size of 8 mm × 7.8 mm. The periphery has a length of 1.8 mm, dominated by the size of buffers. Also shown is the "left" pixel of a double column. The other half is essentially mirror-symmetric, with a common vertical bus in between. The analog section with the bump pad is in the left part of the pixel, as far away as possible from the bus with its frequent digital activity.

The transition from DMILL to a 0.25 μm process brought a number of advantages. The smaller feature size is only one of them and it was deliberately not fully exploited. The translation permits roughly a factor 6 increase in device density even when the requirements of the radiation-tolerant layout, like ring gates and guard rings, are taken into account.[13] Instead of consequently using minimal layout rules and the highest device count, the design emphasized good yield and compatibility with multiple vendors. It was still possible to reduce the pixel size from 150 μm × 150 μm to 100 μm × 150 μm. In view of the cluster sizes, no advantages were expected from reducing the pixel size further. The size of the periphery, on the other hand, was reduced significantly, from 2.8 mm to 1.8 mm, despite an increase in buffer capacity.

Five metal layers and finer trace pitches permit significant further improvements, although the use of the top metal layers is severely restricted by fill factor limitations and the need to completely shield the sensor from crosstalk.

While the DMILL chip was never intended to be operated in the 4 cm layer at full design luminosity, the improvements allowed by the 0.25 μm technology made this possible without a fundamental redesign of the readout architecture.

Last but not least, the power consumption of the chip went down by a factor of 4. The PSI46 consumes $30\,\mu W$ per pixel.

4.1. Analog section

The smallness of the input capacitance of the pixel makes the intrinsic noise of a charge-sensitive amplifier almost irrelevant. More important than the analog quality of the amplifier is the speed needed to separate the LHC bunch crossings. Maximizing speed requires maximizing transconductance for a given current and minimizing parasitic capacitance where it slows down the amplifier.

The amplifiers are push–pull stages. The transconductance of p- and n-transistors adds up and is higher than in other configurations. The input transistors operate in the weak inversion regime, where the transconductance depends only on the drain current I_D, and g_m/I_D is maximal. A push–pull amplifier is in principle quite susceptible to power supply ripple. However, with the on-chip regulators the system is sufficiently robust. The working point of the amplifiers is in the middle between the power rails and independent of voltage drops across the chip because the analog power rails are kept symmetric and separate from other power rails. This minimizes systematic variations of thresholds and feedback transistors.

The charge-sensitive preamplifier has sufficient gain for signals of a few thousand electrons when the integrating feedback capacitance is a few fF. For larger values, the gain is reduced but the integration is faster (or the same speed is achieved with lower current). In terms of the total power, a solution with a rather large feedback of $C_f = 20$ fF and an additional gain stage was preferred to a single-stage design with small C_f.

The preamplifier input is DC-coupled to the sensor pixels. Its feedback must absorb the expected sensor leakage current of $10\,nA$ per pixel. Active and passive feedback networks have been considered. The value R_f of a passive feedback must be small enough to absorb the leakage current without causing an offset of more than $\sim100\,mV$. On the other hand, it must be large enough to keep the shaping time $(R_f C_f)$ bigger than the amplifier rise time, so as to avoid loss of signal. Both requirements are fulfilled by a $\sim M\Omega$ resistor. Passive feedback was therefore chosen because of its simplicity and robustness. The resistor is implemented as a weak p-transistor[c] operating in its linear region. The working point of the push–pull stage is at one transistor threshold voltage above analog ground, such that the feedback resistor

---

[c]Such weak n-transistors are not possible in the radiation-hard layout.

is properly biased when the gate is near ground. Both gate and substrate potential are made adjustable to control the resistance of the feedback resistors.

The second stage ("shaper") has the same push–pull design and is AC-coupled to the preamplifier. The AC coupling serves a dual purpose. It removes offsets caused by leakage current and it is part of the gain stage. The gain is the ratio of coupling capacitance to feedback capacitance.

The shaper output is connected to the comparator and the sample-and-hold circuit. A buffer in front of the sampling capacitor reduces the load of the shaper output (not shown in Fig. 6), which is critical for keeping the input to the comparator fast. The comparator threshold is adjustable with an eight-bit DAC. Production-related random variations of transistors lead to pixel-to-pixel signal threshold variations, but good threshold uniformity is essential for obtaining a low global threshold without driving a large number of pixels into saturation. Additional four-bit DACs in each pixel can compensate for these variations and make fine adjustments for individual pixel thresholds. The untrimmed threshold dispersion in the $0.25\,\mu\text{m}$ chip is approximately 450 electrons, which can be reduced by trimming to 80 electrons.

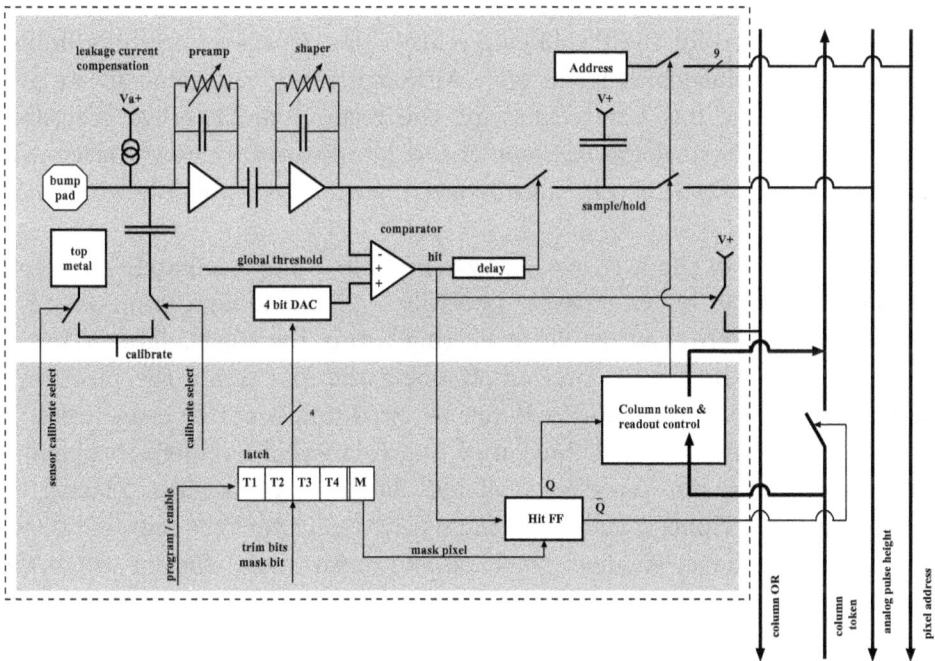

Fig. 6.   Schematic view of a pixel cell.

The analog pulse height information is stored for later readout on the sample-and-hold capacitor. There is no bunch crossing clock in the pixels and the sampling point is simply given by a delayed version of the comparator output. Nonlinearities related to this and other effects are compensated for off-line. Only a modest accuracy equivalent to four bits is needed, because charge fluctuation dominates the resolution obtained by interpolation.[23] Saturation of the amplifier occurs near two times the average signal of a minimal ionizing particle. Larger signals occur frequently, but are not useful for interpolation because they are due to long range delta electrons.

The finite rise times in the amplifier and comparator lead to time walk. Larger signals are detected earlier than smaller ones. The peaking time is slightly larger than the bunch crossing separation, which means that not all pulses will reach the detection threshold within the same bunch crossing. This determines the effective threshold for useful signals. For a nominal threshold of 3000 electrons the effective threshold is 3800 electrons with an analog current of $5\,\mu A$ per pixel.

4.2. *Readout architecture*

Minimizing the logic inside pixels and the number of bus lines in the pixel array was essential for the original DMILL layout, and the readout architecture developed for the PSI43 chip reflects this. It was successfully implemented and verified in a beam test. Although those requirements are less stringent for the $0.25\,\mu m$ technology, the readout architecture described here is essentially unchanged. Some of the data losses that were consciously accepted in the DMILL chip were reduced when it was possible with small modifications.

The latency of the first level trigger in CMS is approximately $4\,\mu s$, corresponding to 150 bunch crossing intervals of 25 ns. At high luminosity the probability of more than one hit occurring within the trigger latency in the same pixel can be several percent. Keeping the hits inside the pixel during the latency would require a buffer to avoid unacceptable dead times. It would also require the distribution of bunch crossing numbers throughout the chip to allow the association of hits and bunch crossings. Placing all buffers in the periphery, outside of the sensitive region of the sensors, allows a simpler (and hence smaller) pixel cell and makes more efficient use of the buffers because they can be shared by more pixels. The data rate into the buffer should not exceed more than a few percent of the transfer capability, to keep waiting times of pixel cells small. Simulations show that grouping two columns of pixels leads to manageable rates per buffer. The readout chip is therefore organized into double columns of pixels that operate essentially independently. The pixels communicate the detection of hits over a wired OR. No clock or bunch crossing numbers need to be distributed over the pixel array.

The periphery synchronizes the wired OR with the LHC clock and latches the current bunch crossing number in a time stamp buffer whenever a hit is found. The state of the pixels that have a hit at that time is frozen and their data are subsequently collected. A token passing from pixel to pixel controls the transfer. Even though empty pixels are skipped relatively rapidly (asynchronously), finding the next pixel with a hit can take longer than one bunch crossing. Hit pixels therefore must resynchronize with the periphery. The handshake reduces the transfer speed to 20 MHz. However, in most cases the number of hits per column is small and the total transfer time has a significant contribution from skipping empty pixels anyway.

The data transmitted and stored in the data buffers are the pixel (row) address and the analog pulse height. Marker bits in the data buffers keep track of the association between time stamps and hits. Maintaining this synchronization under all conceivable circumstances is crucial. A flaw that could cause the loss of a marker when an event was triggered while the buffer was just overflowing was discovered during high rate test beam and fixed in the production version of the PSI46. High rate beam tests were an important part in the development cycle of the readout chip, because the effect of random hit patterns on a chip with data-driven architecture cannot be tested fully in the laboratory.

Hits and time stamps are stored in the buffers during the latency of the trigger. If no trigger signal arrives when a time stamp expires, the corresponding buffer cells are again available for new hits. The buffers overflow when more than 12 time stamps or more than 32 hits occur during the trigger latency, in which case some data are lost. If a time stamp is triggered, the double column stops data-taking and waits for readout. More details about the detector readout are given in Sec. 6.

4.3. *Control*

During data-taking, the readout chip is controlled by the bunch crossing clock and the trigger. All control and configuration data are transmitted by differential lines similar to LVDS at 40 MHz. As is usual in CMS, a trigger signal is three clock cycles long and can encode different commands. In addition to the first level trigger this is used for soft resets and the injection of calibration signals. Configuration data use a serial protocol similar to I2C but operating at 40 MHz. The fast transfer was chosen in order to keep the detector setup times short and to permit frequent reloading of the configuration in case single event upset (SEU) should make this necessary.

SEU was originally assumed to be a problem but is now expected to be much less serious. A simple protection scheme using a capacitive load in the feedback of the storage cells has reduced the vulnerability by two orders of magnitude. The SEU cross section of such protected storage cells has been measured to be $\sigma = 2.6 \times 10^{-16}$ cm^2 for 300 MeV/c pions. Less than one such storage cell per second is expected to be affected at full LHC luminosity.

Pixel thresholds and masks are the largest part of the configuration data. Chipwide DAC settings and control registers constitute only 1% of it. A direct readback of the configuration is not foreseen. DAC adjustments can be verified using the chip header in the readout data (Sec. 6).

A lot of adjustability has been retained for internal operating points and amplifier bias currents. Adjusting the 26 DACs complicates detector operation but it ensures enough flexibility to compensate for process variations and to adapt to ambient temperature variations or radiation effects. Most of the DACs control the analog front-end and the analog readout chain. Others control supply voltage regulators and the delay and magnitude of calibration signals. The DACs are referenced to an on-chip, temperature-compensated, band gap reference.

5. Modules

The modules of the pixel barrel are shown in Fig. 7. A full module consists of two rows of eight readout chips connected to one sensor. This module size permits efficient use of the 4″ wafers used by the sensor vendor.

The sensor elements at the chip boundaries have twice the normal width, to avoid dead regions between readout chips. Pixels in the corners are four times larger. The readout chips extend 0.8 mm beyond the sensor edge, allowing access to the wire bond pads. Wire bonds step up from the readout chip to a thin high density interconnect (HDI) glued onto the back-side of the sensor. Base strips are glued underneath the readout chips. They allow mounting of modules on the barrel mechanics with small screws and carry no components. The base strips are made of silicon nitride with a thickness of 250 μm. Silicon nitride has excellent mechanical properties and a coefficient of thermal expansion that matches very well that of silicon. The 4160 bump bonds per chip have sufficient strength to hold the chips in place and no further mechanical parts are needed.

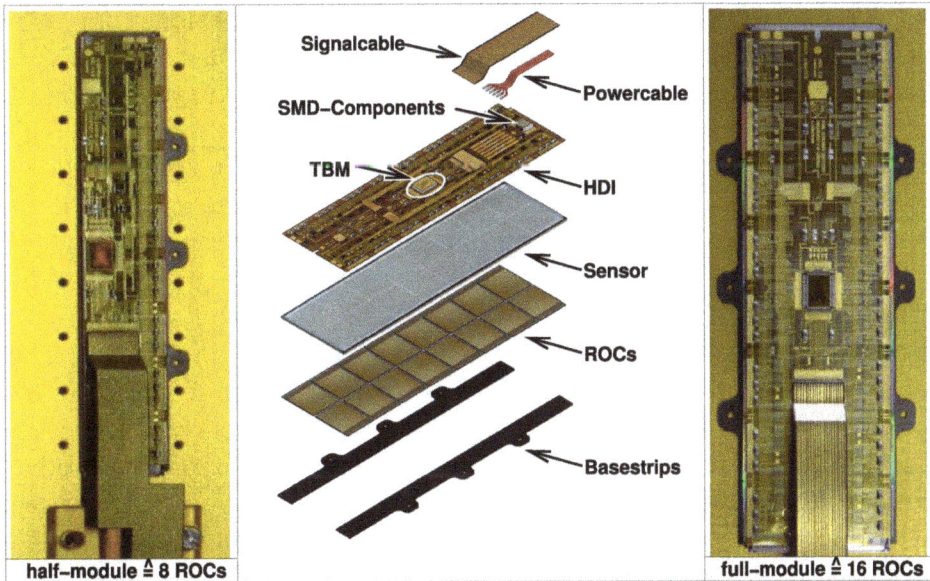

Fig. 7. Components of the CMS pixel barrel modules. The full module (right) has 16 readout chips (ROCs) and a size of 66 mm × 26 mm. Half-modules (left) have only one row of ROCs and minimize the insensitive region near the vertical split of the detector.

Half-modules with only one row of eight readout chips are needed at the boundary of the barrel half-shells to avoid a gap (Fig. 10). The three barrel layers consist of 672 modules and 96 half-modules.

Power, control signals, and readout are distributed over the three-layer HDI. The power and ground layers at the bottom reduce coupling of the fast switching signals into the sensor. All digital signals on the HDI are differential with 300 mV amplitude. Unterminated lines and low power drivers minimize the power consumption connected with signal distribution. A module controller ASIC, the TBM,[24] receives all external control signals and distributes them to the readout chips. It also coordinates the readout of the 16 readout chips, as described in Sec. 6.

Signals to and from the module and the readout are transferred over a two-layer Kapton/copper compound cable with 21 traces and 300 μm pitch. Unacceptable crosstalk from the digital signals into the analog readout on the cable has led to the use of a very low voltage swing protocol using only 80 mV. Low voltage power and detector bias are brought over six copper-coated aluminum wires of 250 μm diameter.

A completed full module has the dimensions $66.6 \times 26.0 \text{ mm}^2$. The power consumption of such a module is $\sim 2\text{ W}$. It weighs 2.2 g plus up to 1.3 g for cables. The cable length varies between 5 cm and 40 cm, depending on the position of the module in the barrel.

The material of the pixel barrel amounts to 5% of a radiation length in the central region. Sensors and readout chips contribute one third of the material, while the support structure and cooling fluid contribute about 50%.

Each disk of the endcap detector has 24 wedge-shaped segments or blades. The blades extend from 6 cm to 15 cm in radius and have a width between 3 cm and 5 cm. Full coverage of the blade area is achieved by an arrangement of rectangular subassemblies (plaquettes) of varying size mounted on two panels, as shown in Fig. 12. Two panels are mounted on either side of a cooling channel such that each one covers the insensitive regions of the other. From the point of view of readout and control, a panel is the equivalent of a barrel module with one TBM and up to 24 readout chips. The four disks are made up of 96 panels populated by 672 plaquettes of five different sizes.

Bump-bonding of sensors to readout chips was done in-house at PSI using indium bumps[8] for the barrel modules and in industry with Pb/Sn bumps for the endcaps.[25] The in-house process was particularly valuable for evaluating test devices and prototypes.

6. Readout

The readout of the pixel data is based on the analog optical links of the CMS strip tracker. Unlike the strip detector data, the pixel data are sparsified, which means that pixel addresses need to be transmitted along with the charge information.

For efficient link usage, the addresses are not binary-coded, but use a set of 6 discrete analog levels (~ 2.5 bits/clock). A hit in a given chip is encoded in a sequence of $5 + 1$ clock cycles: 2 for the double column, 3 for the row, and 1 for the pulse height. Between 8 and 24 chips are serially readout over a common link. The number of chips reflects the expected hit rate in the respective part of the detector.

The serial readout is controlled by a token going in a fixed order from chip to chip. The chip that has the token transmits all hits for a given trigger and then passes the token to the next chip. Each chip adds a header, regardless of whether it actually has hits to send or not. For very low occupancy, this means a significant overhead, but it is efficient at high occupancy (~ 1 hit per chip) and avoids the need for transmitting chip IDs.

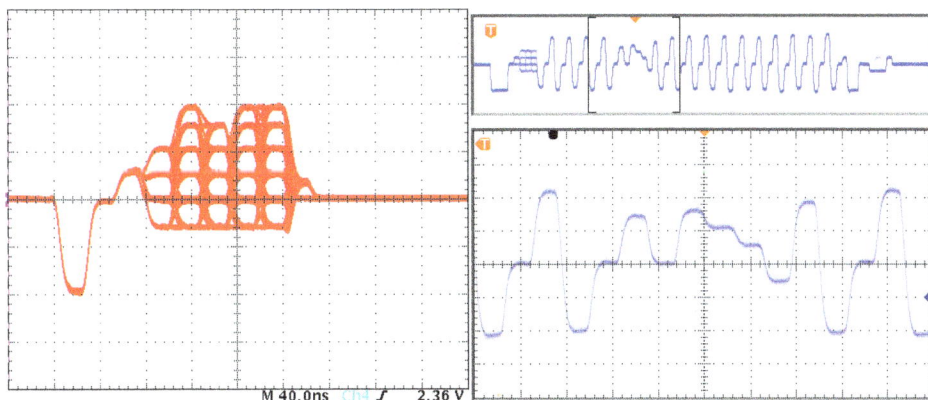

Fig. 8. An "eye diagram" of the readout of a single chip with one pixel hit is shown on the left side. The horizontal axis is time, and the vertical axis shows the amplitude of the chip output. All possible pixel positions are scanned and the oscilloscope traces superimposed. The first three 25 ns cycles are the chip header. It is always the same. The following five cycles encode the column and row of the pixel. The last cycle represents the pulse height, which is constant in this case. The figure on the right shows the readout of a barrel module with a TBM and 16 chips. The TBM header contains an event counter. The fifth chip has one hit; the corresponding section is shown enlarged at the bottom.

Inside the chip, the readout is again controlled by a token that goes from double column to double column, skipping those that are empty. The length of the header, three clock cycles, accommodates skipping through a full chip if necessary. The first two header cycles are used to create an unambiguous pattern identifying chip headers. The third cycle can carry information about DAC adjustment. Figure 8 shows the readout of a single chip with exactly one pixel hit (generated by calibration signals).

Triggers and readout tokens are both counted and hits are readout only when the token number matches the readout number: every token pass collects hits for exactly one trigger. The need for transmitting time stamp information and event-building is avoided, but it must be ensured that exactly one token for every trigger is issued and that there is never more than one token. This is the task of the Token Bit Manager chip (TBM) on the modules/blades. For every first level trigger it sends a token to the first chip and waits until the token returns from the last chip in the chain. It keeps track of triggers arriving while the readout is still busy with a FIFO. The TBM adds a header with an event number and a trailer with status information to each readout. The most important status information during data-taking is token FIFO overflow warning. A loss of a token destroys the synchronization between module and data acquisition, effectively leading to the loss of all subsequent data. The trigger counter is used to verify the correct synchronization. A full readout is shown in Fig. 8.

The TBM can manage two readout groups in parallel and independently. This is used in the barrel modules of the inner two layers where two readout fibers are needed to have enough bandwidth. A single fiber is sufficient for the third barrel layer and the endcap panels. The TBM is configurable by software such that the same module type can be used in all barrel layers.

The output of the readout chip is differential electrical. The TBM serves as an analog multiplexer and the transmission continues over differential lines for distances up to ~1 m to the analog optical hybrid, where a custom-made level translator shifts and amplifies the signal to match the input specifications of the laser driver. The optical signals are received, digitized, and decoded by front-end drivers (FEDs), after some 40 m of optical fiber.

The robustness of the analog-coded addresses was a big concern. To reduce the amount of data, the FED decodes the levels according to pre-defined address level windows immediately after digitization. Offline corrections are not possible; the raw data is accessible only in test runs.

The transmission chain is completely transparent from the readout chip to the FED. The accumulated rise time of all transmission stages had to stay below a few ns in order not to spoil the level identification. Impedance-matched transmission lines were used at all stages to minimize distortions. The output levels can be adjusted by DACs in each chip to reduce chip-to-chip variations. In addition, the level decoding by the FED is calibrated on a per chip basis. Pixel-to-pixel variations were a problem in the DMILL chip but have been eliminated in the $0.25\,\mu$m chip. Baseline drifts of the optical link, mostly due to small temperature variations, are continuously monitored and compensated for in the FED. A typical separation of 26:1 for the full readout chain (Fig. 9) makes the analog-coded address transmission robust and reliable. The trigger number and status information in the TBM header and trailer use a similar encoding.

The noise added by the analog transmission to the pulse height information corresponds to 300 electrons, which is negligible for the position resolution.

7. Mechanics and Installation

Relatively frequent access for replacement and beam pipe bake-out has always been a part of the concept of the CMS pixel detector. Mechanical structure and insertion procedure are designed to permit extraction and

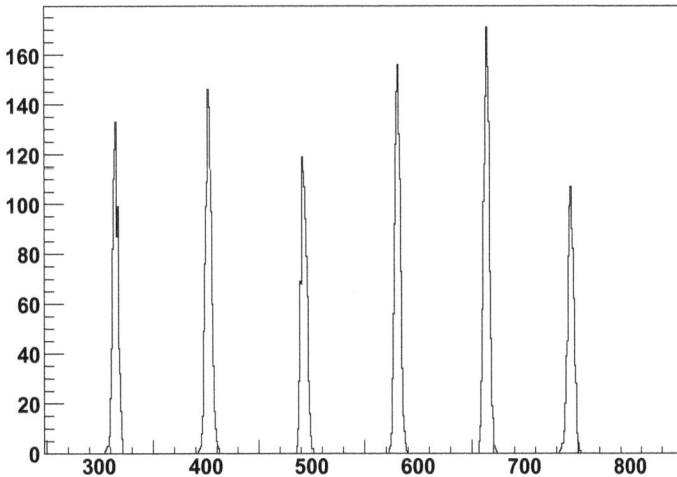

Fig. 9. Address level histogram as received by the FED. The horizontal axis is in ADC units corresponding to the analog level. Robust decoding is possible with the RMS-to-distance ratio of 26:1.

reinsertion of the detector within a relatively short time and in the presence of the LHC beam pipe.

The pixel system is installed inside a cylindrical volume with a diameter of 42 cm and a length of 5.6 m formed by the inner wall of the CMS strip tracker. The beam pipe is cylindrical in the central region, followed by a conical section. It is fixed to the CMS detector with vertical supports at $z = \pm 88$ cm.

In order to permit installation of the pixel detector in the presence of the LHC beam pipe, the whole pixel system is split vertically and mounted on wheels that are guided by grooves in the floor and ceiling of the support tube. The grooves for the left and right halves are parallel in the center and diverge near the end of the tracker volume to avoid collisions with the conical section of the beam pipe.

One half-shell of the barrel mechanics[26] is shown in Fig. 10. The barrel structure consists of the aluminum cooling tubes connected by thin carbon fiber blades. The blades are glued alternating to the inside and outside of the tubes such that adjacent blades have a small overlap. Eight modules are mounted on each blade; four are readout to the $+z$ side and four to the $-z$ side. The cooling tubes are attached to cooling manifolds embedded in the bulkheads. Wheels for installation are mounted at the top and the bottom of the bulkheads and two carbon fiber rods between the wheels on the $\pm z$ side protect the barrel from horizontal forces during insertion. The structure

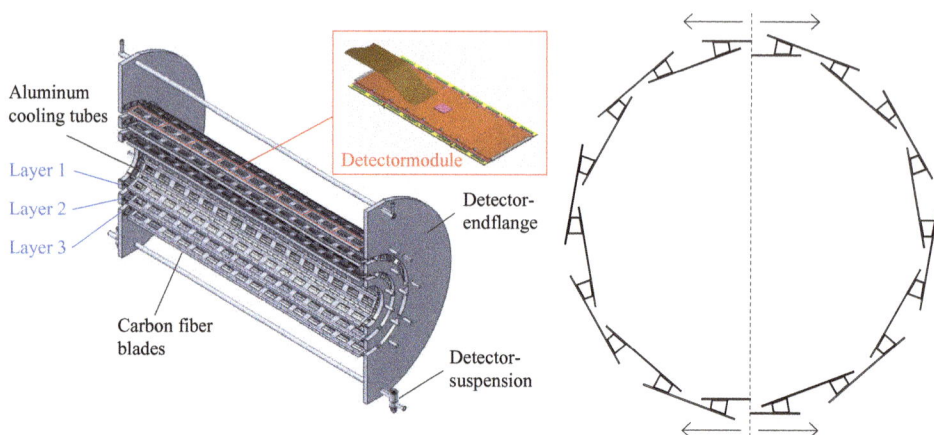

Fig. 10. Schematic view of the innermost pixel barrel layer. The modules are mounted with alternating orientation (inside/outside) onto carbon fiber blades held by trapezoidal cooling tubes. Half-modules allow joining the two shells without a gap.

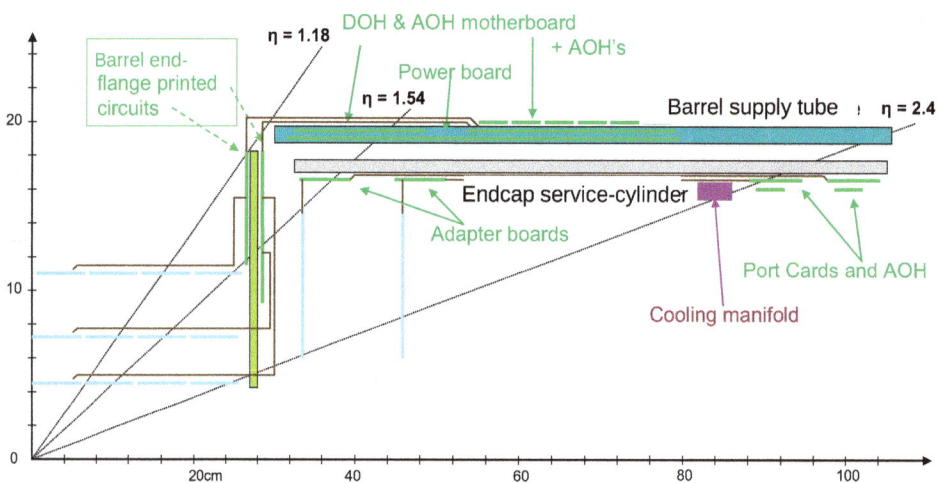

Fig. 11. Schematic view of one $r - z$ quadrant of the pixel detector.

can be separated into three layers for the mounting of the modules. So-called supply tubes with a length of 2.2 m and a thickness between 1 cm and 3 cm bring all services from patch panels outside the tracker volume to the barrel bulkheads. The volume inside of the supply tube is used by the pixel endcaps (Fig. 11).

No access to the detector is possible in its final position; all connections must be made before insertion. All connections between barrel and supply

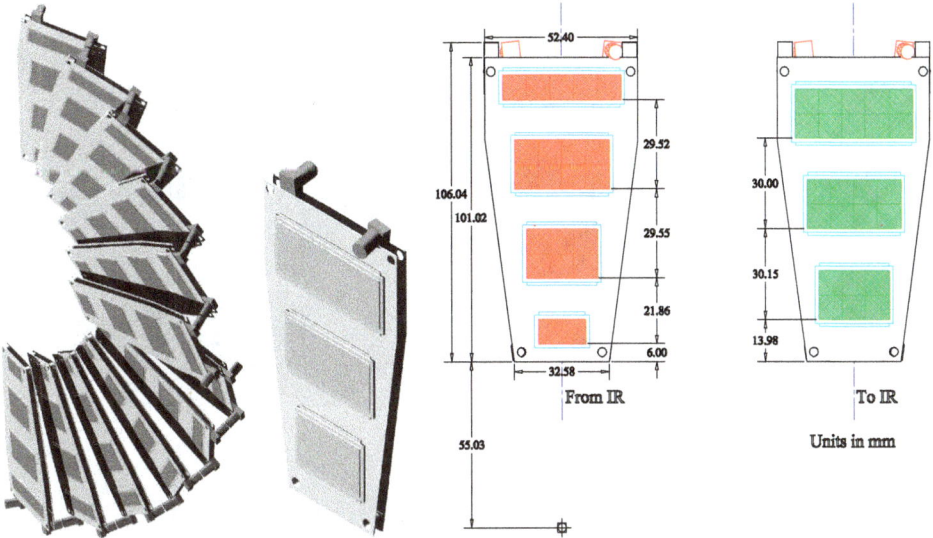

Fig. 12. Endcap detector geometry. The left figure shows one half-disk with 12 blades. A blade consists of a cooling channel with detector panels on either side. The gaps between sensor plaquettes on one panel are covered by plaquettes on the second panel.

tubes must be flexible, because the angle changes by a few degrees during insertion. The cooling connections are made with fiberglass-reinforced silicone tubes. All electrical cables have enough slack to provide the required flexibility. The electrical/optical conversion is done on the supply tube and no optical fibers go to the barrel.

The endcap detector is divided into four independent half-cylinders for installation reasons. Two half-cylinders are installed from each z side on a rail system similar to that of the barrel detector. The blades of the endcap detectors are mounted onto aluminum cooling channels that define the turbine geometry (Fig. 12). Connections to the patch panel are made via service cylinders, on which also the optical links and auxiliary electronics are mounted. The endcap detector is inserted after the barrel into the volume inside the barrel supply tube.

8. Power

The nominal operating voltage of the $0.25\,\mu\text{m}$ process is 2.5 V. The digital logic including the asynchronous parts is still sufficiently fast when operated above 2 V. A similar supply voltage is needed for the discriminator and all parts of the analog readout chain with the exception of the preamplifier

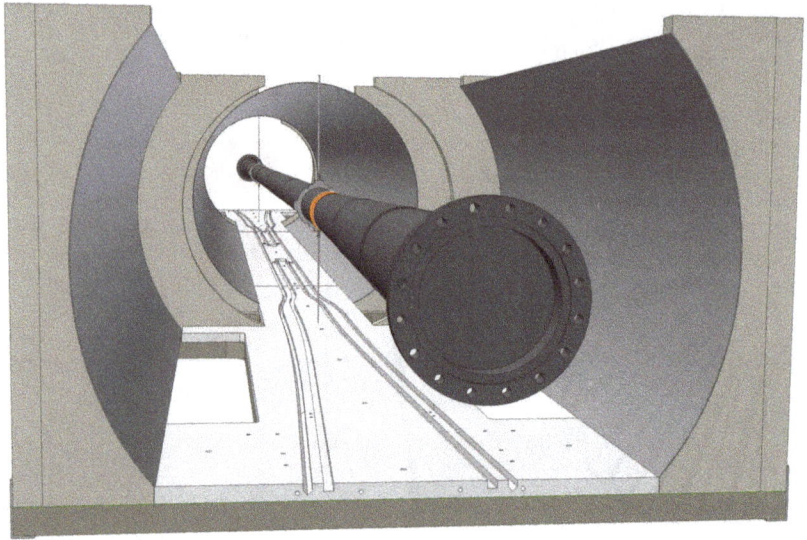

Fig. 13. Simplified drawing of the installation volume with beam pipe. The outermost pair of grooves is for the pixel barrel. The barrel is inserted from one side. The inner pairs of grooves are for the pixel endcaps, which are inserted from both sides. The volume shown has a length of 5 m.

and shaper. A single external supply voltage of approximately 2.5 V is used to generate these internal voltages with adjustable on-chip regulators. The internal operating voltage of the preamplifier and shaper is ≈ 1.2 V, much lower that that of the digital section. Deriving this from a separate supply (1.5 V) is more efficient in terms of power consumed in the front-end. A separate analog supply helps to isolate the sensitive analog sections from the digital parts. Internally to the chip this isolation is maintained by keeping separated analog and digital nets for power and substrate connections.

Groups of typically 12 modules are powered by a common supply. A voltage drop of ~ 3 V over the 40 m of cable is compensated for using sense wires at the end of the supply tube. The remaining voltage drop between that point and the modules varies between 100 mV and 400 mV, depending on the length of the module cable and the power consumption of the module. The on-chip regulators provide well-defined internal operating voltages and reject ripple on the power lines up to 150 kHz.

While the analog power consumption is very constant, the digital power consumption has fast transients and varies substantially with time due to

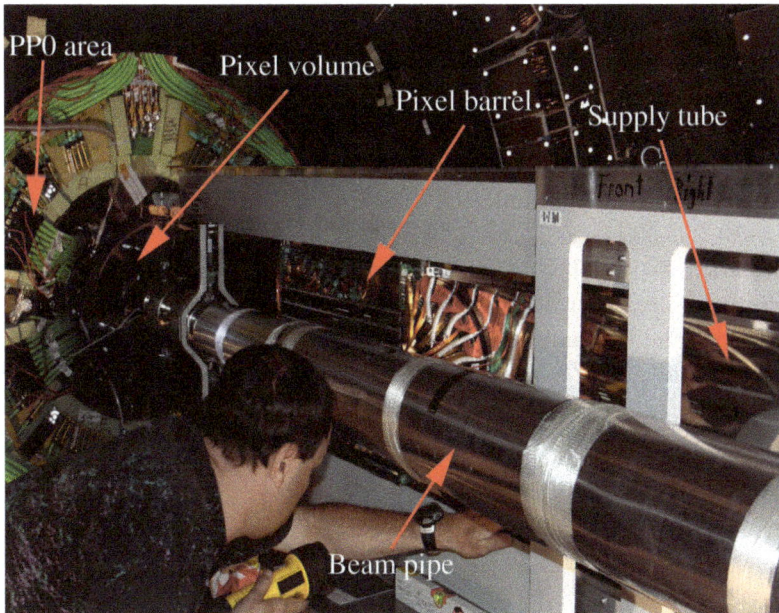

Fig. 14.   Installation of a pixel barrel shell into the CMS detector.

the data-driven architecture. During gaps in the bunch structure of the LHC that last up to $3\,\mu s$, the power consumption drops up to 30%. Every read-out chip has two external $1\,\mu F$ SMD capacitors on the HDI that absorb such variations and avoid large voltage fluctuation or inductive effects on the cables. Crosstalk among readout chips and modules coupling through the power connections was found to be small compared to chip internal crosstalk.

9. Conclusion

The challenges of precise tracking near the interaction point of the LHC required a huge step in detector technology. New levels of radiation, track rate, and track density demand a finely segmented pixel detector with a powerful readout architecture and careful design for radiation hardness. The CMS pixel detector has been installed and commissioned in 2008 after more than 10 years of development and construction time. It achieves excellent resolution in the $r-\phi$ and z direction, with almost-square pixels and interpolation based on charge sharing. Low occupancy and three-dimensional space point measurements give the pixel detector a unique role in the tracking of the CMS experiment.

References

1. M. Della Negra *et al.* CMS: The Compact Muon Solenoid — letter of intent for a general purpose detector at the LHC. CERN-LHCC-92-03.
2. CMS physics: technical design report. CERN-LHCC-2006-001.
3. S. Cucciarelli, M. Konecki, D. Kotlinski and T. Todorov. Track reconstruction, primary vertex finding and seed generation with the pixel detector. CERN-CMS-NOTE-2006-026.
4. B. Henrich, W. Bertl, G. K. and R. Horisberger. Depth profile of signal charge collected in heavily irradiated silicon pixels. CERN-CMS-NOTE-1997-021.
5. B. Henrich and R. Kaufmann. Lorentz-angle in irradiated silicon, *Nucl. Instrum. Methods A* **477** (2002) 304–307. doi: 10.1016/S0168-9002(01)01865-4.
6. R. Horisberger. Design requirements and first results of an LHC adequate analog block for the CMS pixel detector, *Nucl. Instrum. Methods A* **395** (1997) 310–312. doi: 10.1016/S0168-9002(96)00000-9.
7. M. Lechner. Development of a radiation-hard pixel analog block for the CMS vertex detector and search for the rare decays $\Phi \to \eta e^+ e^-$ and $\Phi \to \pi^0 e^+ e^-$ at CMD-2. DISS-ETH-12866.
8. C. Broennimann *et al.* Development of an indium bump bond process for silicon pixel detectors at PSI, *Nucl. Instrum. Methods A* **565** (2006) 303–308. doi: 10.1016/j.nima.2006.05.011.
9. CMS: tracker technical design report. CERN-LHCC-98-06.
10. M. Barbero *et al.* Design and test of the CMS pixel readout chip, *Nucl. Instrum. Methods A* **517** (2004) 349–359. doi: 10.1016/j.nima.2003.09.043.
11. M. Dentan *et al.* Final acceptance of the DMILL technology stabilized at TEMIC/MHS. Given at 4th Workshop on Electronics for LHC Experiments (LEB 98), Rome, Italy, 21–25 Sep. 1998.
12. W. Snoeys *et al.* Layout techniques to enhance the radiation tolerance of standard CMOS technologies demonstrated on a pixel detector readout chip, *Nucl. Instrum. Methods A* **439** (2000) 349–360. doi: 10.1016/S0168-9002(99)00899-2.
13. P. Fischer. Design considerations for pixel readout chips, *Nucl. Instrum. Methods A* **501** (2003) 175–182. doi: 10.1016/S0168-9002(02)02029-6.
14. H. C. Kastli *et al.* Design and performance of the CMS pixel detector readout chip, *Nucl. Instrum. Methods A* **565** (2006) 188–194. doi: 10.1016/j.nima.2006.05.038.
15. D. Kotlinski. Status of the CMS pixel detector, *JINST* **4** (2009) P03019. doi: 10.1088/1748-0221/4/03/P03019.
16. L. Caminada and A. Starodumov. Building and commissioning of the CMS pixel barrel detector, *JINST* **4** (2009) P03017. doi: 10.1088/1748-0221/4/03/P03017.
17. A. Kumar. Commissioning of the CMS forward pixel detector, *JINST* **4** (2009) P03026. doi: 10.1088/1748-0221/4/03/P03026.
18. R. Adolphi *et al.* The CMS experiment at the CERN LHC, *JINST* **0803** (2008) S08004. doi: 10.1088/1748-0221/3/08/S08004.

19. G. Bolla *et al.* Design and test of pixel sensors for the CMS experiment, *Nucl. Instrum. Methods A* **461** (2001) 182–184.
20. J. Kemmer *et al.* Streifendetektor. Patentoffenlegungsschrift DE 19620081 A1.
21. G. Lindström, M. Ahmed, S. Albergo, P. Allport, D. Anderson, L. Andricek *et al.* Radiation-hard silicon detectors — developments by the RD48 (ROSE) collaboration, *Nucl. Instrum. Methods A* **466** (2001) 308–326.
22. T. Rohe *et al.* Fluence dependence of charge collection of irradiated pixel sensors, *Nucl. Instrum. Methods A* **552** (2005) 232–238. doi: 10.1016/j.nima.2005.06.037.
23. CMS Collaboration, CMS tracker. Technical Design Report LHCC 98-6 (CERN, Geneva, Switzerland, 1998).
24. E. Bartz. The 0.25-mu-m token bit manager chip for the CMS pixel readout. Prepared for 11th Workshop on Electronics for LHC and Future Experiments (LECC 2005), Heidelberg, Germany, 12–16 Sep. 2005.
25. P. Merkel. Experience with mass production bump bonding with outside vendors in the CMS FPIX project, *Nucl. Instrum. Methods A* **582** (2007) 771–775. doi: 10.1016/j.nima.2007.07.089.
26. C. Amsler *et al.* Mechanical design and material budget of the CMS barrel pixel detector, *JINST* **0904** (2009) S05003. doi: 10.1088/1748-0221/4/05/P05003.

Chapter 3

THE HYBRID TRACKING SYSTEM OF ATLAS

Leonardo Rossi

Istituto Nazionale di Fisica Nucleare, Sezione di Genova
33 Via Dodecaneso, Genova, 16146, Italy
leonardo.rossi@ge.infn.it

The central tracker of the ATLAS experiment is built using both silicon and gaseous detectors immersed in a 2 T solenoidal magnetic field. To better match the topology of the tracks emerging from the proton–proton collisions, the tracker is separated into a central barrel part (measuring below pseudorapidity $|\eta| \approx 1.2$) and two end-caps (measuring from $\approx$1.2 to 2.5). Different technologies are used at different radii to optimize the cost–performance ratio. The innermost part, immediately surrounding the beam pipe and up to $\approx$15 cm, is made up of silicon pixels for best pattern recognition and maximal radiation resistance. The intermediate region (radii from 30 to 60 cm) uses microstrip detectors and provides excellent space resolution over a large area. The outer layer (radii from 60 to 95 cm) is made up of a large number of small diameter drift tubes (straws) which provide good space resolution in the track bending plane and greatly contribute to pattern recognition with multiple measurements. The transition radiation detection capability of this gaseous detector also helps in electron identification.

1. Introduction

In 1984, at the time of initial studies for a general purpose detector for LHC physics,[1] the operation of a central tracker was considered problematic already at the luminosity of a few 10^{32} cm^{-2}s^{-1}. A huge R&D effort on fast radiation-hard detectors for LHC physics was then promoted by CERN under the control of a dedicated scientific committee.[2] This effort addressed all the critical detector issues for multi-TeV high luminosity colliders, but it has been instrumental in pushing the key tracking technologies to cope with unprecedented particle densities and radiation doses. The design luminosity of the LHC increased to 10^{34} cm^{-2}s^{-1} by the time the Letter of Intent to build ATLAS was written,[3] but a realistic and powerful tracker could already be proposed then.

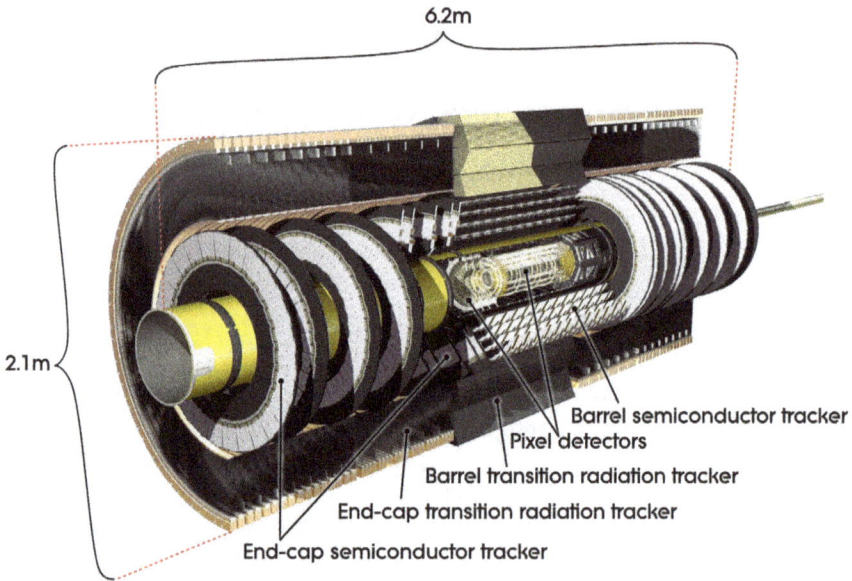

Fig. 1. Artist's view of the ATLAS inner detector. The ID consists of a set of precision silicon tracking detectors (pixel and silicon microstrips (or semiconductor tracker, SCT)) covering the rapidity region $|\eta| < 2.5$ and extending up to a radius of $\approx$60 cm and a matrix of straw tubes of the transition radiation tracker (TRT) for larger radii. Space accuracies in the bending plane are of $O(15\,\mu m)$ in the silicon part and $O(150\,\mu m)$ in the TRT.

The ATLAS inner detector (ID) was later detailed in a Technical Proposal[4] and has now been built and installed in ATLAS, where it has been operated for several months, collecting cosmic rays, and is finally ready to record collision data. The layout of the ID is shown in Fig. 1.

This chapter is organized as follows. The main guidelines and requirements at the root of the ATLAS central tracker design will be illustrated and discussed in Sec. 2; a description of the tracker will follow in Sec. 3, in sufficient detail to make the reader understand the main issues, but referring to already-published results where this is justified. A few key challenges will then be discussed in more detail in Sec. 4, with particular emphasis on what has been learnt during the construction and commissioning phases. Conclusions and lessons for the future will be the subject of Sec. 5.

2. Design Issues

The optimization of the ATLAS central tracker has been part of a more general process of optimization of the ATLAS detector. In particular, for the tracker, one should consider geometrical boundary conditions, environmental constraints and performance requirements. Most of the constraints and requirements are indeed correlated and have performance implications for (and from) other parts of ATLAS. It is anyhow useful, even if oversimplified, to single them out and discuss them.

2.1. *Geometrical constraints*

The geometrical constraints are strongly influenced by the magnetic field and by the electromagnetic calorimeter. A solenoidal magnetic field has been chosen to optimize the track momentum measurement accuracy in the central rapidity region. In ATLAS the superconducting solenoid is between the tracker and the calorimeter and it has windings extending for 5.8 m. It provides a 2 T magnetic field within ±2 m from the center, as shown in Fig. 2. This allows good track momentum measurement, as illustrated in more detail later in this section, with minimal dead material in front of the calorimeter.

The electromagnetic calorimeter immediately follows the solenoid and has a cost which depends heavily on its inner radius. ATLAS has chosen a lead–liquid argon sampling calorimeter to guarantee linear behavior, stability of the response over time and radiation tolerance; moreover, its implementation (accordion geometry) allows a very high granularity and a very good hermeticity. The dimension of the calorimeter has a natural scale, the Moliere

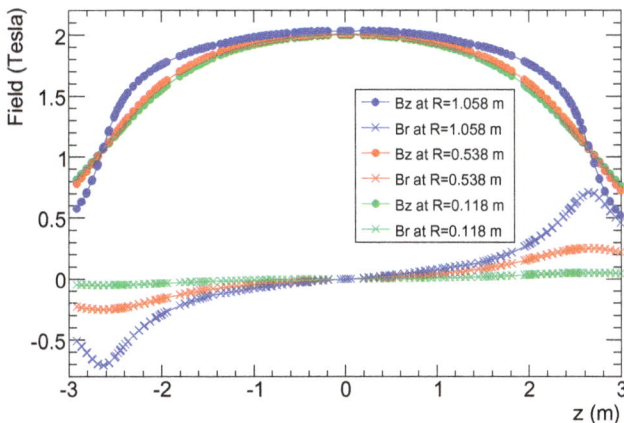

Fig. 2. R and z dependence of the radial (B_r) and axial (B_z) magnetic field components in the inner detector cavity, at fixed azimuth. The symbols denote the measured axial and radial field components, and the lines are the result of the fit to the data.

radius ρ_M, which is strongly correlated with the transverse dimension[a] of a shower and has to be low enough to separate showers (ρ_M is $\approx$4.8 cm in the ATLAS electromagnetic calorimeter). The conflicting requirements of shower separation and minimal cost have been considered together with the tracker performance and gave an optimized calorimeter inner radius of $\approx$1.3 m. The technique chosen for the calorimetric measurements allows optimal integration with the superconducting solenoid. The very first part of the calorimeter (preshower detector) is in fact placed inside the cryostat, uses the cryostat wall as part of the radiator and helps in correcting for the energy loss in the coil. From these considerations, it was agreed to place the cryostat wall at 115 cm, with the coil at 123 cm, followed by the cold preshower and the electromagnetic calorimeter.

The length of the cavity of the cryostat is then determined by the η range which needs to be covered ($\pm$2.5 units of rapidity). A length of 5.4 m was finally chosen in order to have enough B-field integral for the p_T measurement at $\eta = 2.5$.

The inner radius of the tracker is limited by the LHC beam pipe, which has a radius of 3.6 cm, including the thermal insulation. This insulation is mandatory, in order to protect the inner layer of the tracker during the bakeout operation necessary for optimizing the vacuum in the beam intersection region.

2.2. *Environmental constraints*

The external environmental constraints are related to the operation of the accelerator. The tracker has to cope with a high integrated radiation dose and a large instantaneous rate of tracks.

The absorbed radiation dose is the mean energy deposited per unit mass, taking into account all energy loss mechanisms. Ionization is the dominant energy loss mechanism in the central tracker, but nonionizing energy loss is also considered in order to understand detector and electronic damage effects. The radiation dose is given in units of Gy/yr, where one year corresponds to 8×10^{15} inelastic proton–proton collisions (this assumes an inelastic cross-section of 80 mb, a luminosity of 10^{34} cm^{-2}s^{-1} and a data-taking period of 10^7 s). Comparisons of the calculated ionizing dose in the ID between two simulation programs, FLUKA[5] and GCALOR,[6] show agreement within a

---

[a]The Moliere radius, ρ_M, is the radius of a cylinder containing on average 90% of the electromagnetic shower's energy deposition. It is a characteristic constant of a material.

factor of 2. Doses in the tracker volume range from $O(10^5)$ Gy/yr at a radius of 5 cm to $O(10^3)$ Gy/yr at a radius of 100 cm and have required optimization of the design of the detectors and proper choice of materials, especially for the innermost layers. The underlying design criterion is to guarantee a 10-year operation of all detectors, with the exception of the innermost layer.[b]

The natural time quantum at the LHC is 25 ns, i.e. the time interval between two bunch crossings (BCs). This is the time resolution that a detector must have for optimal performance. Worse time resolution (up to two or three BCs) is acceptable at the cost of a higher hit occupancy. The number of charged particles in the tracker acceptance at nominal luminosity is $O(10^3)$ per BC. The speed and granularity of the detectors to be used in the tracker must comply with the numbers above.

The choice of detector technologies is primarily driven by radiation hardness, high rate performance and granularity. This imposes the use of semiconductor detectors at small radius while allowing the use of gaseous detectors at large radius. The rule of thumb for efficient pattern recognition with a low ghost rate is that each layer should have an average occupancy below 1%. This is easily satisfied at nominal luminosity by pixel detectors even at the lowest practical radius of 5 cm (assuming that each pixel sensor is 50 μm wide and 400 μm long); it is satisfied by microstrips above a radius which depends on the strip geometry (30 cm for 80-μm-wide and 12-cm-long strips) but cannot be satisfied by a gaseous detector like the straw tubes [the cross-section of a barrel straw tube is $O(3 \times 10^3$ mm$^2)$], even at radii exceeding 60 cm, where straws are expected to see occupancies well above 30% at a 1 GHz collision rate. To overcome this limitation one must counterbalance the occupancy with the amount of correlated information which can be derived by a set of contiguous detectors. This is the choice of ATLAS, with typically 36 points measured on track segments. The "equivalent" space resolution of a set of N contiguous measurements scales with $1/\sqrt{N}$, giving, with 36 points, a measurement accuracy ($\approx$20 μm)[c] in the bending plane comparable to a silicon microstrip layer. This measurement is relatively robust, as it does not depend appreciably on local inefficiencies.

One must finally note that there are operational constraints which are related to the instantaneous and integrated doses. For instance, fast

---

[b]There was no technical solution for a high granularity tracking detector withstanding 1 MGy. A detector at 5 cm radius must then be replaced after few years of operation at nominal luminosity.

[c]This assumes that the systematic effects can be mastered within this accuracy.

electronics requires more current to function, which immediately translates into increase of material both to bring power in through cables and to take power out through a cooling system. Another example is the need to operate silicon sensors below 0°C to minimize the effect of long term annealing and therefore maximize the lifetime under irradiation[7]; this increases the power load on the cooling system and also forces one to define a dry sealed environment around the silicon detectors to avoid condensation and the related problems (corrosion, electrical discharges, etc.).

2.3. *Performance requirements*

The tracker should measure precisely and efficiently all the charged particles emerging from the proton–proton collision with a transverse momentum above 1 GeV and within $|\eta| < 2.5$. This means both reconstructing tracks inside jets (at an angular distance down to 2 mrad) and contributing to the precise measurement of the track momenta,[d] which is especially important in the case of electrons and muons. Granularity is the most important parameter for efficiently and unambiguously reconstructing tracks inside jets, while the track momentum measurement addresses more specifically the issue of spatial accuracy.

To determine the track curvature a minimum of three measurements is needed in the plane where the bending due to the B field occurs. This does not rely on assuming that the primary vertex point is known, since there is no guarantee that the track originates at the collision point and, indeed, some tracks bringing relevant physics information are expected not to come from the primary vertex. As no layer can realistically be expected to achieve 100% efficiency, four measurement planes is the practical minimum. The momentum measurement resolution can be calculated analytically.[8] In the hypothesis of $N + 1$ equally spaced detectors, all measuring with the same space accuracy σ in a uniform magnetic field of strength B,

$$\frac{\delta p_t}{p_t} = \frac{p_t \sigma}{0.3BL^2} \sqrt{A_N}, \tag{1}$$

---

[d]Another very important role of the tracker is to reconstruct the position of the p–p collision and tag for secondary vertex activity, especially to identify b jets and tau decays. The role of the pixel detector is of paramount importance for precise vertex determination, and therefore this issue is treated by Einsweiler in another chapter of this book.

where

$$A_N = \frac{720N^3}{(N-1)(N+1)(N+2)(N+3)}, \tag{2}$$

the units are GeV, tesla and meter, and p_t is the component of the track momentum perpendicular to the magnetic field direction.

This formula illustrates that the most important parameter for the momentum resolution is the lever arm L (i.e. the length of the track segment in the plane orthogonal to the magnetic field), followed by the field strength B. The number of measurement points enters the formula only under a radical sign. The determination of the number of measuring points depends more on the requirement of robustness of the pattern recognition than on momentum accuracy.

The measurement of a space point along a track trajectory will perturb the measurement itself. A unit charge particle with momentum p traversing a thickness x of a material with radiation length X_0 will be deflected by multiple Coulomb scattering by nuclei according to the relation

$$\theta_{\rm rms} = \frac{0.0136}{\beta p} \sqrt{\frac{x}{X_0}}, \tag{3}$$

where $\theta_{\rm rms}$ is the deflection angle in rad projected on the measurement plane and p is the particle momentum in GeV.

The multiple scattering smears the position resolution and correlates the position measurements.

The error due to multiple scattering dominates up to a momentum of

$$p = \frac{0.0136}{\beta \sigma} \sqrt{\frac{x C_N L^3}{X_0 A_N}}, \tag{4}$$

where $C_N \approx 1.3$ is a coefficient which depends weakly on the number of measuring planes.[8]

Assuming 10 measurement layers, of 3% X_0 thickness each, uniformly distributed over 1 m inside a 2 T magnetic field and measuring with 20 μm accuracy, the multiple scattering would dominate up to $\approx$50 GeV.

Minimization of the radiation length of the tracker is very important in order to make full use of the tracker intrinsic accuracy at lower momenta and to limit early electromagnetic showering, which deteriorates the liquid argon calorimeter energy resolution.

In ATLAS the muon spectrometer has a very good stand-alone momentum measurement performance, thanks to the air core superconducting toroidal magnets and the large area high resolution drift chambers. The contribution of the ID to the muon momentum resolution is important only below 100 GeV or in the transition region between the barrel and the forward system (i.e. $1.1 < |\eta| < 1.7$), where the combined effect of low bending power, additional material and a limited number of measurements deteriorates the stand-alone muon performance.

The electron energy will be measured by the liquid argon electromagnetic calorimeter and the unique, most important contribution of the ID to the electron measurement is the determination of the charge sign. This is crucial for measuring charge asymmetries arising from decays of possible heavy gauge bosons (W' and Z'). In order to have similar performance in electron and muons decay modes, the momentum resolution of the tracker alone should be

$$\frac{\delta p_t}{p_t} = 0.05\% p_t \oplus 1\%. \tag{5}$$

This defines the ID momentum measurement specifications and, consequently, the space accuracy requirements in the given magnetic field.

Background rejection is also heavily based on the tracker performance through the E/p ratio (energy measured in the calorimeter and momentum in the tracker). This ratio should be ≈ 1 to reduce the rate of pions faking an electron signal to below the irreducible background from real electrons from physics signals. Identification of photons also depends heavily on the tracker performance and should, in principle, be based on the absence of a track signal prior to the calorimeter energy deposit.

In practice the situation is more complex, as the amount of material in the tracker will cause a sizeable fraction of photons to convert (and a similar fraction of electrons to start showering) still inside the tracker. Sophisticated software tools are then necessary in order to deal with those cases and offline-correct the measurements accordingly.

Additional ID contributions to the lepton sector are the independent electron identification through transition radiation (made possible by the adoption of a straw system able to generate and detect transition radiation x-rays) and the isolation criteria to be applied around a lepton candidate.

The cross-correlation between the tracker and the electromagnetic calorimeter and the muon spectrometer requires that the tracks be extrapolated within the expected accuracy to guarantee good matching. This does not constitute an additional requirement on the track accuracy. The calorimeter spatial accuracy in the precision coordinate is $>6\,\mathrm{mm}/\sqrt{E}$ and there is no significant additional rejection power in track shower matching once the E/p cut is implemented. To reach the muon spectrometer a muon has to cross $\approx$100 radiation lengths (mostly due to electromagnetic and hadronic calorimeters). This will make it lose $\approx$3 GeV and deviate from its original direction because of multiple scattering (30 mrad at 5 GcV) by much larger amounts than the intrinsic resolution of the tracker as set by the $\delta p_t/p_t$ requirement.

In general the tracker plays a crucial role in momentum measurement and reduces the background through matching of independent information. It adds robustness to measurements, which is particularly important in case of discoveries.

3. The ATLAS Central Tracker

This chapter will illustrate how the design issues presented in Sec. 2 have been implemented in the ATLAS tracker. The reader can find a more detailed description elsewhere.[9]

The central tracker layout is shown in Fig. 3, It is inside a cylindrical envelope of length 7024 mm and radius 1150 mm, within a solenoidal magnetic field of 2 T (see Fig. 2).

The ID is subdivided in two parts: the precision tracking detectors, pixels and semiconductor tracker (SCT), for radii below $\approx$60 cm, and the transition radiation tracker (TRT) above. The precision tracking part is made up of silicon detectors with different diode implantation patterns and covers the region $|\eta| < 2.5$. The TRT is a gaseous detector which extends only up to $\eta = 2.0$ and is made up of 4-mm-diameter drift tubes closely packed together and interleaved with transition radiation material. All detectors are arranged in a barrel region as concentric cylinders around the beam axis and in two end-cap regions as disks perpendicular to the beam axis. The η coverage of the barrel region depends on the radius at which the layer is located and it varies from $|\eta| < 2.7$ of the innermost pixel layer (b layer) to $|\eta| < 0.75$ of the outermost TRT layer. All the detectors, with the notable exception of the pixel b layer, are designed to sustain a 10-year operation lifetime. Each subdetector has been designed to have some "stand-alone" track finding capability; this allows one to cope with local faults and add robustness to the system.

In the following a reminder of the pixel detector's main characteristics will be given together with a more detailed description of the SCT and the TRT.

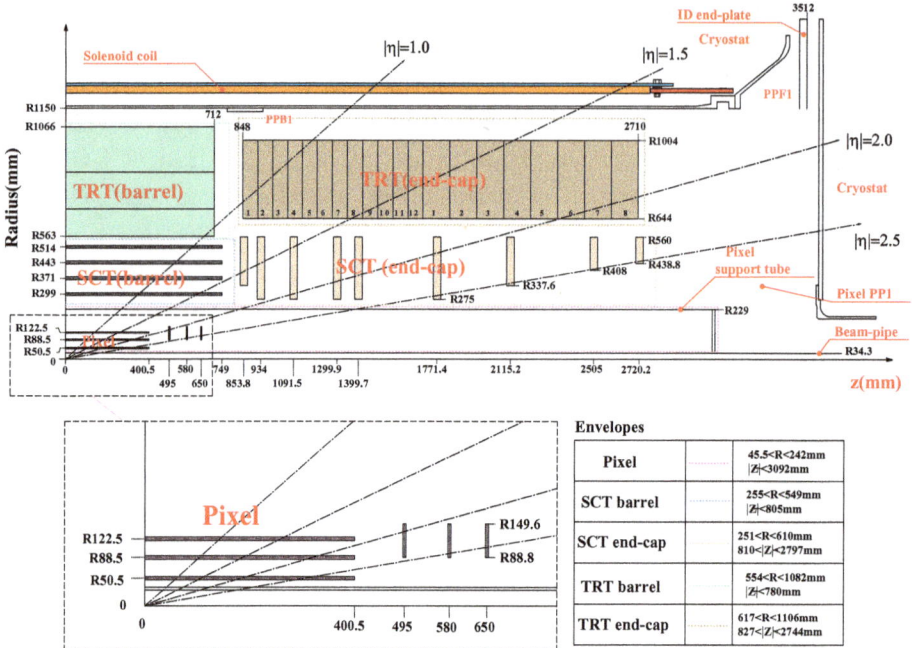

Fig. 3.   Plan view of a quarter section of the ATLAS inner detector, showing each of the major detector elements with its active dimensions and envelopes. The labels PP1, PPB1 and PPF1 indicate the patch panels for the ID services.

3.1. *Pixels*

The pixel detector is made up of 1744 modules covering three barrel layers and three disks per side. It has a total of ≈80 million individual sensing elements.

Each module is made up of one sensor and 16 readout chips, and has a $6.04 \times 1.64 \, \mathrm{cm}^2$ active area. All modules are equal and are built using n-doped sensor tiles containing 47,532 n^+ pixel implants. The use of n^+ implants allows partially depleted operation after type inversion, should this be needed at the end of the detector lifetime. The large majority (≈98%) of the pixels cover a surface of $50 \, \mu m \times 400 \, \mu m$; the remaining ones are bigger, to allow for a continuous sensitive area also in between the front-end chips. Having rectangular pixels implies better resolution in the plane where the track sagitta (and therefore the particle momentum) is measured. Every pixel is read out through an amplifier, followed by a discriminator which detects when the pulse exceeds an adjustable threshold. The time resolution is below 25 ns, as needed to unambiguously associate the pixel hits with a given LHC beam–beam interaction.

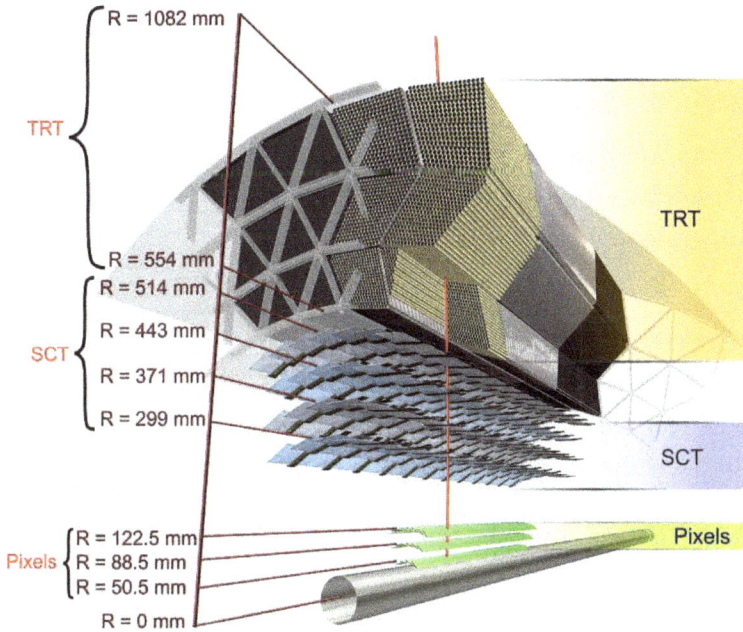

Fig. 4. Artist's view of the barrel inner detector traversed by a charged track of 10 GeV p_T and $\eta = 0.3$ (only the active elements are shown). The track traverses successively the beryllium beampipe, the three cylindrical silicon pixel layers with individual sensor elements of $50 \times 400\,\mu m^2$, the four cylindrical double layers (one axial and one with a stereo angle of 40 mrad) of barrel silicon microstrip sensors (SCT) of pitch $80\,\mu m$, and approximately 36 axial straws of 4 mm diameter contained in the barrel transition radiation tracker modules within their support structure.

Figure 4 shows an artist's view of the barrel layout of the ID and how the pixel is located within the central tracking system.

The barrel part is built using staves, which are ≈2 cm wide and ≈80 cm long local supports arranged in a turbine geometry to allow for hermetic track coverage. The tilt angle has been chosen to counterbalance (at least in part) the effect of the Lorentz force and therefore to minimize charge sharing between pixels. This choice favors robustness of operation toward space resolution, which can be improved with charge interpolation. Thirteen modules are placed on each stave facing the interaction point. The central pixel detector is made up of three barrel layers of average radii 5.0, 9.8 and 12.2 cm, built with, respectively, 22, 38 and 52 staves. The innermost layer is of special importance, as it determines the impact parameter resolution (i.e. the accuracy of the tracks extrapolated to the interaction vertex).

The forward part is built using sectors of 8.9 cm inner radius and 15.0 cm outer radius. Each sector has six modules (three in front and three at the back) and the hermeticity is obtained overlapping the front and back modules. Eight sectors make up a disk, and six disks make up the entire pixel forward system. Three pixel layers are always crossed by all particles emitted within ±2.5 units of rapidity; modules are superimposed by a few millimeters to facilitate alignment with tracks.

Figure 5 shows an artist's view of the forward layout of the ID; this helps in understanding the location of the pixel in the forward tracking system and will later be referenced, together with Fig. 4, to also describe the rest of the ID.

The ATLAS pixel project is characterized by a high degree of standardization meant to ease the production process: all modules are equal; all staves and all sectors are equal.

The pixel detector can be installed independently once the rest of the ATLAS tracker is in place. This facilitates repairs and upgrades, which nevertheless remain a formidable challenge given the complexity of the device and the access restrictions after irradiation.

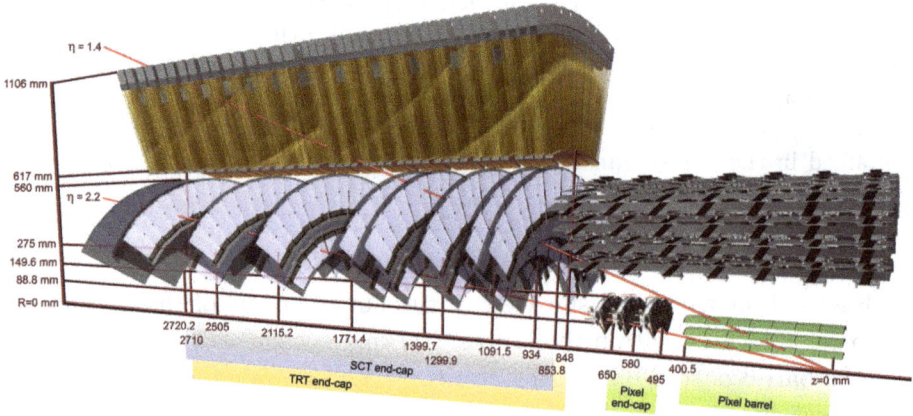

Fig. 5. Artist's view of the end-cap inner detector traversed by two charged tracks of 10 GeV p_T and $\eta = 1.4$ and 2.2 (no services are shown). The track at $\eta = 1.4$ traverses successively the beryllium beampipe, the three cylindrical silicon pixel layers with individual sensor elements of $50 \times 400 \, \mu m^2$, four of the disks with double layers (one radial and one with a stereo angle of 40 mrad) of the end-cap silicon microstrip sensors (SCT) of pitch 80 μm, and approximately 40 radial straws of 4 mm diameter contained in the end-cap transition radiation tracker wheels. The track at $\eta = 2.2$ traverses successively the beryllium beampipe, only the first of the cylindrical silicon pixel layers, two end-cap pixel disks, and the last four disks of the end-cap SCT.

3.2. *Semiconductor tracker*

The SCT[9, 10] consists of 4088 sensor modules: 2112 located over 4 equally spaced barrel layers and 1976 over 18 disks (9 per side). The total number of sensing elements is ≈6 million.

Each module is made up of two strip sensors mounted back to back on a thermally conductive baseboard; they are tilted by 40 mrad and are wire-bonded to a double-sided electronics hybrid. The tilt angle is chosen to minimize the number of ghost hits (when associating the information of the two sensors) still keeping a z resolution well below a millimeter. The module baseboard is built from anisotropic thermal pyrolytic graphite with an in-plane thermal conductivity of about 1500 W/mK at 20°C, which provides a good thermal path between the source of the heat (the modules) and the sink (the cooling pipe). A blowup sketch of a module is shown in Fig. 6.

The sensors are single-sided p-in-n microstrip detectors, implanted on 285-μm-thick high resistivity silicon wafers. Each sensor has 768 AC-coupled readout strips resulting in 1536 readout channels per module. Modules are all the same in the barrel and feature an 80 μm pitch. Four different shapes at different radial distances are instead needed to match the end-cap geometry. The strip pitch also changes with the radius and varies between 57 μm on the innermost sensors and 90 μm on the outermost ones. The main parameters of the SCT modules are shown in Table 1.

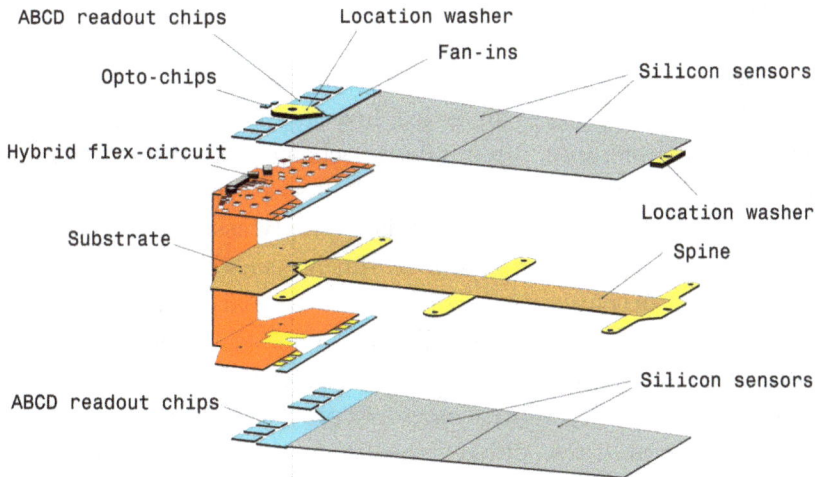

Fig. 6.   Exploded view of the different components of an end-cap module, including the high thermal conductivity spine, the polyimide hybrid and the readout ASICs.

Table 1. SCT barrel and end-cap module specifications.

Parameter	Description	
Strips	2×768 active strips	± 20 mrad stereo rotation
Nominal resolution	$17 \, \mu$m	in-plane lateral $(R - \phi)$
	$580 \, \mu$m	in-plane longitudinal $(z$ or $R)$
Module active length		
• barrel	126.09 mm	
• outer end-cap	119.14 mm	(radius 438.77–560.00 mm)
• middle end-cap	115.61 mm	(radius 337.60–455.30)
• short middle end-cap	52.48 mm	(radius 402.82–455.30 mm)
• inner end-cap	59.1 mm	(radius 275.00–334.10 mm)
Hybrid power consumption	5.5–7.5 W	
Sensor power consumption	Up to 460 V bias	<1 W at $-7°$C

The strip pitch was chosen to optimize two-track separation and to provide a silicon strip occupancy below 1.0% at full luminosity. Under the same luminosity condition the probability of having ghost hits is ≈30% per track. The rise time of the front-end (F–E) electronics (20 ns) was designed to associate all hits with a unique bunch crossing; a "1" is stored in the F–E memory if the signal is above a threshold which is adjustable channel by channel. The readout is performed via the hybrid, a multilayer flexible Cu/polyimide circuit which carries 12 binary ABCD[11] readout chips, 6 on each side. The hybrid is also required to have good thermal conductivity in order to remove the heat generated by the chips (up to 7.5 W/module). Data are then read out at a 100 kHz rate through an optical link once a level-1 trigger is received. To increase fault tolerance, a redundancy scheme is implemented to reroute signals in case a single chip or a readout link stops functioning.

As for the rest of the tracker, a detailed quality assurance (QA) including database traceability was implemented during construction at any level of integration. The procurement of 25% additional parts allowed to complete construction and integration and still left a number of spares used to build an ≈10% test system. This has proven to be very useful for exercising new software and newcomers. Module assembly proceeded in clean rooms of class 10,000 or better at a few different assembly sites. The assembly process started by aligning and gluing pairs of sensors onto the baseboard, using a thermally conductive epoxy. The readout chips were assembled to the hybrids with electrically conductive glue. Once populated and tested the hybrids were

mounted onto the sensor–baseboard sandwiches. The last assembly step is
the ultrasonic wire-bonding of the signal and supply lines of the sensors to
the hybrids.

Requiring noise occupancy below 5×10^{-4} at $1\,fC$ threshold, less than
1% bad channels and absence of microdischarge below $350\,V$ (with a current
limit of $4\,mA$ at $500\,V$) gave a production yield slightly above 90%.

The 2112 barrel modules are supported by four concentric low mass cylinders (see Fig. 4) designed to be extremely stable to both temperature and
humidity variations, and to long term creep.[12] These cylinders are made from
three-layer $(0°, +60°, -60°)$ high modulus carbon fiber skins of $<200\,\mu m$
total thickness over a carbon-fiber-reinforced plastic (CRFP) honeycomb
core to form a $6\,mm$ sandwich. The cylinder ends are closed with CRFP
flanges, incorporating holes that are machined to high precision and serve as
mechanical reference in the integration process. The modules are mounted
directly on the cylinders without intermediate local supports and tilted by
$11°$ (to compensate for the Lorentz angle effect). Absence of local supports
imposes concentration of the integration effort in very few labs.

Each end-cap is made up of nine disks surrounding the beam axis; the
layout of one end-cap is shown in Fig. 5. Each disk has up to three rings
of modules: outer, middle and inner,[e] for a total of 1976 modules of four
different types (see Table 1). The modules in each ring overlap in azimuth to
avoid gaps, and the rings overlap in radius (when seen from the interaction
region) so that particles always cross four detector layers. There are, nevertheless, small gaps in the acceptance in the barrel–end-cap transition region
where only three measurements are available. This loss is caused by the barrel services being routed between the end of the barrel and the beginning of
the end-cap.

Cooling is provided by evaporation of a fluorocarbon (C_3F_8) which
expands in an aluminum tube thermally coupled to the modules after having
passed through a capillary. This system is in common with the pixel and will
be described in more detail in Sec. 4.

The SCT has been operated with the rest of ATLAS in autumn 2008.
Two million cosmic rays have crossed the detector, allowing the precise mea-
surement of its efficiency. This has turned out to be very high, as shown in
Fig. 7 for the four barrel layers.

---

[e]It is not necessary for all disks to have outer, middle and inner rings of modules
to ensure full coverage.

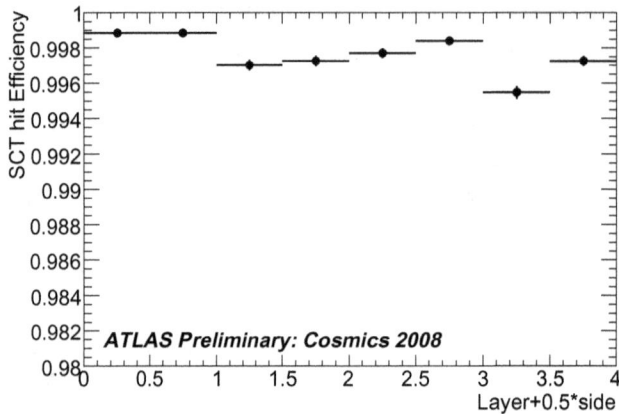

Fig. 7. The unbiased hit efficiency for each layer of the SCT barrel as measured with cosmic ray tracks. Each layer has two sides, and a hit is a cluster with at least one strip hit on one side. The efficiency is defined for expected hits only, taking into account that a few percent of modules are inactive for known reasons (typically 1% in the barrel). Only tracks with an incidence angle of less than 40° to the normal have been used.

3.3. Transition radiation tracker

As the TRT is peculiar to ATLAS and justifies the word "hybrid" in the title of this chapter, I will spend a few more words on it and, in particular, on the basic detector element.

Transition radiation[13, 14] (TR) is emitted by a relativistic charged particle when it is crossing a boundary between regions with different indices of refraction, n_1 and n_2. The particle will polarize the medium and the electric field will suddenly change when the refraction index changes; to restore continuity photons must be emitted. The photon flux depends on the Lorentz factor γ and on the number of boundaries crossed. The photons are emitted at an angle of $1/\gamma$ to the track direction (therefore very much collimated with the charged tracks in all cases of interest) and have energies ranging from 5 to 15 KeV. Abundant generation of TR photons requires many n_1, n_2 interfaces (e.g. many plastic foils). Efficient detection of TR photons requires high Z sensors (e.g. Xe-filled proportional counters). The detector described below optimizes both tracking and sensitivity to TR.

The TRT[9, 15] is made up of 372,032 proportional drift tubes of 4 mm inner diameter and filled with a Xe-rich gas mixture. The tubes' diameter is a compromise between speed of response, number of ionization clusters, and operational stability. The length of the tubes (or straws, as they are more commonly called) is different if they are used in the barrel or in the end-cap, but all other characteristics are the same. The straws in the barrel are 144 cm long and parallel to the beam (see Fig. 4), while the straws in the end-cap are shorter (37 cm) and oriented radially (see Fig. 5).

The straw tube wall is made up of two layers of a 35-μm-thick Kapton film wound with a pitch of about 13 mm on a precisely tooled mandrel at a temperature of $\approx$260°C. A special film coating technology (using Al and graphite-loaded polyimide) was developed and then applied on the Kapton tape to obtain the desired mechanical and electrical properties of the final straw. This coating guarantees both good electrical conductivity and protection from cathode etching, which may be caused by microdischarges or other erosive effects.

The Kapton film has poor mechanical properties. Addition of four 1000-filament carbon fiber bundles glued on the outer skin of the straw increases rigidity and minimizes the effect of environmental factors (T, humidity). For instance, the reinforced straws have a thermal expansion coefficient of $2 \times 10^{-6}\,\mathrm{K}^{-1}$, an order of magnitude lower than the nonreinforced straws. A total of 1.5×10^5 160-cm-long straws were produced and then cut to length for the barrel and for the end-cap use. The deviation from straightness is less than 0.3 (1.0) mm for the end-cap (barrel) straws.

The addition of the reinforcing carbon fiber bundles also reduces the cathode resistivity to $\approx$20 $\Omega\,\mathrm{m}^{-1}$, thus improving signal propagation. An electrical connection of the outer and inner conductive surfaces of the straw is made to avoid loss of conductivity in case of damage to the inner tube conductive layer. The anode wire is 30-μm-diameter tungsten plated with an $\approx$0.6 μm gold layer. In order to ensure gas gain uniformity, deviations from wire diameter and circularity are less than 1% and 2%, respectively. The resistance of the anode wire is 60 $\Omega\mathrm{m}^{-1}$ and the wires are strung to 70 g (this gives a sagitta of 15 μm, due to the gravitational force on the barrel straws). The barrel anode wires are electrically separated in the middle by a glass joint[f] and are read out from both ends to reduce occupancy.

The gas mixture in the straws is Xe(70%) (for efficient x-ray conversion), CO_2(27%) (for high drift velocity) and O_2(3%) (for photon quenching and increased operation stability) circulated at a flow which guarantees about one volume change per hour. The exit gas is recovered, cleaned and recirculated

---

[f]The first nine radial layers have two glass joints per wire. These straws are blind in the center and active only over the last 31 cm on each side. This is done to further reduce occupancy.

and the amount of Xe lost in this process is $\approx 5\,1/day$, a very low value for such a large system. The straw cathodes are typically operated at a high voltage of 1530 V, corresponding to a gas gain of 2.5×10^4 and a maximum drift time of $\approx 50\,ns$. The use of a high Z' gas like Xe is dictated by the need to convert within the active volume of the detector the TR photons which are generated by particles with $\gamma > 1000$ when passing through the polyethylene foils (in the end-cap) or polyethylene/polypropylene fibers (in the barrel) which are placed in between straw layers. It is important that Xe does not fill this "in between" region, otherwise the TR photons would deposit their energy outside the straws. For that reason (and to extract the heat produced by the straws themselves) the radiator region is flushed with CO_2.

The TR material interspersed in the ATLAS TRT allows one to have a converted x-ray in about 20% (30%) of the barrel (end-cap) straws for a high momentum electron. As the energy loss by ionization is typically 2 keV (versus the 5–15 keV of the TR photons), the TR signal can easily be discriminated, allowing it to contribute to electron (and, more in general, to any high γ particle) identification. An example of the response of the TRT to high momentum muons is shown in Fig. 8.

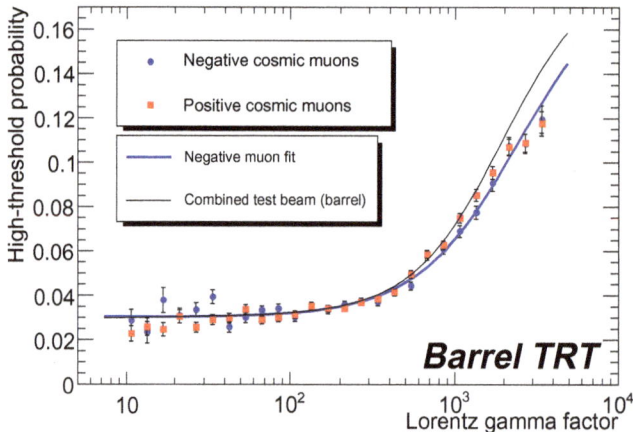

Fig. 8.   Turn-on of the production of TR photons as a function of gamma as measured by the barrel TRT for the tracks of cosmic particles (muons) during cosmic data-taking of the ATLAS detector in October 2008. On the y axis the probability of a high-threshold (i.e. >6 keV) hit (indicator for a TR photon conversion) is given. The data points are shown for both muon charges (positive — red dots; negative — blue dots) and are compared to the results obtained in the ATLAS combined test beam in 2004 (black line).

The readout of the 424,576 channels requires recording the drift time (with a 1 ns rms, corresponding to 130 μm) and being able to discriminate if the pulse is coming from the ionization signal (therefore with energy deposited in the gas $E_d > 0.2$ keV) or from the conversion of a TR photon (therefore $E_d > 6$ keV). A double threshold system is therefore implemented and sampled with a 3.125 ns binning. A fast baseline restorer network ensures that the pulse baseline level is reached within 25–50 ns, depending on signal amplitude. The analog signal processing and digitization is performed in the front-end electronics; the information is stored locally and transferred through 40 Mbit twisted-pair LVDS[g] links.

The barrel TRT consists of 52,544 axial staws organized in three cylindrical rings, each made up of 32 identical and independent modules of 144 cm length housed in a precisely machined CFRP support structure [see Fig. 9(a)]. The average straw spacing is 6.8 mm to leave adequate space for the 15 μm polypropylene/polyethylene fiber radiators. In this configuration 7 TR photons convert on average in the active gas for 20 GeV electrons.

The end-cap TRT consists of two sets of wheels, each containing eight planes of radially oriented straws [see Fig. 9(b)]. Each wheel contains 6244 straws placed on eight successive layers separated by 8 mm (or 15 mm) along the beam direction. The space in between two successive layers is filled with radiator foils (more foils at larger rapidity). A typical track will fire 32–45 straws (with 10–15 TR photon hits for a high momentum electron).

A gas proportional detector (like the TRT) is intrinsically less stable than an ionization detector (like the SCT), as there is an additional gain in the system due to the proportional amplification factor in the gas mixture. This factor depends on the electric field, the temperature and the gas mixture. All those parameters need to be monitored so as to avoid gain variations exceeding 20%.

4. Some Key Challenges

This section briefly describes three challenges which are relevant to the successful operation of the ATLAS central tracker: the silicon cooling system, the detector alignment and the material (mostly X_0) minimization. Indeed, there are many more challenges which had to be faced in order to realize such a complex detector, but these are the issues which have turned out to be more relevant at system level.

---

[g]Low voltage differential signaling.

(a)

(b)

Fig. 9.   (a) Photograph of one quarter of the barrel TRT during integration. The shapes of one outer, one middle and one inner TRT module are highlighted. The barrel support structure space frame can be seen with its triangular substructure. (b) Photograph of a four-plane TRT end-cap wheel during assembly. The inner and outer carbon fiber rings can be seen, as well as the first layer of straws and the first stack of polypropylene radiator foils beneath.

4.1. *Evaporative cooling system*

The cooling system must extract the heat produced in the silicon detectors and related electronics ($\approx$20 kW in the pixel volume and $\approx$40 kW in the SCT volume) and should allow stable operation of the silicon sensors below 0°C so as to minimize the effect of reverse annealing. The choice of ATLAS is of an evaporative system with an inert fluorocarbon coolant fluid (C_3F_8) evaporating at −20°C (i.e. at a pressure of 2 bars).[h] The main motivations for preferring an evaporative cooling system to a monophase system are: the higher heat transfer coefficient between the cooling fluid and the cooling tubes, the smaller temperature gradients along the cooling channels, and the

---

[h]The evaporation temperature can be adjusted by changing the evaporation pressure. A coolant evaporation temperature from −25°C to +10°C is possible with the present layout. The sensor temperature is $\approx$15°C higher than the coolant evaporation temperature in the standard operating condition.

smaller size of (and the absence of thermal insulation around) the cooling channels.

The evaporative system's functionality is similar to that of a standard industrial direct expansion cooling system (like the refrigerator everybody has at home). The fluid is delivered in liquid phase at room temperature from the condenser to the 0.6–0.8-mm diameter capillaries located immediately before the detector structures (staves in the case illustrated in Fig. 10). The fluid expands through the capillaries and then remains in saturation conditions (boiling) along the cooling circuit on the detector structures. A heater, located at the exhaust of each circuit of the detector structures, evaporates the residual liquid and raises the temperature of the vapor above the cavern dew point. The gaseous C_3F_8 is then brought back to the compressor and finally to the condenser. Recuperative heat exchangers (HEX) between the inlet liquid (warm) and the return fluid (cold) are implemented to increase the efficiency of the thermodynamic cycle, by decreasing the vapor quality at the inlet of the detector structures and hence the required flow. The cold parts of the system are all inside the solenoid bore and within a dry nitrogen environment.

The system is made up of 204 independent channels (88 for the pixel, 44 for the SCT barrel and 36 for each SCT end-cap) and has been operated stably, after some initial problems related to leaks and to heater electrical connections, for more than 2000 h (see Fig. 11), with only a-few-% downtime. The thermodynamic performance of the system agrees with the expectations;

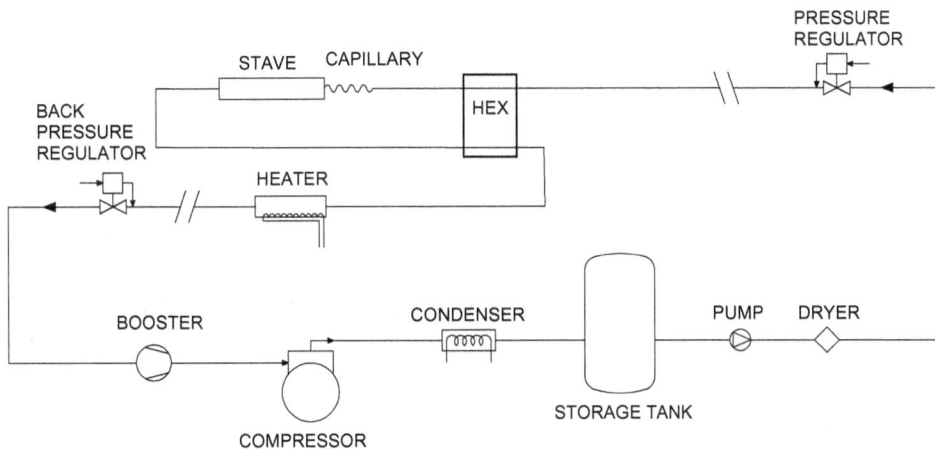

Fig. 10. Simplified scheme of the ATLAS evaporative cooling system (see text for details).

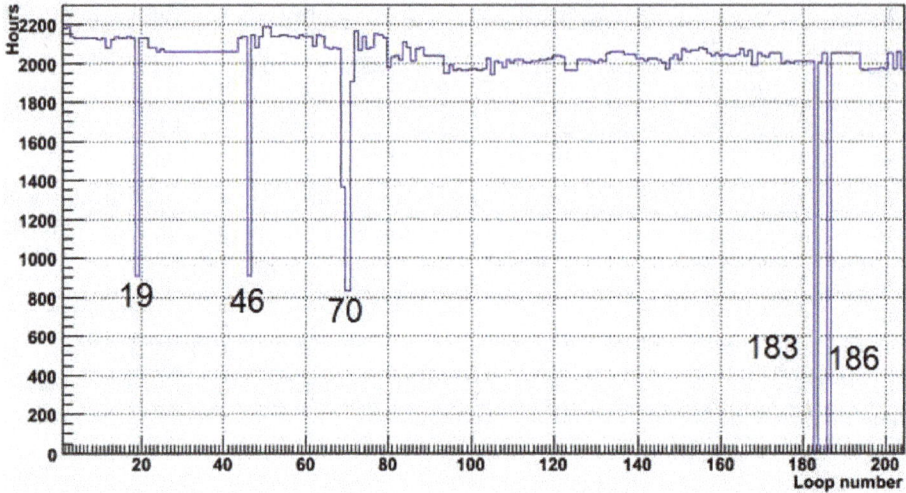

Fig. 11. Number of hours of operation for the 204 channels of the ATLAS evaporative cooling system in autumn 2008. Two SCT channels (183 and 186) were off because of leaks; three pixel channels (19, 46 and 70) have been switched off midway through the run to upgrade them with new diagnostic tools.

more details can be found elsewhere,[16] together with a full description of the cooling plant.

As part of the tracker (silicon) has to operate cold and part (TRT) has to operate at room temperature the environment around each subsystem has to be set and controlled independently. Each volume is filled with the appropriate dry gas (CO_2 for the TRT and N_2 for the silicon sub-systems) and surrounded by active thermal barriers to guarantee thermal neutrality and avoid condensation.

4.2. *Alignment*

The precise determination of the module positions is the crucial element for assure optimal performance of the ID tracking. ATLAS aims to achieve the high precision in coordinate determination by performing alignment using information from the reconstructed tracks. The alignment algorithms are based on minimizing the hit residuals, i.e. the distances between the hits on the modules and the fitted track positions at each module. To obtain track parameter uncertainties within 20% of the intrinsic resolution, silicon module positions must be known to an accuracy of 10 μm in the most sensitive coordinate. The track-based alignment algorithms require a sufficiently precise knowledge of the initial module positions to converge on accurate and precise final values. This is provided by mechanical and optical surveys during the construction of the modules, as well as during initial commissioning while the detectors are placed in their final location. The ATLAS SCT is also equipped with an interferometric system,[17] important as a cross-check and to eventually follow up on short term displacements. Optimal algorithm performance is obtained by combining information from data sets sensitive to different types of misalignment, including cosmic rays and beam haloes, with information from tracks from collisions.

Alignment constants will be calculated on a daily basis using $O(10^6)$ high p_T tracks from collisions. 200k cosmic muons taken over a period of several days in autumn 2008 have been used to obtain the preliminary results shown in Fig. 12. These results indicate that the knowledge of the original geometry is sufficient to have the alignment process converging well enough to come very close to the nominal resolution already with a reduced set of data. It is also worth mentioning the absence of significant time variations of the alignment constants over days.

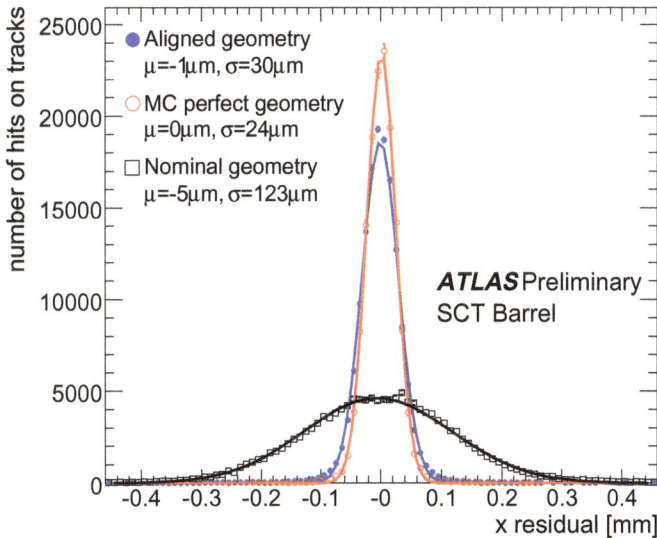

Fig. 12. Residual distribution in x, integrated over all hits-on-tracks in the SCT barrel for the nominal geometry and the preliminary aligned geometry. The residual is defined as the measured hit position minus the expected hit position from the track extrapolation. Shown is the projection onto the local x coordinate, which is the precision coordinate. Tracks are selected to have $p_T \geq 2\,\text{GeV}$ and an impact parameter to the nominal collision point smaller than $50\,\text{mm}$ across the beam and $400\,\text{mm}$ along the beam. Data are fitted with a single Gaussian.

4.3. *Material minimization*

As already mentioned, the ATLAS central tracker must cope with an unprecedented harsh environment. High speed, large channel density and high radiation dose all call for large power and therefore large mass (necessary for bringing the power in and for taking the heat out). This results in an overall weight ($\approx$4.5 tonnes) and material budget of the ID (shown in Fig. 13) much larger than for previous tracking detectors.

As the tracker X_0 limits the quality of some physics measurements planned by ATLAS (like $H \rightarrow \gamma\gamma$ or $H \rightarrow eeee$), huge efforts have been made to minimize it. Low X_0 alternative solutions have always received priority in the design of the construction phases. While this has been successful for the mechanical supports thanks to the know-how transferred from the space industry, some problems arose for the electrical and cooling services.

For instance, the aluminum-on-Kapton power and signal cables connecting the SCT end-cap to the first patch panel had to be changed to a more standard copper-on-Kapton solution after discovery of cracks in the conductor lines due to manipulation. This has happened in a late phase of the detector integration despite the careful prototype qualification tests.

The thin wall aluminum pipes of the cooling system required more than one iteration to solve all the fabrication and galvanic corrosion problems. Low mass hydraulic connections also turned out to be critical and required a lot of skill in the integration phase of all the silicon subdetectors. After installation in the experimental hall, about 1% of the silicon channels cannot be operated

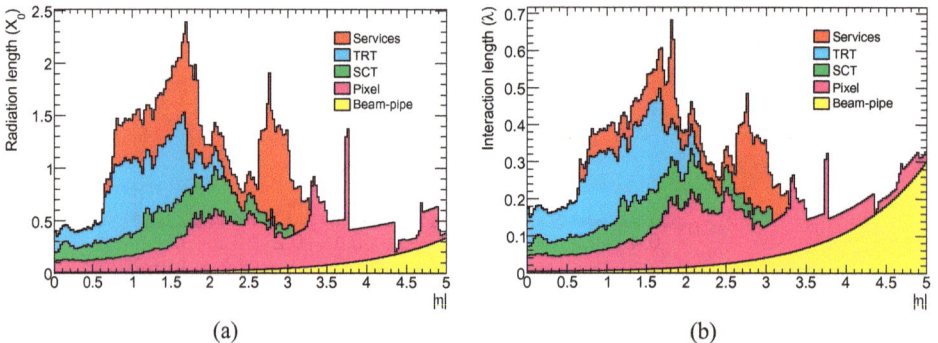

Fig. 13. Radiation length X_0 (a) and interaction length λ (b) as seen from a particle at the exit of the ATLAS inner detector envelope, including the services and thermal enclosures. The distribution is shown as a function of pseudorapidity $|\eta|$ and averaged over ϕ. The breakdown shows the contributions of different ID components.

because of cooling leaks. This is, at present, the largest individual source of inefficiency in the ATLAS tracker, but it can probably be reduced after verification that mildly leaky loops can be operated without problems.

5. Conclusions and Lessons Learnt

The ATLAS tracker has been installed in the experimental hall, carefully tested in stand-alone mode and then commissioned with the rest of ATLAS. This has allowed verification of its stable operation over a few months and within the performance requirements. Several million cosmic muons have been recorded, allowing careful verification of the alignment procedure and *in situ* confirmation of the many laboratory and test beam measurements made over a decade. All the information available indicates that the tracker is ready to efficiently collect p–p collisions in late 2009.

The design of the ATLAS tracker started ≈15 years ago and a lot of tests and engineering studies have been necessary in order to move from the conceptual design to a project able to survive a decade close to an LHC collision point. The time to define all the details of the detector, including the qualification of the many partners (laboratories and industries), has been significantly longer than the construction time itself. The length and complexity of the project has also implied that new concepts or techniques could not always be included in the final implementation, due to lack of long term experience sometimes enhanced by conservatism. The integration, installation and commissioning of such a large and diversified detector has also required more time and effort than initially planned for.

The main lesson learnt is that the most significant difficulties came from the noncore technologies (cooling system, electrical and optical connectivity) and noncore activities (high level integration, installation). This clearly indicates that the brainpower of the collaboration behind this detector was devoted to the study of the key detector elements more than to the optimization of the system aspects. This strategy finds its explanation not only in the difficulty of the challenge, but also in the cultural bias of the physicists involved. It is *a posteriori* clear that an additional engineering overview in the early stage of the project would have facilitated stable operation and easy maintenance.

References

1. *Proc. CERN–ECFA Workshop* (Lausanne–Geneve, 21–27 Mar. 1984). CERN Report 84–10 (5 Sep. 1984).
2. http://cdsweb.cern.ch/collection/DRDC%20Public%20Documents?ln=itas=

3. ATLAS Collaboration. Letter of Intent for a general-purpose pp experiment at the Large Hadron Collider at CERN. CERN/LHCC/92-4.

4. ATLAS Collaboration. Technical Design Report of the ATLAS Inner Detector. CERN/LHCC/97-16 (Vol. 1) and CERN/LHCC/97-17 (Vol. 2) (1997).

5. A. Ferrari, P. R. Sala, A. Fasso and J. Ranft. FLUKA: A multi-particle transport code (program version 2005). CERN Report CERN-2005-010 (2005).

6. C. Zeitnitz and T. A. Gabriel. The GEANT–CALOR interface and benchmark calculations for Zeus calorimeters, *Nucl. Instrum. Methods A* **349** (1994) 106.

7. R. Wunstorf *et al.* Investigation of donor and acceptor removal and long term annealing in silicon with different boron/phosphorus ratios, *Nucl. Instrum. Methods A* **377** (1999) 228.

8. R. L. Gluckstern. Uncertainties in track momentum and direction, due to multiple scattering and measurement errors, *Nucl. Instrum. Methods* **24** (1963) 381.

9. ATLAS Collaboration (G. Aad *et al.*). The ATLAS Experiment at the CERN Large Hadron Collider, *JINST* **3** (2008) S08003.

10. A. Clark *et al.* Status of the ATLAS SCT, *Nucl. Instrum. Methods Phys. Res. A* **579** (2007) 580.

11. F. Campabadal *et al.* Design and performance of the ABCD3TA ASIC for readout of silicon strip detectors in the ATLAS semiconductor tracker, *Nucl. Instrum. Methods A* **552** (2005) 292.

12. A. Abdesselam *et al.* The integration and engineering of the ATLAS semiconductor tracker barrel, *JINST* **3** (2008) P10006.

13. V. L. Ginzburg and I. M. Frank. *J. Phys. USSR* **9** (1945) 353.

14. V. L. Ginzburg and I. M. Frank. *J. Exp. Theor. Phys.* **16** (1946) 15.

15. T. Akesson *et al.* Status of design and construction of the transition radiation tracker (TRT) for the ATLAS experiment at the LHC, *Nucl. Instrum. Methods A* **552** (2004) 131.

16. D. Attree *et al.* The evaporative cooling system for the ATLAS inner detector, *JINST* **3** (2008) P07003.

17. S. M. Gibson *et al.* Coordinate measurement in 2-D and 3-D geometries using frequency scanning interferometry, *Opt. Laser Eng.* **44** (2005) 79.

Chapter 4

THE ALL-SILICON STRIP CMS TRACKER:
MICROTECHNOLOGY AT THE MACROSCALE

M. Mannelli

CMS Experiment, CERN
CH-1211 Genève 23, Switzerland
Marcello.Mannelli@cern.ch

The CMS silicon strip tracker is the largest device of its kind ever built, and is the first instance of a large collider detector relying exclusively on this all-silicon technology for tracking. With an instrumented surface of over $200\,\mathrm{m}^2$, and over nine million readout channels, it is two orders of magnitude larger than the silicon vertex detectors of the LEP experiments, and 30–40 times larger than the silicon inner detectors of the CDF and D0 experiments at the Tevatron. It makes use of microelectronics technology, but deploys it at the macroscale.

   The photograph below shows the completed CMS silicon tracker, ready for installation in the experiment, side by side with the silicon strip vertex detector of the OPAL experiment at LEP. Approximately a decade and

The CMS silicon strip tracker ready for insertion in CMS, standing next to the OPAL silicon strip vertex detector.

more than two orders of magnitude in scale separate these two silicon strip detectors.

In the following we motivate the choice of an all-silicon tracker for the CMS experiment, and discuss the reasons for the major design choices that define the system.

1. Introduction and Overview

In keeping with the spirit of this book, this is not a technical overview of the CMS tracker, or a detailed discussion of its expected performance. These can be found in Refs. 1, 2, 3 and 9. Rather, this chapter is a narrative discussion on the primary considerations that drove the underlying design choices that define the CMS tracker, and on some of the factors that proved crucial to its success. The electronics and optical link systems, key aspects of the CMS tracker, are described in other chapters of this book and therefore not discussed here.

The design of the CMS experiment reflects a shift away from an earlier skepticism regarding the potential for useful tracking at hadron colliders, in particular at very high energy and luminosity, and toward the recognition of the fundamental importance of precision tracking in a high magnetic field for an all-purpose discovery experiment. The CMS tracker is designed to provide a powerful tool for the identification and precise measurement of isolated leptons, photons, and jets over a large energy range, in conjunction with the CMS experiments 4 T solenoid magnet, and its calorimeter and muon systems. This was one of the key choices that made CMS what it is, namely a compact muon solenoid detector for the LHC. In addition, the inner pixel layers of the CMS tracker, described elsewhere in this book, allow for efficient identification of b jets over a large energy range.

Figure 1 shows one of the possible flagship physics channels at the LHC: the decay of a Higgs boson into a pair of Z bosons, one of which then decays into a pair of electrons, shown in green, and the other into a pair of muons, shown in red. The same event is also shown with the expected pileup of about 20 interactions in a bunch crossing at the design luminosity of 10^{34} cm^{-2}s^{-1}. It is evident that robust pattern recognition is essential for reliable tracking.

In the early days of thinking of the design for LHC trackers, the prejudice remained that the outer tracking region would be used for pattern recognition, with the reconstructed tracks then being extrapolated and correlated with the corresponding hits in the inner region of the tracker for vertex reconstruction. In view of the difficult environment of the LHC, with

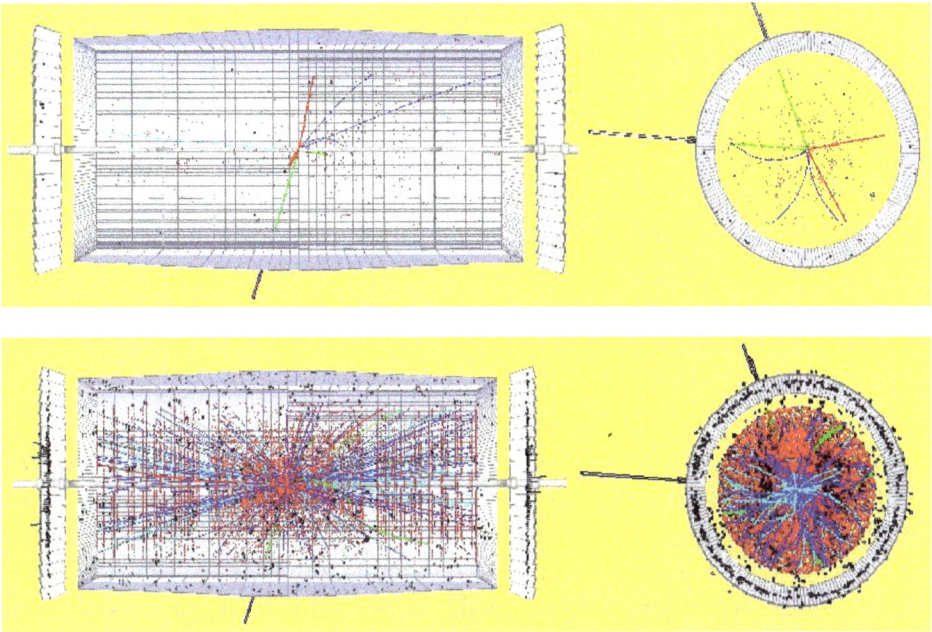

Fig. 1. The decay of a Higgs boson into a pair of Z bosons, one of which then decays into a pair of electrons, shown in green, and the other into a pair of muons, shown in red. The same event is also shown with the expected pileup of about 20 interactions in a bunch crossing at the design luminosity of $10^{34}\,\mathrm{cm}^{-2}\mathrm{s}^{-1}$.

some 20–25 minimum bias events superimposed in each bunch crossing at the design luminosity of $10^{34}\,\mathrm{cm}^{-2}\mathrm{s}^{-1}$, it was assumed that individual hits would inevitably be confused by overlapping tracks, and that robust pattern recognition would require a large number of measurements along track trajectories in the outer tracking region.

In contrast to that, the choice taken for the CMS tracker was to rely on the use of a relatively small number of sensor layers, each able to provide robust and high resolution measurements, thanks to a fine pitch and low cell occupancy, from the innermost region dedicated to vertex reconstruction to the entire tracking volume.[4] These requirements could be met by a set of inner pixel layers, complemented by an all-microstrip tracker. This was a courageous choice at the time, but one that can now be seen as far-sighted.

For the innermost vertex layers, pixel systems were proposed for the region below about 20 cm in radius, due essentially to their ability to provide full three-dimensional track and vertex determination. However, due to the considerable challenges and uncertainties faced by instrumenting this very difficult region with an ambitious technology still in its infancy, the inner pixel layers were initially given a lower priority for the tracker project. As discussed in the corresponding chapter of this book, these challenges were in fact surmounted, and a three-layer pixel system now sits at the heart of the CMS tracker.

At that time, in the early 1990s, there were two microstrip technologies with the potential to meet the performance requirements of the LHC environment, and with the potential for large scale deployment. They were silicon strip sensors and microstrip gas chambers (MSGCs).

Even so, the choice of sensor technology for the region between about 20 cm and 60 cm in radius was fraught with uncertainty at the time. Whereas silicon sensors constituted the best-understood potentially viable technology at the time, they had yet to be shown to function adequately in the very high radiation environment close to the interaction vertex, and a vigorous program of R&D was underway to establish their viability and optimize their design. At the same time as different design options for silicon sensors were being investigated, more exotic semiconductor materials for sensor fabrication were also being pursued within the larger community, such as diamond and gallium arsenide.

In order to maintain the overall number of channels at a manageable level, a few planes of strip sensors were envisaged, at radial intervals of approximately 10 cm. A strip pitch of order 100 μm and a length of order 10 cm were sufficient to provide both adequate resolution in the bend plane and sufficiently low cell occupancy for robust measurements. Stereo layers were proposed for providing additional information in the r–z plane. This alone implied the need for deploying silicon strip detectors for tracking at a scale of at least more than an order of magnitude larger than what had previously been achieved.

Although the rapid falloff in charge track density and radiation exposure eased the requirements for the surrounding region of the tracker, from about 60 cm up to the outer radius of approximately 110 cm, the sheer scale required provided a formidable challenge. In the early 1990s, when these ideas were taking shape, using silicon strip sensors for the entire tracking volume appeared out of reach: an all-silicon strip tracker would have required some two orders of magnitude more silicon sensors than any previous experiment had deployed at that time, and this raised obvious questions of production capacity and cost.

At that time, the alternative technology of choice for the outermost region of the CMS tracker was that of MSGCs, an innovative and promising new technique for gaseous detectors with a resolution and granularity similar to that achievable with silicon strip sensors. Interest in the MSGCs was primarily driven by cost considerations. Even though the technology was in its infancy, it was thought that the production process at a large scale would prove to be simpler, and thus less expensive than the production of silicon strip sensors.

By the second half of the 1990s, proof of principle of the performance potential of the chosen technologies had been established, and by 1998 these considerations had led to a Technical Design Report for the CMS tracker[1] with three inner silicon pixel layers, four inner silicon strip layers and six outer MSGC layers covering the central region up to $\eta \sim 1$, with end-cap disks extending the coverage to $\eta < 2.5$.

However, the step from successful proof of principle to the complete engineering of the two systems, and the viable production and deployment of both of these technologies on an unprecedented scale, remained a formidable challenge. By the end of 1999, all the elements had been put in place to establish the feasibility of an all-silicon strip tracker, and it was realized that moving from two technologies to a single one was a unique opportunity to concentrate all available effort on a reduced set of problems. It was on these grounds that the decision was taken to rebaseline the CMS tracker project as an all-silicon strip tracker, the first of its kind.[2]

In the following, we discuss how the performance requirements drove the basic design and layout of the CMS all-silicon strip tracker, and then review some of the key design features of the silicon strip sensors and modules, which are used throughout the tracker, and motivate the choices made. We then summarize and comment on the salient features of the expected performance of the CMS tracker, before concluding with a few remarks on some of the lessons learnt during this project.

2. Performance Requirements and Layout of the CMS Silicon Tracker

Much of the physics of interest at the LHC decays into, or is otherwise evidenced by intermediate vector bosons and their subsequent decays into high Pt isolated electrons or muons. It is therefore natural to require an experimental resolution on the mass of Z bosons decaying into electron or muon pairs, comparable to or smaller than the natural width of the Z. Such a resolution is also well matched to the discovery of new narrow resonances decaying directly into lepton pairs, such as the SUSY Higgs, over a large range of SUSY parameters. In addition, it allows for good electron and muon charge discrimination up to Pt of $\sim$2 TeV, corresponding to the approximately 4 TeV limit of discovery, for example of a new heavy Z' boson.

Whereas for Pt larger than about 15 GeV the electromagnetic calorimeter provides the primary precision measurement of the electron energy, CMS relies on the tracker for high precision muon momentum measurements over the full accessible range of Pt. A primary requirement for the CMS tracker was therefore the efficient identification of isolated high Pt leptons, and the momentum reconstruction of isolated muons with an indicative precision of $\Delta Pt/Pt \sim 1\%$ at 100 GeV Pt and 10% at 1 TeV Pt.[4]

The very large 4 T solenoid magnet at the heart of the CMS experiment is largely motivated by this requirement. To set the scale, the transverse momentum kick given to a charged track traversing a magnetic field is approximately 300 MeV/m*T. As a result of the 4 T field then, within the ~110 cm useful radius of the tracking volume, a track of Pt ~ 100 GeV will have an ~1.9 mm sagitta, and a track of Pt ~ 1 TeV a sagitta of ~190 μm.

If we consider a layout with equally spaced layers of strip sensors, and assuming a digital position resolution of approximately pitch/sqrt(12), then a resolution of $\Delta Pt/Pt \sim 0.1*Pt$ (Pt in TeV) would for example require 12 layers of sensors with an 80 μm pitch. Taking into account the improved resolution expected through the measurement of charge sharing on adjacent strips, a pitch of ~100 μm should in fact be sufficient.

Efficient and robust high Pt lepton track reconstruction with a relatively small number of layers requires a low cell occupancy, to ensure that the probability of any given hit being disrupted by an overlying track is sufficiently small. With this in mind, a target cell occupancy of order 1–2% was set, throughout the tracker volume.

In order to establish effective isolation criteria, and also to allow for effective jet reconstruction and tau-jet and b-tagging, efficient and clean inclusive track reconstruction, over a broad momentum range going down to at least ~2 GeV, is required, with a target efficiency of 90% or better. The high granularity and low cell occupancy target set above is also well suited to this task.

Figure 2 shows the strong radial dependence of the charged track density inside the CMS tracker volume, at the nominal 10^{34} cm^{-2}s^{-1} luminosity. In the absence of a magnetic field, the charged track density simply falls off as $1/r^2$. It can be seen that the charged particle rate is dominated by tracks with Pt < 1 GeV. The presence of the strong magnetic field curls up low momentum tracks and, as a result, it can be seen that the charged track density is increased below about 60 cm, and correspondingly decreased above 60 cm, with respect to the zero field situation. This further accentuates the radial dependence of the charged track density.

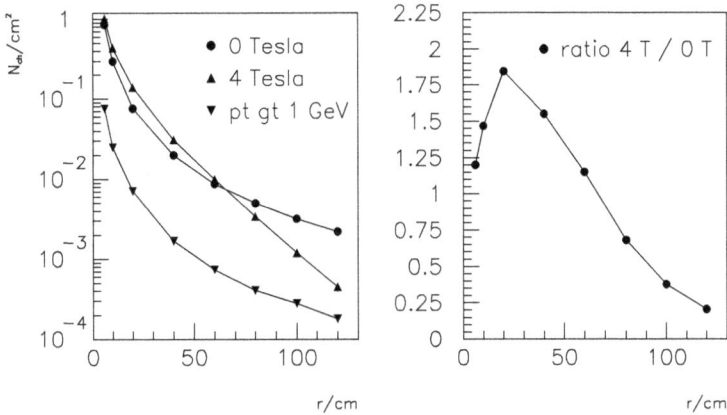

Fig. 2. The primary charged particle density per cm^2 at $\eta = 0$, for 20 minimum bias events superimposed (in units of %), from the CMS tracker TDR.[1]

The net result is a charged track density for each 25 ns bunch crossing ranging over three orders of magnitude, from about 10^{-3}/mm^2 at a 20 cm radius to about 10^{-4}/mm^2 at a 60 cm radius, and down to about 10^{-5}/mm^2 at a 100 cm radius. In practice, the interaction of particles with the material in the tracker contributes an additional source of occupancy, which increases with the radius and dampens substantially the radial dependence of the resulting hit density. With this in mind, cell sizes in the range of about 20–100 mm^2 are well matched to the 1–2% occupancy target.

Since the charged particle flux is dominated by very low Pt tracks, which cross the sensors at a shallow angle, the transverse cell size is effectively given by the active sensor thickness, rather than by the pitch. With a thickness ranging from 300 μm to 500 μm for the silicon sensors below and above 60 cm radius respectively, strip lengths of 10–20 cm will provide cell sizes in the desired range.

In addition to the precise measurements in the transverse plane necessary for determining the Pt of charged tracks, coarser measurements in the nonbending plane are also required. For these, a precision of order 1 mm is sufficient to ensure efficient matching with the inner pixel layers, and for extrapolating tracks to the electromagnetic calorimeter with a precision well within a single crystal. A stereo arrangement with a 100 mr angle was chosen to provide these measurements.

The layout of the CMS tracker is shown schematically in Fig. 3.

Given the large angular region covered by the CMS tracker, which extends up to $\eta \sim 2.5$, it was natural to use a barrel arrangement in the central region, complemented by disks in the forward regions.

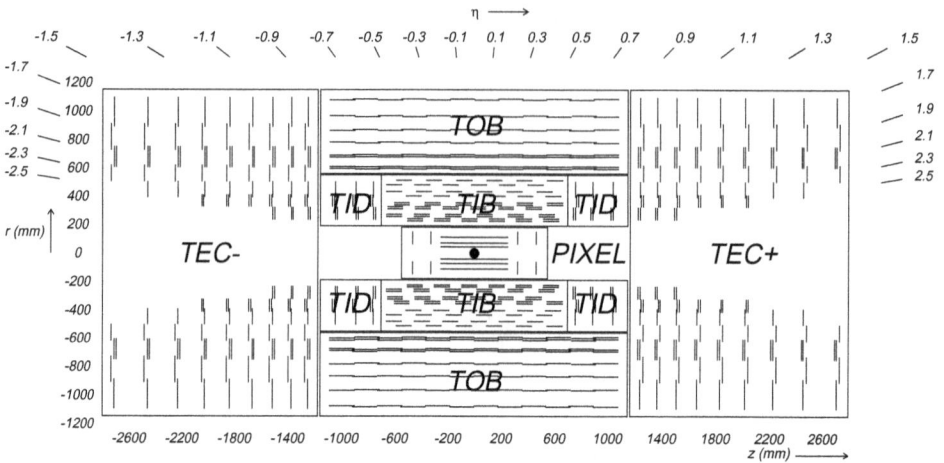

Fig. 3.   A schematic illustration of the layout of the CMS silicon tracker.[3]

The central barrels were subdivided in a tracker inner barrel (TIB) covering the region from 20 cm to 60 cm in radius, and a tracker outer barrel (TOB) covering the region above 60 cm radius, which allowed for separately optimizing the sensor and module geometries in these two radial regions.

In the TIB sensors of approximately 6 cm width and with a strip length of approximately 11 cm were used, whereas in the TOB sensors of approximately 9 cm width were paired and electrically daisy-chained within a module to provide an effective strip length of about 18 cm.

The radial interval between barrel layers was kept approximately constant at about 10 cm throughout the tracker, so that there are four layers in the TIB and six layers in the TOB. The inner two layers of each TIB and TOB incorporate stereo modules, with sensors placed at a 100 mr stereo angle, to provide the required nonbending plane measurements.

A natural transition point between barrel and disk geometries is at around 45°. In order to simplify the mechanics, and the arrangement of services in the transition region, it was decided to impose a constant length for each of the four TIB layers and six TOB layers. The transition between barrel and end-cap geometries is necessarily disruptive to the tracker measurements and, in order to provide a wide window of clean acceptance, the length of each TIB and TOB was set such that their coverage reaches approximately 45° at their respective outer radii and extends well beyond that at the inner radii.

As a result, the TIB is substantially shorter than the TOB. This required the addition of tracker inner disks (TIDs) to extend the coverage longitudinally, and provide a common transition to the tracker end-cap (TEC) disks.

There are nine TEC disks on each side of the CMS tracker, with seven rings of modules arranged in an annular fashion spanning the region from about 20 cm to 110 cm in radius. The sensors are wedge-shaped, with radial strips projecting toward the beam-line in the r–phi plane. The TEC disks also include stereo modules at approximately the same radii as in the TIB and TOB.

Unlike in the barrel region, the radius of the last measurement point varies as tracks span the TEC disks in η. The number and longitudinal position of the TEC disks were chosen to ensure a minimal lever arm similar to that of the penultimate TOB layer, of almost 100 cm. Similarly to TIB and TOB, there is a transition from single sensor modules with strip lengths

Fig. 4. (a) Putting the finishing touches to a TIB half-shell; (b) inspecting a complete TIB assembly.

Fig. 5. (a) The complete TOB is shown, with its services integrated on the TOB end-flange and the inner surface of the tracker support tube. (b) One of the two TIB assemblies is shown, after insertion inside the TOB.

around 10 cm below about 60 cm to modules with a pair of electrically daisy-chained sensors with an effective strip length of around 20 cm above 60 cm.

Below are a few photographs of the TIB, TOB and TEC tracker subdetectors, taken at various stages of their assembly and integration.

Although obviously different in many respects, the mechanical structures shared a number of features. The structures were arranged to ensure a viable level of modularity, by introducing substructures on which sets of modules could be assembled, and which integrated all the services required to fully operate and test each completed substructure independently. In the case of the relatively compact TIB, these substructures consisted of half-shells, each covering half the length of the TIB, and disks were used as the substructures for the TID. In the case of the TOB, ladderlike "ROD" substructures were used, with the analogous substructure for the TEC being pie-shaped "petals." These are shown in Figs. 6 and 7, respectively.

After careful testing, the TIB half-shells were directly assembled together, whereas for the TOB and TEC the RODs and petals respectively were assembled into framelike support structures. At the periphery of each the TIB, TOB and TEC manifolds and connectors provided a transition between the cooling and electrical services internal to the subdetectors, to pipes and

(a)   (b)

Fig. 6.   (a) Putting the finishing touches to a TEC disk; (b) a complete TEC assembly, with all its services and ready for insertion into the tracker support tube.

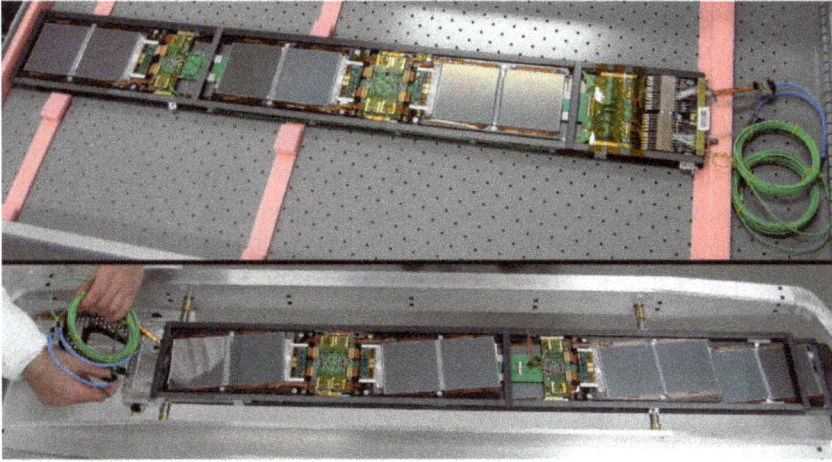

Fig. 7. Top: An assembled double-sided rod, showing the modules with sensors oriented along the r–phi view. Bottom: The opposite side of the same ROD, showing the stereo modules.

Fig. 8. The front and back sides of a TEC petal.

cables linking these to a patch panel outside the tracker volume. This transition was an important one for the rational integration of the subdetectors but, as will be seen later, it plays a dominant role in determining the material budget of the tracker.

3. Design Choices for the Silicon Strip Sensors and Modules of the CMS Tracker

At the heart of the CMS tracker are some 25,000 silicon strip sensors, arranged in approximately 17,000 modules. The silicon strip sensors and modules for the CMS tracker had to satisfy stringent performance requirements, in a harsh radiation environment, and also to be suitable for very large scale and consistently high quality production. In the following, we motivate the key design choices, which ensured that the CMS tracker silicon strip sensors and modules met these requirements.

3.1. *Radiation hardness requirements*

The high design luminosity of the LHC presents a very challenging radiation environment for the detectors. The charged and neutral hadron flux inside the tracker volume has been simulated in detail, and the results are summarized in Table 1.

Table 1.   The expected hadron fluence and radiation dose in different radial layers of the CMS tracker, in the central region, for an integrated luminosity of $500 \, \text{fb}^{-1}$, representative of 10 years of nominal LHC operation. The fast hadron fluence is a good approximation to the 1 MeV neutron equivalent fluence, used to calculate radiation damage.[3]

Radius (cm)	Fluence of fast hadrons $(10^{14} \, \text{cm}^{-2})$	Dose (kGy)	Charged particle flux $(\text{cm}^{-2}\text{s}^{-1})$
4	32	840	10^8
11	4.6	190	
22	1.6	70	6×10^6
75	0.3	7	
115	1.2	1.8	3×10^5

Whereas the charged hadron flux falls off sharply with the radius, the neutron flux has a much weaker radial dependence. At a radius of 20 cm the charged hadron flux dominates by over an order of magnitude. By about 60 cm the overall hadron flux has fallen by about a factor of 4, driven by the fall in the charged hadron flux, which has decreased to a level similar to the neutron flux. The overall hadron flux then decreases by only about a factor of 2, from 60 cm to 100 cm.

It will be seen that this radial dependence of the hadron flux offered an opportunity to separately optimize the sensor design in the regions below and above an ∼60 cm radius, and to provide for the use of longer strips in the outer region of the tracker, where the reduced charged track density allows the channel occupancy targets to be met with larger cell sizes.

3.2. *Choice of substrate type*

The first — and most crucial — issue is the charge collection efficiency of the two types of charge carriers: electrons and holes. The operational conditions under which efficient charge collection is achieved have a fundamental impact on the design requirements for the sensors, and indeed on some important aspects on the overall system.

The standard material for sensor fabrication is high resistivity n-type bulk, fabricated with the so-called float zone process.

The simplest sensors use heavily doped p++ implanted strips to provide a p–n junction on which an electric field can be applied and charges generated by tracks crossing the sensor, in this case holes, can be efficiently collected. A heavily doped n++ back-side implant is used as a charge injection barrier. Such single-sided p-on-n sensors require lithographic processing on the p++ strip implant side only (single-sided process). They are therefore potentially suited to large scale production and deployment.

In more sophisticated, double-sided sensors, the n++ back-side implant is also segmented in order to be able to collect electrons in addition to holes. In this case, lithographic processing is required on both sensor surfaces (double-sided process), and the processing of the back-side is further complicated by the need for additional p structures, which are required to ensure the isolation of the charge collecting n++ electrodes (p-stop and/or p-spray). These additional structures required to isolate the n++ electrodes also increase the capacitive load of these electrodes, and therefore result in higher electronic noise.

Radiation damage substantially affects the sensor characteristics. The three main effects are changes in donor and receptor concentration in the bulk, an increase in leakage current, and increased concentration of trapped charges at the silicon to silicon-dioxide interface, at the surface of the sensors.

It had long been known that the most visible radiation damage effect on the silicon sensors, aside from a linear increase in the leakage current, is to increase the concentration of donor-like sites, which will cause an initially n-type bulk to rapidly undergo "type inversion" to become an effectively p-type bulk. As a result of this, the p–n junction of the sensor could be expected to move from the p++ strip implants to the n++ back-side implant. This raised serious doubt as to the feasibility of efficient hole collection with the p++ strips, and was an early argument in favor of electron collection with n++ electrodes. An additional argument in favor of electron collection is their higher mobility, which raised the prospect for a lower trapping rate.

Those considerations, coupled with the extremely high fluences at radii of 10 cm and below, drove the choice of so-called n-on-n sensors for the inner pixel layers. These sensors, in which the n++ electrodes on an (initially) n-type bulk are used to collect electrons, have indeed demonstrated the ability to efficiently collect charge at these very high fluences. However, this comes at the price that a costly double-sided process is required to produce what is effectively a single-sided sensor. While this is viable on the scale of the inner pixel layers, which have about $1\,m^2$ of silicon, it would have been a very costly solution for surfaces one or two orders of magnitude larger.

An interesting alternative might have been to use n++ strips on a p-type bulk. While this would not have removed the need for p-stops and the associated complications, it would have allowed the collection of electrons with single-sided sensors. At the time, however, adequate p-type wafers were unavailable and this was not a viable option.

In this context, a crucial set of measurements in 1995–1996 clearly showed that efficient hole collection on the p++ strips is indeed possible at radii above 20 cm, provided that the sensors can be operated at a sufficiently high bias voltage, of up to 400 V.[5] This is to be compared with most previous systems, for which the bias voltage was not required to exceed 100 V, and was typically substantially lower than that.

While on the one hand this demonstrated the potential viability of the simpler p-on-n sensors to instrument the strip tracker, on the other hand it set unprecedented requirements on the quality and robustness of the sensors, to ensure reliable operation at such high bias voltages and notwithstanding the effects of radiation damage.

As a result, an intensive program of R&D was focused on optimizing both the design and detailed choices of materials for radiation-hard p-on-n sensors, in view of very large scale fabrication and deployment of these sensors, and robust and reliable operation at very high bias voltage.

3.3. *Coupling to readout electronics and its implications for the sensor technology*

An early design choice was to arrange for AC coupling of the sensors to the readout chip. This choice avoided the need for the readout chips to maintain the strips at ground voltage supply, and to sink the relatively large leakage currents required to do so. The option of producing separate ad-hoc coupling capacitors and biasing structures for each strip was briefly considered, and discarded. Instead, these were integrated directly onto the sensors.

This was a widely debated choice at the time. On the one hand, it removed what many felt was a potential source of possible difficulties and electronic performance degradation of the front-end readout system; on the other, it made substantial demands on the sensor technology. In the end, it was felt that the technology for integrating the coupling capacitors and bias structures onto the sensors was sufficiently mature, and that this was the best choice for the overall robustness of the system.

In order to ensure efficient charge collection from the readout electrodes, the coupling capacitors must be substantially larger than the strip capacitance. Coupling capacitors directly integrated onto the sensors can cover the full length of the strips. This is a major advantage over coupling capacitors produced on an ad-hoc structure, external to the sensor, as these are subject to severe space constraints. The coupling capacitors are realized with thin oxide layers separating the strip implants from the readout metal electrode. Since these oxide layers can be thin, coupling capacitances more than an order of magnitude larger that the implant strip capacitance can be produced in this way. The major challenge associated with this technique is that a single pinhole through the dielectric layers forming the coupling capacitors will disable the readout strip and, depending on the details of the readout chip design, may also adversely affect the performance of neighboring channels.

Almost all strip sensor designs feature a bias ring which surrounds the sensor active surface, and which is maintained at ground potential. The bias ring is itself surrounded by one or more guard rings, which are usually kept floating, and which isolate the active area of the sensor from leakage current around the sensor periphery. The strips themselves are biased to ground through a high impedance connection to the bias ring. Two types of structures were studied to provide this high impedance bias to the strips.

The most economical biasing structure can be implemented simply by designing the gap region between the ends of the strip implants and the bias ring such that a "punch-through" mechanism will effectively provide a high impedance connection between them. An electrode placed on top of the oxide layer above the gap region may also be used, to control the punch-through behavior of the device. This technique is attractive because, while it involves some design optimization, it requires no additional processing steps to produce.

Such structures had been used successfully before, for example in the silicon strip vertex detectors of some of the LEP experiments.

For heavily irradiated sensors, the bias structures must compensate for a high level of leakage current, and may in addition be affected by radiation damage. To distinguish between these two effects, the characteristics of punch-through structures of heavily irradiated sensors were compared with those of nonirradiated sensors in which comparable levels of leakage current were induced. It was found that the punch-through characteristics of the two sets of sensors exhibited the same leakage current dependence, with no evidence of radiation damage directly affecting this behavior.

However, earlier results from the CDF experiment had shown that punch-through biased strip sensors showed anomalously high increase in shot noise with radiation.[7] In light of this, measurements of shot noise as a function of leakage current were performed on this set of punch-through biased sensors. The results confirmed the anomalous shot noise leakage current dependence, but found that this was the same for the irradiated and nonirradiated sensors. The anomalous shot noise therefore could be ascribed to an intrinsic property of current flow through the punch-through structures, rather than the direct consequence of radiation damage to the punch-through structures.

Nonetheless, this behavior ruled out the simple punch-through structure as a viable biasing technique for the LHC.

This drove the choice for biasing with integrated polysilicon resistors, connecting the strip implants to the bias ring implant. The bias resistance was chosen to be 1 MΩ, sufficiently higher than the expected input impedance of the readout chip to ensure efficient charge collection.

3.4. *Optimization of sensor design and parameters*

Having defined the sensor technology, attention then turned to determining the dependence of the critical sensor characteristics on the design parameters and substrate type. Systematic studies were carried out, using test wafers with multiple geometries designed so that the dependence of sensor characteristics on each of the parameters under scrutiny could be clearly demonstrated, over a range of the strip pitch from 60 μm to 280 μm. In addition to geometric design parameters, different substrates types were also studied, to examine the influence of the choice of crystal lattice orientation relative to the sensor plane, of the substrate resistivity and of the substrate thickness on the sensor characteristics and performance.

In the following, we motivate the detailed choices for the geometry of the strips, the crystal lattice orientation of the silicon wafers, as well as the resistivity and thickness of the wafers. Finally, we motivate our choice for moving from what was the de facto standard of 4″ diameter wafers at the time to production on 6″ wafers.

An important sensor characteristic is the capacitance that the readout strips present to the input of the readout chip, as this plays crucial a role in determining the electronic noise. The dependence of electronic noise with input capacitance for the CMS tracker readout chip (APV25) is shown in Fig. 9.

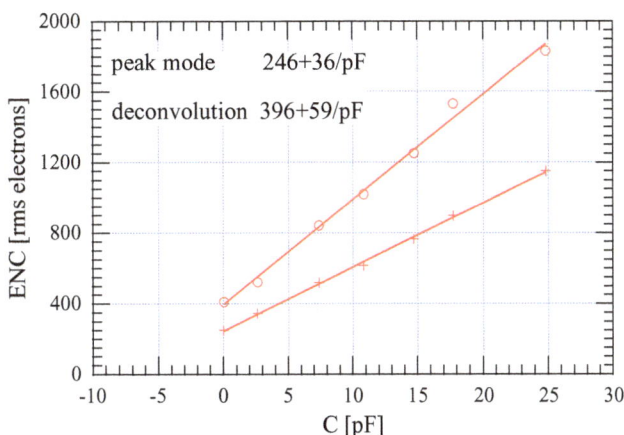

Fig. 9. The electronic noise of the CMS tracker APV25 readout chip, as a function of the input capacitive load. The APV25 is designed for operation in deconvolution mode, to ensure clean discrimination of signals from each LHC bunch crossing, which are separated by 25 ns.[2]

The measurements on the multigeometry test sensors produced some unexpected results, which are summarized in Figs. 10 and 11.[8]

The first result, shown in Fig. 10, was that the total strip capacitance depends only on the ratio of the strip implant width to the strip pitch (w/p), and is independent of the strip pitch and sensor thickness, at least as long as the sensor thickness is larger than the strip pitch. The total strip capacitance is composed of the strip capacitance to the backplane and to the neighboring strips (interstrip). The observed behavior, which contradicted the prevailing expectation that larger strip pitches would translate to larger strip capacitance, is due to a subtle interplay of the strip backplane and interstrip capacitance, whose dependence on the strip pitch and on the substrate thickness exactly offset each other, as shown in Fig. 11. A similar behavior can be seen as the thickness of the sensor is varied, for a fixed value of the w/p ratio.

This implied that, provided a uniform value for the w/p ratio was maintained, all the sensor geometries under consideration could be expected to have similar capacitive noise behavior, with a simple linear dependence on the strip length. That this is indeed the case can be seen in Fig. 12, where the noise performance of the different silicon strip module types of the CMS tracker is shown, as a function of the strip length.

The second result concerned the observed differences in the strip capacitance of sensors produced on substrates with the sensor plane cutting

M. Mannelli

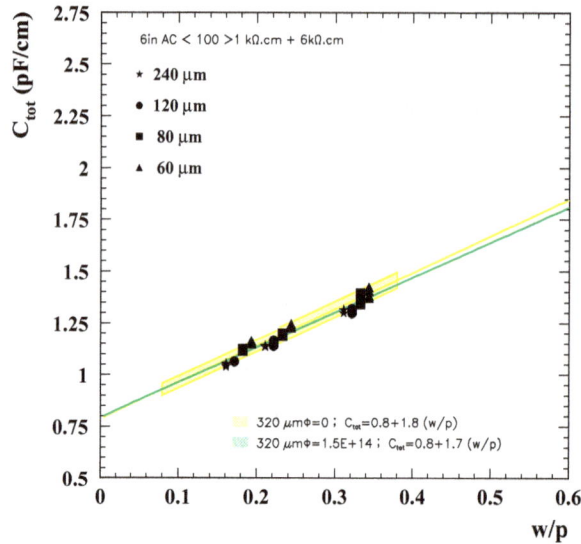

Fig. 10.   Total capacitance (C_{tot}) of sensors with the strip ranging from $60\,\mu$m to $240\,\mu$m, as a function of the w/p ratio, manufactured on substrates with $\langle 100 \rangle$ crystal lattice orientation. The symbols show the measurements after irradiation equivalent to a nominal 10 years of LHC operation, whereas the yellow band shows the range of values measured on sensors prior to irradiation.[8]

diagonally through the crystal lattice $\langle 111 \rangle$, compared to identical sensors produced on substrates with crystal orientation orthogonal to the sensor plane $\langle 100 \rangle$.

For mainly historical reasons, most strip sensors to date had been fabricated on $\langle 111 \rangle$ substrates, and this was therefore the default choice also for the LHC trackers. However, for sensors fabricated on the $\langle 111 \rangle$ substrate, there was considerable evidence that the overall strip capacitance, in particular the interstrip capacitance, was substantially increased by irradiation, due to the effect of radiation-induced trapped charges at the interface of the bulk silicon and the surface silicon-dioxide layer.

This result was indeed confirmed. The increased strip capacitance was found to be most significant for bias voltage below the full depletion voltage, but the increase remains significant, even at much higher bias voltage, well above the depletion voltage. In addition, the strip capacitance showed clear signs of hysteresis, which would further complicate the sensor operation.

On the other hand, no such radiation-induced increase in strip capacitance was found for the sensors produced on the $\langle 100 \rangle$ substrates, as can be

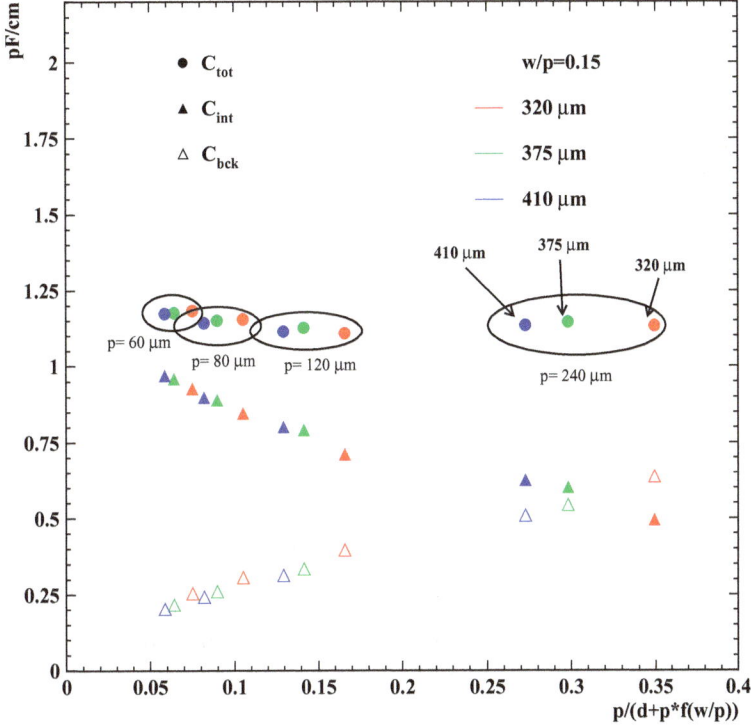

Fig. 11. Backplane capacitance (C_{bck}), interstrip capacitance (C_{int}) and the sum of the two (C_{tot}) measured in sensors with the pitch (p) ranging from $60\,\mu$m to $240\,\mu$m, and of thickness (d) in the range from $320\,\mu$m to $410\,\mu$m, but with the same value for the width-to-pitch ratio (w/p).[8]

seen from Fig. 10. This effect has been ascribed to the cleaner cleavage of the lattice at the surface of the sensor, compared with the worst-case situation provided by the $\langle 111 \rangle$ substrate.

This result, and the fact that all subsequent scrutiny found no adverse effects, motivated a break with the past practice, and the choice of $\langle 100 \rangle$ substrates for the silicon strip sensors of the CMS tracker.

As previously mentioned, an essential condition for efficient charge collection with p-on-n sensors after irradiation is the ability to operate at very high bias voltages, well above the full depletion voltage. This imperative motivated many studies aimed at optimizing guard ring structures surrounding the sensor active area, as well as the detailed geometry of the readout strips. Two important results drove the detailed design choices aimed at ensuring optimal high voltage performance.

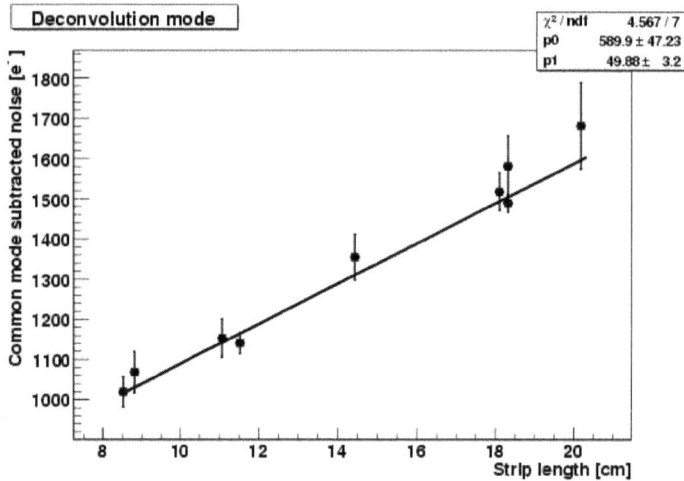

Fig. 12.   The measured noise performance for the different geometries of the CMS silicon strip tracker modules. The strip pitch ranges from 80 mm to 240 mm, and the sensor thickness is 320 mm for the short strip modules and 500 mm for the long strip modules. All geometries, however, have a constant value for the w/p ratio of 0.25. It can be seen that, as a result, the noise does indeed scale with the strip length, and is independent of the strip pitch and sensor thickness, as expected.[3]

The first result concerned the use of metal electrodes wider than the strip or guard ring implants (metal overhang). This arrangement removes the highest fields, at the edges of the strips, from the silicon bulk and into the silicon-dioxide layer at the surface of the sensor. Since the silicon dioxide has a breakdown voltage that is typically at least an order of magnitude higher than that of the silicon bulk, it was surmised that this arrangement would improve the high voltage stability of the sensors. Not only was this expectation confirmed, it was also found that the effect of the metal overhang on the strip capacitance was small enough to be essentially negligible.[8] These results motivated the use of a metal overhang for the strips and guard rings of the CMS tracker silicon strip sensors.

Our studies also confirmed the expectation that high voltage stability is best for geometries with a large w/p ratio, whereas strip capacitance and thus electronic noise are minimized for small values of the w/p ratio. A value of 0.25 was chosen for the w/p ratio, which was shown to be safely above the minimum required to ensure stable operation at up to 600 V, and yet small enough to result in an acceptable strip capacitive load and electronic noise.

Having settled on these parameters, and optimized the sensors for a low capacitive load, to minimize electronic noise, and stable high voltage operation, to maximize charge collection efficiency, the remaining choices concerned the substrate resistivity and thickness.

As discussed above, there is a steep fall of charged track density with the radius in the tracker volume. One consequence of this is that larger cell sizes, and in particular substantially longer strips derived by electrically daisy-chaining two sensors together, may be used in the outer part of the tracker compared to the inner part, while preserving cell occupancies within the target range. On the other hand, longer strips result in a large capacitive load and electronic noise, which must be compensated for by a higher collected charge in order to preserve the signal-to-noise performance (S/N). As can be seen from Fig. 12, with a 1.2 pF/cm capacitive load the electronic noise grows from $\sim$1'100e$^-$ for $\sim$10-cm-long strips, to $\sim$1'600e$^-$ for $\sim$20-cm-long strips.

The obvious way, and indeed the only way, to increase the collected charge in a silicon sensor is to increase the thickness of the sensor substrate. However, for heavily irradiated sensors the sensor thickness will be limited by the corresponding increase in the bias voltage required for efficient operation of thicker sensors, for which the target limit was 400 V. Clearly, this limit depends on the fluence to which the sensors are expected to be exposed, to over the lifetime of the detector.

Whereas the overall hadron fluence decreases by a factor of approximately 4 from the inner radius to approximately the center of the CMS strip tracker, from 60 cm to the 110 cm outer radius of the tracker, the hadron fluence decreases only by about a factor of 2.

Taken together, these considerations motivated the study of charge collection for sensors of different thickness for the inner and outer regions of the CMS strip tracker, with 60 cm being the approximate dividing line between these two regions.

As a result of these studies, we finally settled on a substrate thickness of $\sim$320 μm for the inner "thin" sensors, resulting in a nominal charge of $\sim$24'000e$^-$, and 500 μm for the outer "thick" sensors, resulting in a nominal charge of $\sim$40'000e$^-$. The 500 μm thickness was also chosen because, in addition to providing sufficient charge, it constitutes a more broadly adopted industrial standard, and it was expected that this would ease the very large scale production of the thick sensors.

The resistivity of the substrate plays an important role in the evolution of the bias voltage required for efficient sensor operation with irradiation. This parameter was therefore also separately optimized for the thin and thick sensors deployed in the regions above and below $\sim$60 cm in radius.

M. Mannelli

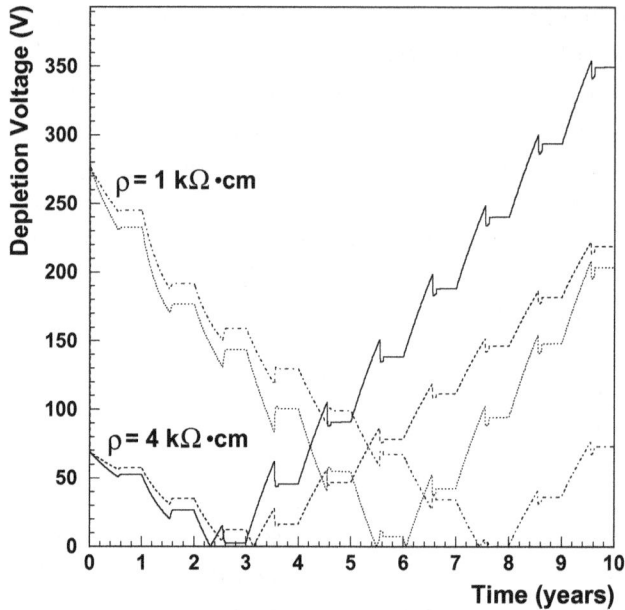

Fig. 13. The predicted evolution of the depletion voltage with time for barrel layer 1, at a 22 cm radius, for two different initial values of resistivity. The worst-case scenario of fluence 1.5 times higher than nominal is also shown.[1]

The voltage required to deplete the sensor is proportional to the square of the thickness and inversely proportional to the resistivity. In order to ensure efficient charge collection, over the full thickness of the sensor, these must be operated at a bias voltage above the full depletion voltage.

For the thin sensors, exposed to the largest fluence, a relatively low value of resistivity was chosen. While this resulted in a high initial bias voltage, it also delayed the onset of type inversion of the substrate from n-type to p-type, and reduced the bias voltage required for efficient charge collection at the end of the operational lifetime of the detector.

This is illustrated in Fig. 13. The depletion characteristics of a highly irradiated sensor are far more complex than this. It remains nevertheless true that, in the case of highly irradiated sensors, these must be operated at bias voltages well above full depletion to achieve efficient charge collection, and that the simplified model on which Fig. 13 is based provides a useful guideline.

For the thick sensors, instead, the expected fluence is expected to be too low to induce type inversion of the substrate, and a low initial resistivity would result in a very high initial bias voltage.

3.5. *Moving from 4″ diameter to 6″ silicon sensor production lines*

Finally, the usual standard for fabrication of silicon strip sensors was based on 4″ diameter wafers, which provide a useful sensor surface of up to about $40\,cm^2$. While this was well matched to the relatively small sensor sizes best suited to vertex detectors at small radii, it presented several difficulties for the CMS strip tracker. In terms of sheer numbers, the over $200\,m^2$ of active silicon strip sensors envisaged for the CMS tracker would have required the production, testing and subsequent manipulation and integration into modules of some 50,000 sensors.

By the late 1990s, a trend toward setting up 6″ wafer silicon sensor production lines was beginning to take shape. The use of larger area wafers would make possible a corresponding reduction in the number of wafers required to produce the sensors for the CMS tracker, thereby considerably reducing the production volume, and associated projected costs. In addition, at least one well-established producer of silicon sensors had demonstrated the ability to maintain a high yield for the larger sensors that could be fabricated on the 6″ wafers. This allowed the production of large area sensors, in the range 80–90 cm^2, better matched to the geometric constraints of our layout and the need to cover very large surfaces, without a substantial cost penalty, and would correspondingly reduce the number of sensors to be manipulated and assembled into modules.

By 1999, in view of the above, we chose to base the production of the silicon sensors for the CMS tracker on 6″ wafers. Even so, the CMS tracker required the production of about 19,000 thick sensors, 6500 thin sensors, and their subsequent assembly in approximately 17,000 modules — a truly enormous task by any measure.

3.6. *Design considerations for the CMS silicon strip modules*

The function of the modules is to integrate all the required elements, including sensors, front-end readout hybrids, and pitch adapters, onto a support frame, to create a low mass, self-contained unit, which can be accurately positioned on the supporting mechanical structures and efficiently cooled.

The modules vary in detail, reflecting the different geometries required by the layouts of the TIB/TID, TOB and TEC subdetectors. In particular, the TIB/TID modules, and the modules for the inner rings of the TEC modules, all have a single thin sensor, whereas the TOB modules, and those for the outer TEC rings, all house a pair of electrically daisy-chained thick sensors.

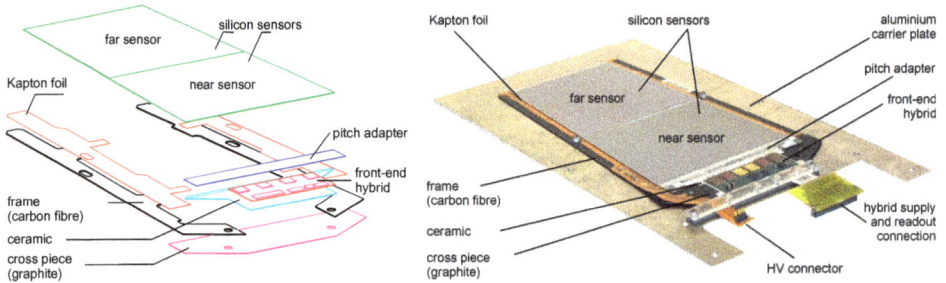

Fig. 14.   Left: An exploded view of a module housing two sensors. Right: An assembled TEC ring 6 module, mounted on a temporary carrier plate.

In addition, whereas the TIB and TOB sensors are all rectangular, the TID and TEC sensors are wedge-shaped.

All module types, however, share a number of basic design features, which were driven by the underlying functional requirements and the need for a design suitable for large scale, automated and high quality assembly. A schematic view of a TEC module, and its various components, are shown in Fig. 14. All modules in the CMS tracker share the basic features illustrated in this figure.

Beyond integrating the various components on a rigid support, the most important requirement for the module was to ensure the correct functionality of the silicon sensors. This required taking the necessary precautions to allow operation with a bias voltage as high as 600 V, avoiding undue mechanical stress on the sensor and, most crucially, ensuring the thermal stability of the sensor.

The first two requirements were a matter of detailed design optimization and choice of materials and adhesives used for the module and its assembly (but, remember, the devil is in the details).

The third requirement drove much of the design not only of the module itself, but also of the design and integration of the cooling circuits into the support mechanics, and of their coupling to the modules.

The issue of the thermal stability of a heavily irradiated silicon sensor is illustrated in Fig. 15.

The bulk leakage current of a silicon detector has an approximately exponential dependence on temperature, varying by a factor of 2 for every approximately 7°C change in temperature. Not only does irradiation increase the bulk leakage current by orders of magnitude, but it also gives rise to the need for high bias voltage operation to ensure efficient charge collection. As a result, at the end of the 10 years of LHC operation the silicon sensors will themselves dissipate substantial power and be subject to self-heating. If the

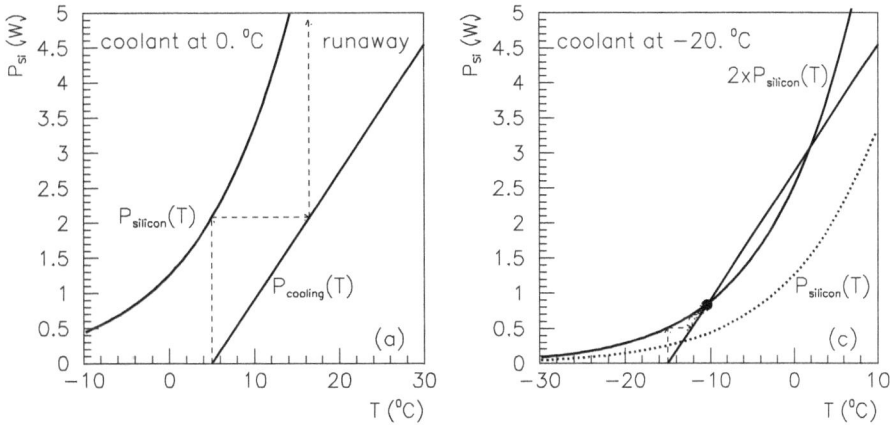

Fig. 15. The silicon sensor temperatures for a representative prototype module design, for two different cooling fluid temperatures. The curves show the temperature dependence of power dissipation of the silicon sensor, with a 500 V bias voltage, reflecting the underlying temperature dependence of the silicon sensor leakage current. The straight lines show the power that the cooling system removes for a given temperature of the silicon. Measurements were taken at several points on the silicon sensor: the line corresponding to the worst-case position is shown. The situation shown in (a), with a cooling fluid temperature of 0°C, is unstable: the sensor temperature will continue increasing unchecked (thermal runaway). The situation shown in (c), with a cooling fluid temperature of −20°C, is stable, even assuming twice the nominal power dissipation for the silicon sensor.[1]

resulting self-heating exceeds the capacity for cooling the sensor, a thermal runaway process will ensue and the sensor will rapidly be disabled.

For a well-optimized design, the cooling capacity can only be improved with the addition of material, which in turn degrades the performance of the detector. On the other hand, operating the sensors at a lower temperature reduces the current drawn and the resulting power dissipation, thereby easing the problem. The curves of Fig. 15 show how, by lowering the temperature of the cooling fluid, stable operating conditions can be achieved for a given cooling configuration. As a result of this and other, similar studies, the target operating temperature for the silicon sensors was set at −10°C, with the guideline that this should be achieved for a cooling fluid temperature of around −25°C.

3.7. *Automated assembly of the silicon strip modules for the CMS tracker*

The assembly of the module components onto their frame is a challenging task. It requires the accurate dispensing of adhesives, and the manipulation and high precision placement of fragile and costly components, such as the sensors, front-end readout hybrids and pitch adapters. Typical tolerances are in the range of $10\,\mu m$ or better, and excellent reproducibility is essential.

Until the advent of the CMS tracker project, module assembly had typically been the task of highly skilled technicians, performing a set of demanding manual operations, with the help of specially designed jigs and precise coordinate measurement machines.

The CMS strip tracker, however, required the assembly of some 17,000 modules, and this clearly necessitated the development of a new approach. The electronics industry had for some time already made extensive use of automated glue-dispensing machines, and of automated pick-and-place machines for the manipulation and placement of components onto a substrate, for large volume assembly of electronic modules. It was natural to take this as a model for the large volume assembly of the CMS silicon strip tracker modules, and make the transition from a handcrafted to an automated process.

The first obvious step in doing so was a survey of available automated glue-dispensing and pick-and-place machines. This, however, led to the conclusion that none of these was adequately suited to the assembly of our modules. The main problem was the requirement to safely handle the very large area sensors, as well as the relatively bulky hybrids, and to place these with sufficient accuracy. As a result, in 1996 the development of a robotic device ("gantry") specifically suited to the automated assembly of the CMS tracker modules was undertaken.[9]

This was based on a commercially available precision X–Y machine, with a large two-dimensional span. An innovative pneumatic head was developed, capable of picking up and operating a set of tools ranging from syringes for glue-dispensing and pick-and-place tools for manipulating and placing the different module components. A video camera with pattern recognition capability provided the ability to automatically spot targets on the different components, and ensure their accurate placement with respect to a set of nominal fiducial marks on the assembly table.

This development took place in parallel with the module design and, as mentioned before, the suitability of the module for automated assembly using the gantry was a driving factor in many detailed aspects of its design.

Figure 16 shows two photographs of a module assembly gantry in action. Several such gantries, all based on the same core hardware and software, but adapted to the production of specific module types, were distributed among the collaborating institutes, and were successfully used to assemble the approximately 17,000 silicon strip modules of the CMS tracker.

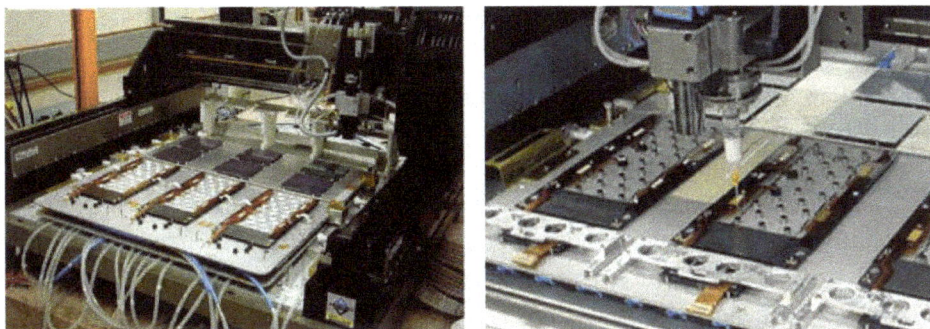

Fig. 16. Two views of the CMS tracker module assembly gantry at work. This custom automated assembly system uses a camera interfaced to a pattern recognition program to survey the various components, and a pneumatic head capable of picking up and operating a set of tools including syringes for glue-dispensing, and pick-and-place tools for manipulating and placing the different module components.

Assembly of the module was, of course, not the only challenge associated with the module production. The reliable bonding of the more than 25 million wire bonds required was also a massive challenge, and was only made possible by highly skilled dedicated teams using state-of-the-art wire-bonding machines at several collaborating institutes.

4. The Performance Potential of the CMS All-Silicon Tracker

The inner pixel system of the CMS tracker plays a crucial role in all aspects of tracking. Indeed, not only does it give the required impact parameter resolution for the tagging of b-jets, but it also provides the initial seeds for track reconstruction, which are then propagated from the inside of the tracker out.

Initiating track reconstruction in the most congested region of the tracker was not immediately intuitive, and ran against previous practice. However, in addition to providing high resolution three-dimensional coordinates, the small pixel cell size results in extremely low cell occupancies of about 10^{-4}, some two orders of magnitude lower than elsewhere in the tracker. The proximity to the beam line also allows the use of the transverse beam spot as a very powerful constraint in reducing the combinatorial complexity for the crucial track-seeding step. These factors make tracking from the inside out by far the most computationally efficient approach. In addition, tracking from the inside out is very advantageous for reconstruction particles, such as pions or electrons, which interact with the material of the tracker as they traverse it.

Muons are reconstructed with an efficiency of better than 99%, over the full acceptance of the CMS silicon tracker. The simulated track parameter resolutions obtained for single muon tracks are shown in Fig. 17.

The middle and right hand side plots showing the impact parameter resolution reflect the performance of the inner pixel layers. From the middle plot it can be seen that a transverse impact parameter resolution of about $20\,\mu m$ is achieved for $10\,GeV$ tracks, over the full range of the acceptance. The right hand side shows that excellent resolution is also achieved in the longitudinal impact parameter.

The left hand plot shows the Pt resolution of the tracker. It can be seen that the Pt resolution in the central region of η up to $\sim$1 is approximately 1.5%, at which point it takes a step up to about 2% until $\eta \sim 1.6$, beyond which it rapidly degrades due to the loss in the lever arm in the forward region, eventually reaching $\sim$7% at $\eta \sim 2.5$.

This is an excellent result, which comes very close to achieving the indicative Pt resolution requirement originally set out in the CMS Letter of Intent.[4] To the extent that it falls somewhat short of reaching the target resolution of 1% at $100\,GeV$, this is substantially due to the effect of multiple scattering in the material inside the tracking volume. The effect of multiple scattering on the resolution in Pt is momentum-independent, and can be directly estimated by examining the simulated Pt resolution for $1\,GeV$ and $10\,GeV$ muons. The multiple scattering is seen to contribute 0.7% to the Pt resolution at $\eta \sim 0$, 1% at $\eta \sim 1$ and almost 1.5% from $\eta \sim 1.3$ and beyond.

Figure 18 shows the simulated Pt resolution for the stand-alone tracker and muon chambers, and for the combined measurement, for a range of muon

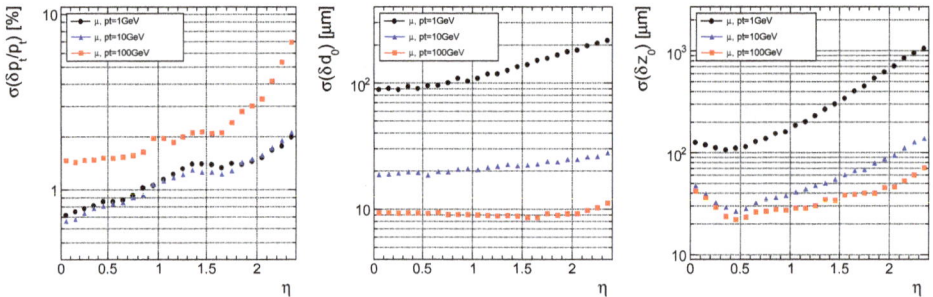

Fig. 17.   Reconstructed track parameter resolutions for single muons of Pt = 1, 10 and 100 GeV: Pt (left), transverse impact parameter (middle) and longitudinal impact parameter (right).[3]

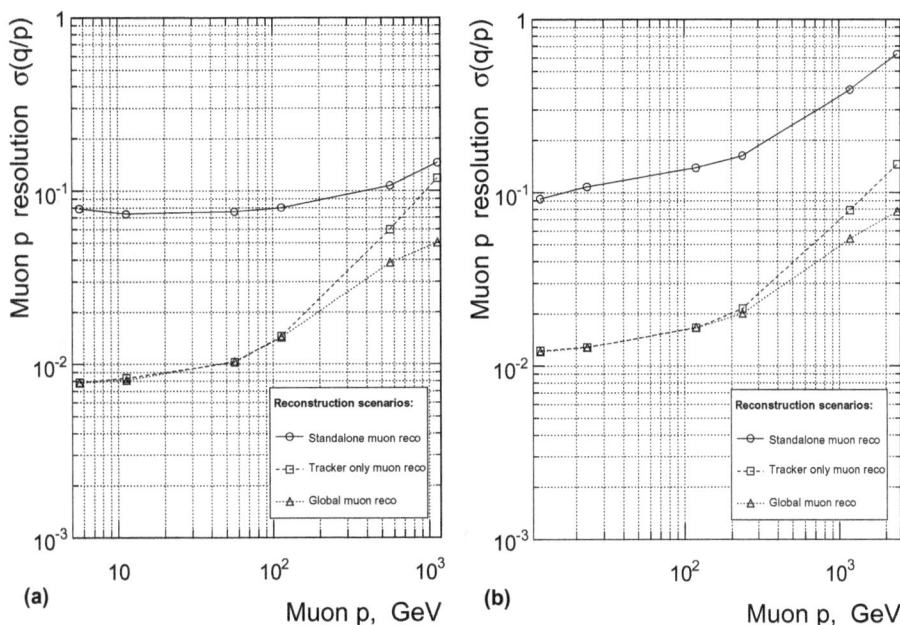

Fig. 18.   Muon momentum resolution as a function of momentum: (a) barrel, $\eta = 0.5$; (b) end-cap, $\eta = 1.5$.[10]

Pt extending up to 1 TeV. It can be seen that the target of 10% resolution for a Pt of 1 TeV is met, and that the tracker plays a key role in the precision determination of muon Pt in the CMS experiment over this entire Pt range, as expected.

The material budget of the CMS tracker as a function on η is shown in Fig. 19. It can be seen that the material budget is between 30% and 40% of a radiation length at η close to 0, but that it then rises steeply beyond that to peak at over 1.6 radiation length at $\eta \sim 1.4$. From the plot on the right, it can be seen that this peak is primarily driven by a very sharp increase in the material budget due to cables and cooling, starting at $\eta \sim 1$, which marks the beginning of the acceptance boundary between the TIB and TOB, and the TID and TEC.

In this region, manifolds and connectors provide a transition between the low mass cooling and electrical services internal to the subdetectors, to rugged pipes and cables linking these to a remote patch panel. The effect of this on the material budget can be directly visualized in Fig. 20.

For smaller objects, such as previous generations of silicon vertex detectors, this transition from "micro" services to "macro" services could usually be displaced outside the detector acceptance. For the CMS strip tracker, due

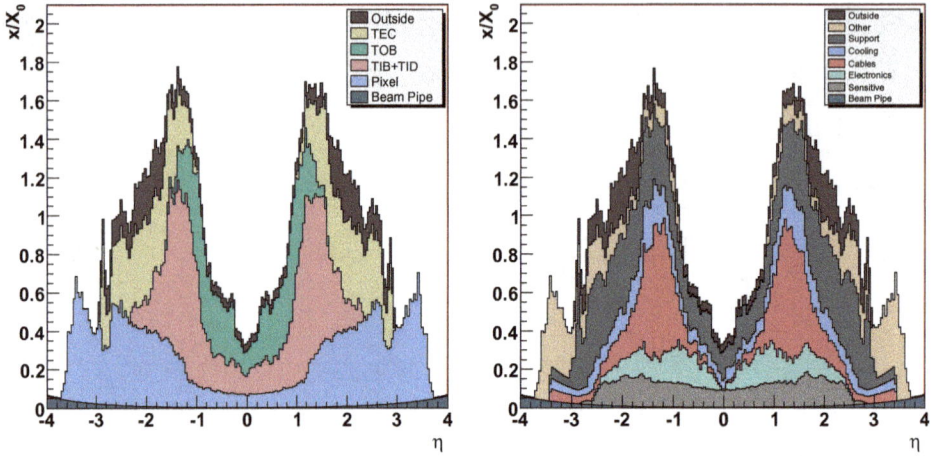

Fig. 19. The material budget in units of radiation length, as a function of η for the different tracker subdetectors (left) and broken down into the functional contributions (right).[4]

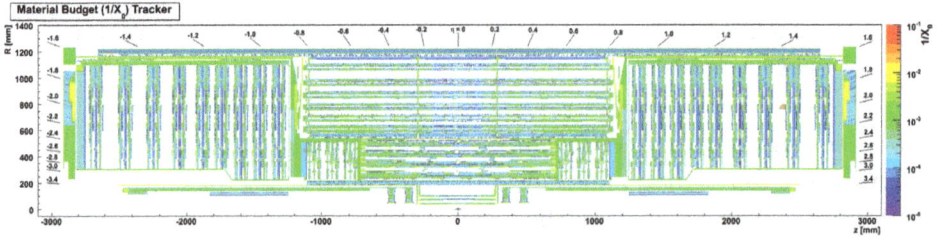

Fig. 20. The density of material for the present CMS strip tracker, in units of radiation length. The main features which can be qualitatively seen from this figure are: the relatively low density of material within the pixel barrel, TIB and TOB; the substantially higher density of material due to pipes and manifolds, and cables and connectors as tracks cross the pixel barrel, TIB and TOB bulkhead regions, where they dominate the tracker material budget; the relatively high density of material on the outer envelope of the tracker, dominated by pipes, cables, and the tracker carbon fiber support tube and thermal screen (Internal Note).

to the scale involved, this transition is internal to the tracking volume and dominates the material budget in the strip tracker. It is essentially the effect of multiple scattering through this transition region which can be seen in the η dependence of simulated muon Pt resolution of Fig. 17, and which affects the pion reconstruction efficiency, as discussed below.

The track reconstruction efficiency for inclusive charged pions is well in line with the original requirements: it is better than 90% for η less than ~1, and remains above 85% all the way to η of 2.5. It is interesting to note that, except inside the central core of very high energy collimated jets, the inclusive track reconstruction efficiency is not limited by the granularity of the tracker, and overlaps. Rather, it is largely driven by the hadronic interaction with the material inside the tracker volume and, in particular, with the services in the transition region between the barrel and end-cap subdetectors.

Initial estimates of the inclusive charged pion tracking efficiency were in fact substantially worse than this, particularly in the regions worse-affected by the material inside the tracker volume. Maintaining high efficiency through these regions was largely made possible by the relatively recent development of an "iterative" approach to tracking. This has also allowed the useful extension of reconstruction to tracks well below 1 GeV, which was previously not considered realistic.

This is but one example of the development of new ideas and algorithms to exploit the intrinsic performance potential provided by the excellent hit resolution and low cell occupancy of the CMS tracker, in ways that were initially not foreseen.

Another powerful example of this is the extensive use of detailed tracker information already at the early stages of the CMS high level trigger (HLT), well beyond what was originally thought possible.[11] This is essentially based on the observation that, particularly with a well-targeted seeding strategy, the quality of the pixel and silicon strips in the CMS tracker is such that pattern recognition and unambiguous track identification are achieved with as few as two or three pixel hits, which are used as track seeds, matched with a further two or three silicon strip hits.

With the implementation of the concept of limiting track reconstruction to geometric regions around candidate objects ("region-of-interest track-ing"), and of stopping the reconstruction of any given track as soon as a given hypothesis can be either verified or excluded ("conditional tracking"), tracking for the HLT becomes at once very fast, and powerful.

5. Summary and Conclusions; Lessons Learnt

By deciding to build an all-silicon strip tracker, the CMS collaboration set itself up to deploy this highly sophisticated microtechnology at an unprecedented scale, in order to realize by far the largest device of its kind to date. This was made possible by pursuing and then capitalizing on a number of timely key developments in this area of technology. These include the access to 0.25 μm CMOS microelectronics technology and the development of low power optical links suitable for analog data transmission, both of which are discussed elsewhere in this book, as well as the availability of a high volume 6″ production line suitable for sensor production, and the development of automated techniques for large volume module assembly.

By early 2007, the integration of the CMS tracker had been completed, at the integration facility constructed at CERN for this purpose. A series of system tests was performed, with approximately 20% of the tracker being operated with the final electronics, readout, power, control and safety systems.

By the end of 2007, the CMS silicon strip tracker had been installed in the CMS experiment, and in the first few months of 2008 it was connected to all of the services and commissioned first in "stand-alone" mode and then with the rest of CMS. In summer 2008, in preparation for the foreseen first run of the LHC, large cosmic ray track data sets were collected with the tracker. By September 2008, the tracker systems (silicon strips and pixels) had been fully commissioned, with more than 98% of all channels operational, and were ready for sustained data taking with beams. Following the LHC incident in the fall of 2008, some eight million cosmic ray tracks were collected by CMS, and the tracker performance fully met the original specifications. The tracker was functional for over 98% of the time during the ~six-week period over which these data were collected.

By the fall of 2008, the tracker measurements of signal-to-noise, efficiency, Lorentz angle and alignment were all at the limit of what could be achieved with cosmic ray tracks and, in particular, the alignment at the module level was better than $30\,\mu$m.

This exceptional performance is the result of many years of dedicated work by the whole CMS tracker collaboration, which includes some 500 physicists and engineers, from about 50 institutes across some 10 countries, with the support of their funding agencies.

The CMS tracker project was a very distributed one, with collaborating institutes throughout Europe and the United States taking on responsibilities for crucial deliverables, ranging from the development and production of single components or entire systems, to participating in the massive task of module production, and finally to integrating and delivering fully qualified subdetectors, ready for installation first into the tracker support tube and then CMS. The development of the tracker software systems, track finding and reconstruction, alignment, calibration and analysis tools was a similarly distributed effort. The complex logistics of such a distributed project required effective coordination and constant attention to detail and quality control.

Looking back at the project, the importance of sound system design and good engineering practice, as well as of a systematic approach to quality assurance, cannot be overstated. In addition, the excellent relationship with, and the commitment of, key industrial partners have been essential to the success of the project.

This has been a project driven by a strong initial vision, of a high precision, highly granular tracker for the LHC, and marked by the will to seize on those concepts that were deemed to provide the best means for building on that vision.

We now look forward to making the best use of this remarkable scientific instrument, and exploring its full performance potential, to study the physics of the LHC.

References

1. The CMS Tracker Project Technical Design Report, CMS/LHCC 98-6.
2. Addendum to the CMS Tracker TDR, CERN/LHCC 2000-2016.
3. The CMS experiment at the CERN LHC, *JINST* **3** (2008) S08004.
4. The Compact Muon Solenoid Letter of Intent, CERN/LHCC 92-93.
5. Charge collection in highly irradiated p-on-n sensors operated at high bias voltage. INFN: Guido.
6. Radiation tests with Foxfet-biased micro-strip detectors, *Nucl. Instrum. Methods A* **418** (1998) 128–137.
7. Radiation damage experience at CDF with SVX, *Nucl. Instrum. Methods A* **383** (1996) 155–158.
8. Investigation of design parameter for radiation-hard silicon micro-strip detectors, *Nucl. Instrum. Methods A* **485** (2002) 343–361.
9. An automated silicon module assembly system for the CMS silicon tracker, CMS Note 2002/005.
10. CMS Detector Performance and Software, Physics TDV, Vol. I, CMS/LHCC 2006-1.
11. CMS Data Acquisition and High Level Trigger, Trigger and DAQ TDR, Vol. II, CMS/LHCC 2002-2026.

Chapter 5

THE ATLAS ELECTROMAGNETIC CALORIMETERS: FEATURES AND PERFORMANCE

Luciano Mandelli

INFN, Sezione di Milan
Via Celoria 16, 20133 Milan, Italy
Mandelli@mi.infn.it

In this chapter it is shown how a sampling electromagnetic calorimeter based on the liquid argon technique satisfies the very demanding requirements of an experiment at the LHC. Section 2 discusses, using a simplified model, the performance that can be achieved in terms of response time, energy resolution and transverse granularity. Section 3 describes how the calorimeters are realized in ATLAS, their segmentation and how from the readout pulses the energy deposited in the calorimeter is computed. The motivations of a presampler detector in front of the calorimeter are also discussed. Section 4 describes how the energy, position and direction of an electron and a photon are computed. Finally, Sec. 5 briefly illustrates the rejection power of the calorimeter against the hadrons and mentions how a Higgs boson signal in the $\gamma\gamma$ channel can already be detected with a luminosity of $10\,\text{fb}^{-1}$.

1. Introduction

The main task of an electromagnetic calorimeter is to measure the energy of electrons and photons and to identify them in the large sea of hadrons. In addition, while the direction of electrons is given by the tracker, the position and direction of photons, at least for the ones without a conversion identified in the tracker, are given by the calorimeter. Electrons and photons are often the key to identifying interesting processes, and an excellent resolution in energy and direction is mandatory for detection of new states (i.e. $Z' \rightarrow e^+e^-$ or $H \rightarrow \gamma\gamma$). We should also keep in mind that the electromagnetic calorimeter is an essential element in the measurement of hadronic showers, being the first calorimeter that particles hit. To reach these goals during the entire life of an LHC experiment ($\geq 10\,\text{yr}$), the calorimeter has to be radiation resistant and stable in time.

Another key quantity for many interesting processes is the missing transverse momentum, a spy of the presence of neutral weakly interacting particles like neutrinos and neutralinos. To reach a good missing transverse momentum resolution it is mandatory that the calorimeters (electromagnetic and hadronic) cover a large solid angle and that the presence of uninstrumented regions is minimized. Particles going in these regions could fake a missing transverse momentum signal. In conventional language the calorimeter has to be hermetic.

Section 2 presents the principles on which liquid argon calorimetry is based. Familiarity with the basic concepts of calorimetry is assumed. Section 3 describes how the liquid argon calorimeters are realized in ATLAS and Sec. 4 how the particle energy and direction are computed. The rejection power against hadrons and the sensitivity of the experiment to the Higgs decay in two photons are briefly discussed in Sec. 5.

2. The Principles of Liquid Argon Electromagnetic Calorimetry

The choice of ATLAS was to design an electromagnetic calorimeter based on liquid argon (LAr) as active medium. This technique was developed starting in the 1970s[1,2] and has already been adopted in detectors operating at Hera and the Tevatron.[3,4]

This technology attracts immediate attention, since being the active medium a monoatomic noble gas it is intrinsically radiation resistant. However, before adopting it for an LHC detector, the proof that the intrinsically slow signal can be handled, that a hermetic detector can be built and that the presence of a cryostat in front of the calorimeter does not significantly deteriorate its performance has to be given.

2.1. *The sampling and the pulse shape*

The choice of LAr as active medium implies that the calorimeter has to be a sampling calorimeter. A sampling calorimeter is built by assembling many layers of passive material; in electromagnetic calorimeters lead is commonly used given its short radiation length ($X_0 = 0.56$ cm), alternated with layers of active material, in our case LAr ($X_0 = 14.0$ cm).

A high energy electromagnetic shower is well contained in $25X_0$. A calorimeter built only with LAr would be $25 \times 14.0 = 350$ cm thick, implying a too-large cryostat.

On the contrary, a layout built with a layer of 2 mm lead and 4 mm LAr has an equivalent radiation length of 1.55 cm, pointing to a calorimeter thickness of 38.7 cm. The cryostat dimensions are now realistic.

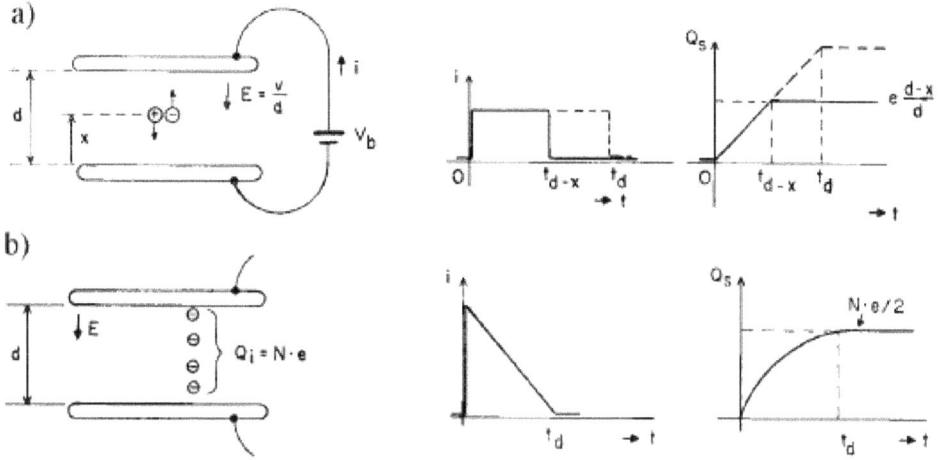

Fig. 1.   Current produced in an ionization chamber: (a) by a localized charge; (b) by a uniform ionization in the gap.[1]

The electrons generated by the electromagnetic shower by ionization in the active material are guided by an electric field up to the anode inducing a current. Each gap of a sampling calorimeter is equivalent to an ionization chamber of thickness d, as sketched in Fig. 1.

It is a simple exercise of electrostatics to compute the current and the charge induced by an electron generated at one point in the gap at a distance d–x from the anode:

$$i = e\frac{v_d}{d} = \frac{e}{t_d}, \quad Q(t) = e\frac{t}{t_d}, \quad 0 \le t \le \frac{d - x}{v_d}, \tag{1}$$

where v_d is the electron drift velocity and $t_d = d/v_d$ is the time the electron takes to cross the full gap. The drift velocity is a function of the electric field E and of the LAr temperature. As an example v_d is $5\,\text{mm}/\mu s$ for a field of $10\,\text{kV/cm}$ and $T = 89\,\text{K}$. Also, the positive argon ions drift to the cathode, but since their drift velocity is about $1/1000$ the electron drift velocity the current induced by them can be neglected in the present LHC program.

The bunch of $e^{\pm}$ produced by a shower crossing a LAr gap generates a uniform ionization and a total negative charge $Q = Ne$. The electrons drifting to the anode induce now a linear current given by

$$i(t) = Ne\frac{1}{t_d}\left(1 - \frac{t}{t_d}\right), \quad Q(t) = Ne\frac{1}{t_d}\left(t - \frac{t^2}{2t_d}\right), \quad 0 \le t \le t_d. \tag{2}$$

Note that $i(0)$ is proportional to the total charge present in the gap and that the current vanishes at $t = t_d$. In a $2\,\text{mm}$ drifting gap (the one used in ATLAS) $t_d = 400\,\text{ns}$.

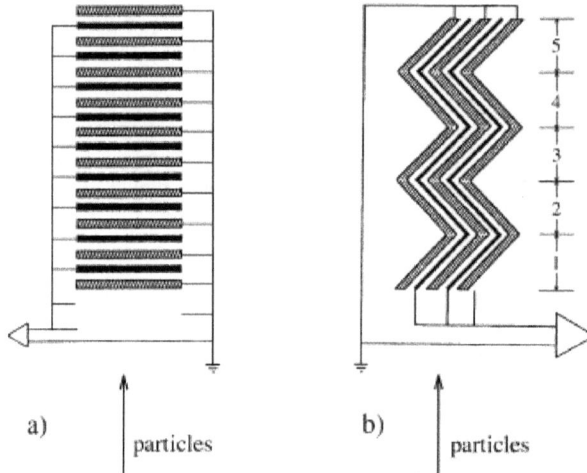

Fig. 2.   Scheme of a sampling calorimeter with: (a) traditional geometry; (b) accordion geometry.

A sampling calorimeter is a sequence of gaps like the one described above and sketched in Fig. 2(a). Several anodes are connected in parallel and the total induced current is measured.

Recalling that the LHC has a bunch crossing each 25 ns and that at the nominal luminosity an average of 25 inelastic interactions are produced at each crossing, a pulse length of 400 ns poses a difficult problem. Suppose that you are interested to measure an event produced at a given bunch crossing. In the calorimeters, pulses of interactions produced up to 16 bunches crossing earlier and 16 bunches crossing later will overlap with the pulses you are interested in.

A solution to the problem is gained by working on two points: granularity and pulse shaping.

2.2. *The longitudinal and transverse development*
of an electromagnetic shower

The energy deposited by a shower in the sensitive medium at a certain depth is proportional to the number of particles at that depth and is given by the simple approximate relation

$$\frac{dE}{dt} = \text{const} \cdot t^\alpha e^{-\beta t}, \tag{3}$$

where $t = x/X_0$ is the depth measured in radiation length, α and β are parameters function of the particle type (e/γ), energy and the material (violating the scaling). At a small depth it increases like t^α while the average particle energy decreases. When the average energy reaches the critical value E_c, the ionization and absorption processes dominate and the number of particles exponentially decreases.

The shower maximum is reached at $t_{max} = \ln\left(\frac{E}{E_c}\right) + c$ ($c = -0.5$ for electrons and $c = +0.5$ for photons), while 98% of the shower is contained in $2.5\, t_{max}$. As an example a shower of a $100\,\mathrm{GeV}$ electron in lead reaches on average the maximum after $9X_0$ and is well contained in $25X_0$.

The lateral size of a shower[5] is due to the multiple scattering of electrons and positrons and characterized by the Molière radius (R_M) of the setup. In a sampling calorimeter $R_M = \frac{21\,\mathrm{MeV}}{E_c} X_0\left(\frac{x+y}{x}\right)$. In our case E_c, X_0 and x are the critical energy, the radiation length and the thickness of the lead, and y is the thickness of the LAr. 95% of the shower energy is contained in a cylinder with a radius equal to $2R_M$. In the layout we took as an example ($2\,\mathrm{mm}$ lead and $4\,\mathrm{mm}$ LAr), we have $2R_M = 9.53\,\mathrm{cm}$. This is the natural length which drives the lateral segmentation of the calorimeter. Since the measurement of the lateral profile of a shower is an important parameter to separate $e^{\pm}/\gamma$ from hadrons, the calorimeter transverse segmentation should be only a fraction of $2R_M$, e.g. $1/3$, giving a design granularity of the order of $3\,\mathrm{cm}$. At $\theta = 90°$ and at $170\,\mathrm{cm}$ from the vertex (the mean depth of a shower in ATLAS) this corresponds to $\Delta\eta = 0.02$. The same granularity should be assumed in azimuth. The segmentation used in ATLAS in the middle longitudinal compartment, where the bulk of the energy is contained, is $\Delta\eta \times \Delta\phi = 0.025 \times 0.025$, over almost all of the calorimeter (Subsec. 3.3).

2.3. *The pileup and the pulse shaping*

At the LHC full luminosity, an average of 25 inelastic interactions with an average of nine particles ($6\pi^{\pm}$ and $3\pi^0$) is predicted to be produced in the central region per unit rapidity. These particles are usually low transverse momentum particles. In 32 bunches this implies $25 \times 9 \times 32 = 7200$ particles and $7200 \times 0.025 \times 0.025/2\pi = 0.7$–$1$ particle hitting a cell of the middle compartment. Suppose now that we have a high transverse momentum e/γ. The pulse generated by the low transverse momentum particle will partially overlap the pulse of the "interesting-in-time particle." This effect is called "pileup." At first glance you could think that it is a small effect, but keep in mind that there could be important fluctuations on the average and that an electromagnetic shower is measured in at least 3×3 cells. Recall also that a measurement of the total transverse energy and of the missing transverse momentum involves all the cells of the calorimeter. The effect of the pileup

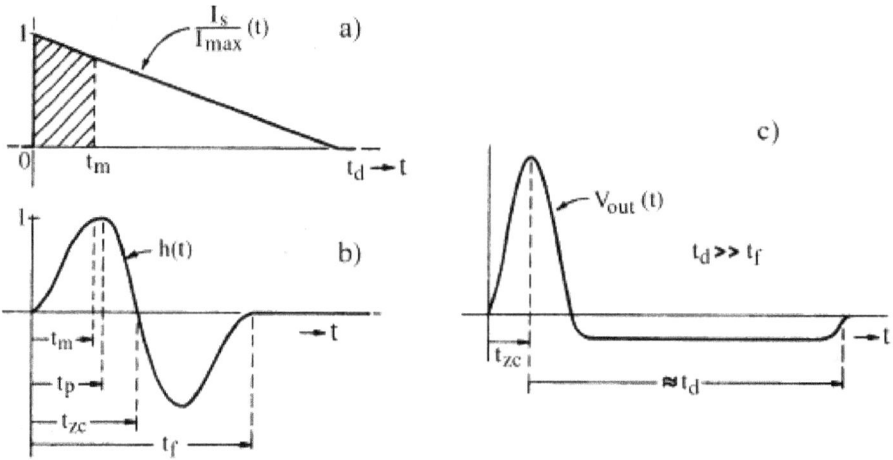

Fig. 3. (a) Current pulse shape generated by a uniform ionization in the calorimeter. (b) Response of the preamplifier-shaper circuit to an impulse pulse. (c) Response of the preamplifier–shaper circuit to the triangular pulse.

should be minimized and well controlled. This target is reached by hardware with the pulse shaping and by software with the optimal filtering.

Shaping: the current after being amplified is differentiated and integrated by a CR–(RC)^2 architecture.

The response time of the preamplifier–shaper system to an impulse has a bipolar shape such as the one shown in Fig. 3(b). The characteristic time $t_p(\delta)$ can be varied. The most important condition for the shaping is that the system impulse response should have zero area, which means that the integral of the pulse over time is zero. When the triangular current signal from the detector is convoluted with the impulse response ($t_p \ll t_d$), the signal has the shape depicted in Fig. 3(c). The maximum of the pulse is proportional to the current at $t = 0$. The pileup modifies the pulse shape and in particular the height and timing of the peak. However, given the bipolar feature of the shaper, on average the pileup pulse does not induce a net shift of the peak amplitude (i.e. energy) but contributes only to the error measurement of the amplitude.

The pile up error increases as $t_p^{1/2}$. However, since the electronic noise increases as $t_p^{-3/2}$ a compromise is found[8] which minimizes the total error. Figure 4 shows the two contributions as a function of t_p for an electromagnetic cluster of 3×3 cells and a luminosity of 10^{33} cm^{-2}s^{-1} and 10^{34} cm^{-2}s^{-1}. At the nominal luminosity of 10^{34} cm^{-2}s^{-1} the minimum is reached for $t_p \sim 25$ ns and corresponds to an error on the energy of $\sim$450 MeV. This error adds up to the calorimeter energy resolution. As an example, for a sampling calorimeter with a sampling term of 10% it is larger than the error due to sampling only for electrons with energy less than 20 GeV. When the calorimeter operates at a different luminosity the optimal filter coefficients (OFCs), discussed in Subsec. 3.6, will act as a digital filter bringing the pileup–electronic noise error to its optimal value.

In other words, what really matters is not the drift time ($\sim$400 ns) but the response time to an impulse ($\sim$25 ns). In this respect we can say that a LAr calorimeter can give a fast response.

An optimum signal-to-noise ratio is obtained if the very fast rising time of the current pulse ($\sim$1 ns; the time taken by the electromagnetic shower to develop) is not deteriorated by the time constant τ_c of the circuit composed of the calorimeter cell capacitance, the connections to the preamplifier and the preamplifier input impedance. In the standard design of a calorimeter such as the one shown in Fig. 2(a), the main contribution to τ_c comes from the leads needed to link together the successive planar electrodes and from the connections to the preamplifier.

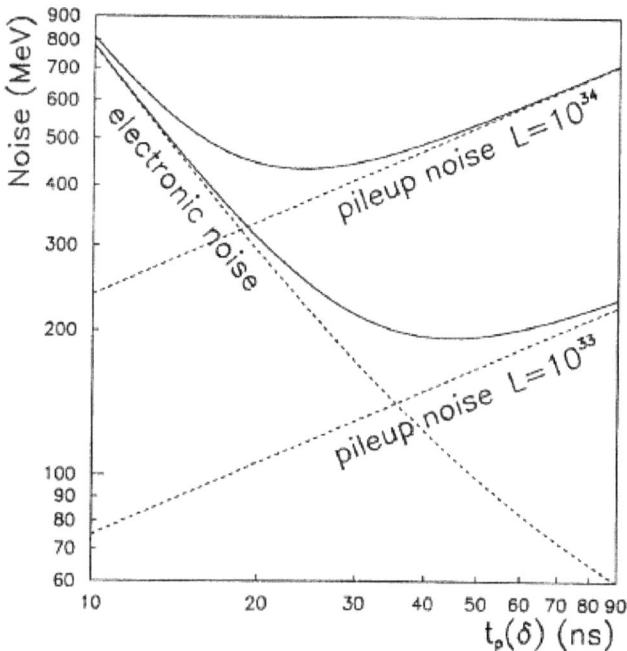

Fig. 4. Electronic and pileup noise versus the amplifier–shaper peaking time t_p to an impulse.

The electromagnetic ATLAS calorimeter benefits from an innovative design [Fig. 2(b)].[7,8] The electrodes are disposed parallel to the particle directions. To avoid channeling the electrodes and the absorber plates are waved in an "accordion" shape [Fig. 2(b)].

The second very important benefit of the accordion design is that signals are picked out from the front and back of the electrodes, avoiding leads and dead spaces. With a proper mechanical design a hermetic detector can be built on a very large solid angle.

2.4. The measurement of the energy deposited in the calorimeter

The maximum of the current pulse is proportional to the ionization charge, i.e. to the energy deposited in the LAr. This energy is only a fraction of the energy deposited in the calorimeter. This fraction is called the "sampling fraction" and is approximated by the expression

$$f_{\text{sampling}} = \frac{E_{\text{LAr}}}{E} = \frac{\left(\frac{dE}{dx}\right)_{\text{LAr}}}{\left(\frac{dE}{dx}\right)_{\text{LAr}} + \left(\frac{dE}{dx}\right)_{\text{Abs}}} \left(\frac{e}{\mu}\right) = f_{\text{mip}} \left(\frac{e}{\mu}\right), \qquad (4)$$

where

$$f_{\text{mip}} = \frac{\left(\frac{dE}{dx}\right)_{\text{LAr}}}{\left(\frac{dE}{dx}\right)_{\text{LAr}} + \left(\frac{dE}{dx}\right)_{\text{Abs}}}. \qquad (5)$$

In the above expressions E is the energy deposited by an e/γ in the calorimeter, f_{mip} is the sampling fraction of a minimum ionizing particle and e/μ is the "electron over muon factor" of the calorimeter. It takes into account the fact that the abundant very low energy photons and electrons present in a shower are preferentially absorbed in the absorber and do not give a signal in the LAr. This factor has been measured in the test beam to be 0.75.

With the very simplified model of the calorimeter as a sequence of 2 mm lead and 4 mm LAr we get

$$f_{\text{mip}} = \frac{2.2 \times 0.4}{2.2 \times 0.4 + 12.73 \times 0.2} = 0.26,$$

$$\left(\frac{dE}{dx}\right)_{\text{LAr}} = 2.2 \,\text{MeV/cm} \quad \left(\frac{dE}{dx}\right)_{\text{Pb}} = 12.73 \,\text{MeV/cm},$$

$$f_{\text{sampling}} = 0.26 \times 0.75 = 0.19.$$

While the quoted value gives the correct order of magnitude, a more precise value must be computed by a detailed simulation of the showers.

Once we know the sampling fraction we have to learn how to compute the energy from the measured current. An approximate answer is given by first principles. Since the maximum of the amplitude is given by the current at $i(0)$,

$$i = Q\frac{1}{t_d} = e\frac{E_{\text{LAr}}}{W}\frac{1}{t_d} = ef_{\text{sampling}}\frac{E_{\text{calo}}}{W}\frac{1}{t_d}, \tag{6}$$

where e is the electron charge, W the ionizing potential of the LAr ($W = 23.6\,\text{eV}$), E_{calo} the energy deposited in the accordion and t_d the electron drift time. In this formula we neglect that a fraction of electrons could be lost either through recombination or through capture by electronegative impurities. With a drift time of 400 ns we get

$$\frac{i}{E_{\text{calo}}} = 3.2\frac{\mu\text{Amp}}{\text{GeV}} \quad \text{or} \quad \frac{E_{\text{calo}}}{i} = 0.312\frac{\text{GeV}}{\mu\text{Amp}}.$$

In conclusion, if we measure the amplitude of the pulse we are able to measure the energy deposited in the calorimeter.

2.5. *The sampling and the energy resolution*

The energy deposited in the active medium fluctuates from event to event, inducing a stochastic error on the measured energy. It can be shown with a simple model of the shower that σ_E/E decreases with the energy as $1/\sqrt{E}$.[5] The fact that the relative error due to statistics decreases with energy makes the calorimeters very useful in high energy experiments. A good parametrization[13] of the stochastic error for a sampling calorimeter is

$$\frac{\sigma_E}{E} = 2.7\%\sqrt{\frac{d}{f_{\text{mip}}}\frac{1}{\sqrt{E}}}, \tag{7}$$

where E is the energy deposited in the calorimeter, d the thickness of the sampling layer in mm and f_{mip} the sampling of a minimum ionizing particle defined in (5).

Assuming the simple model of the calorimeter to be a sequence of 2 mm lead and 4 mm LAr, the expected relative resolution is $\frac{\sigma}{E} = 0.11\frac{1}{\sqrt{E}} = \frac{a}{\sqrt{E}}$. The parameter a is called the sampling (stochastic) term.

To the sampling error one has to add the error due to electronic noise and pileup. Since this error is constant the relative error scales like $1/E$.

Local nonuniformities in the construction, like variations in the thickness of the absorber and LAr gap and the unavoidable limited accuracy with which the particle energy is computed by the software algorithms (see Subsec. 4.1), give an error proportional to the energy and a constant relative error. This error will dominate at high energies. These effects are independent. The total relative error σ/E,

$$\frac{\sigma}{E} = \frac{a}{\sqrt{E}} \oplus \frac{b}{E} \oplus c, \tag{8}$$

is the quadratic sum of the three terms. The energy resolution attained with the accordion calorimeter in the absence of pileup was tested exposing calorimeter modules at an electron beam. After subtraction of the beam energy spread and the noise contribution (measured from random triggers), a sampling term of 10% is obtained.

The constant term is measured reconstructing the electron energy over several calorimeter cells. A value in the range of 0.5–0.7% is reached.

Is this resolution satisfactory for the LHC physics targets? A positive answer comes from a detailed simulation of benchmark reactions like the Higgs decays in two photons. Since the performance of the calorimeter in the ATLAS layout will enter into the discussion, this point will be briefly discussed in Sec. 5.

2.6. Stability with time

The amplitude of the shaped signal is proportional (Subsec. 2.1) to the ionization charge divided by the electron drift time. For stable response of the calorimeter both the charge and the drift time must be kept constant. The drift time, for a given gap size, depends on the electron drift velocity v_d, which is a function of the temperature and the electric field. A variation of $+1°$ around the operating temperature of 89 K gives a variation of -2% in the response. The temperature will be kept constant inside the cryostats and will be monitored with a precision of 10 mK and eventually a correction factor will be applied. The LAr purity must also be monitored, since electrons could be captured by electronegative impurities while drifting to the anode. To avoid poisoning, the calorimeter is built with materials carefully chosen and the LAr purity is monitored. This is done measuring the current induced by radioactive sources placed in the cryostat. In ATLAS the oxygen contamination is better than 0.3 ppm and stable at the % level over one year of operation.

The drift velocity is also a function of the electric field. The variation of the amplitude with the voltage is moderate. For example, the signal is reduced to 77%, decreasing the voltage from 2 kV to 1 kV (barrel). The signal variation is determined experimentally and parametrized in the form $A = aV^b$.

3. The ATLAS Electromagnetic Calorimeters

The above discussion shows that a LAr electromagnetic calorimeter satisfies the conditions required by an experiment at LHC: radiation-hardness, fast response, hermeticity, stability with time. In the present section we will briefly describe how these ideas have been realized in ATLAS. Detailed information is found in Ref. 6 and in the specialized articles of the LAr community.[9−11]

The ATLAS electromagnetic calorimeters are hosted in three cryostats: one barrel and two end-caps. The barrel covers the central region ($|\eta| \leq$ 1.475), and the two end-caps the forward and backward regions (1.375 $\leq |\eta| \leq 3.2$).

The optimization of the calorimeter parameters — sampling, thickness, transverse and longitudinal granularity — is a multivariable problem and is defined with the help of the simulation of e/γ showers and with the study of the response of prototypes to beams of electrons, muons and hadrons. An overview of the system is shown in Fig. 5.

3.1. *The barrel calorimeter*

The barrel calorimeter is hosted in a cylindrical cryostat with the axis along the beam direction. The cryostat length is 6.80 m and the inner and outer radii are 1.15 and 2.25 m. As a natural choice the 2 T solenoid and the

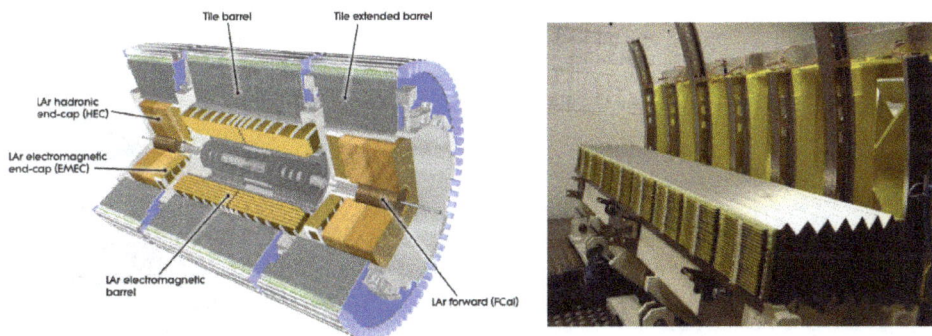

Fig. 5. Left: Overview of the ATLAS calorimeters. The LAr electromagnetic calorimeters are shown in brown. The hadron calorimeters hosted in the end-cap cryostats are shown in dark brown. Right: Photograph of a partly stacked barrel module.

cryostat share the same vacuum vessel at small radii. The electromagnetic showers will develop in a magnetic-field-free region and there will be no spread in azimuth. At $|\eta| = 0$ the material upstream of the presampler is equivalent to about $2X_0$ and varies strongly with $|\eta|$ due to the tracker detector and its services (Fig. 6). The effect of the material in front of the calorimeter on its performance will be discussed in Subsec. 3.4. The barrel calorimeter is a toroid made up of two half-barrels centered at $\eta = 0$, each with a length of 3.2 m. The toroid axis corresponds to the beam axis, and the inner and outer radii are approximately 1.5 and 2.0 m. It is a sandwich of 1024 accordion-shaped absorbers, 2.2 mm thick, made up of lead sheets glued between 0.2-mm-thick stainless steel sheets to give them a good rigidity. For $|\eta| < 0.8$ the lead is 1.53 mm thick, while for $0.8 \leq |\eta| \leq 1.475$ it is 1.13 mm thick to compensate for the increase of the total radiation length and of the thickness of the sampling layer seen by the particles, which would cause a worsening of the stochastic term of the energy resolution. The calorimeter is assembled starting from $16 + 16$ modules. Figure 5 shows a partly stacked module. The electrodes are kept in the middle of a gap by precision spacers made from strips of honeycomb. The drift gap on each side is 2.1 mm. The electrodes, 275 μm thick, are made up of three layers of conductive copper separated by insulating polyimide sheets. The outer layers are kept at high voltage (nominally 2 kV), while the middle one is used for reading the signal through capacitative readout. The granularity of the calorimeter in the longitudinal and transverse direction is reached by etched patterns on the readout layer. The patterns are pointing to the ATLAS center. The drift space is kept constant, increasing the bending angle of the absorbers and the electrodes in the radial direction.

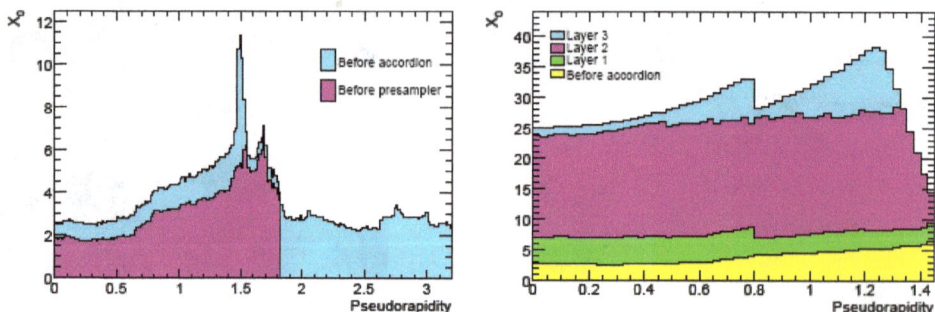

Fig. 6. Amount of material in radiation lengths in front of the calorimeter and the presampler (left). Thickness in radiation lengths of the different barrel calorimeter layers (right).

3.2. *The end-cap calorimeters*

The LAr end-cap calorimeters consist of two wheels with the axis centered on the beam and an internal radius of about 33 cm and an external radius of 210 cm. Each wheel is placed on each side of the barrel and covers the region $1.375 \leq |\eta| \leq 3.2$ partially in the shadow of the barrel. Each wheel consists of two coaxial wheels, the outer wheel covering $1.375 \leq |\eta| \leq 2.5$ and the inner wheel covering $2.5 \leq |\eta| \leq 3.2$. Between the two wheels at $|\eta| = 2.5$ there is a small discontinuity which introduces a very small nonsensitive region. Each outer wheel is assembled from 768 absorbers, and the inner wheel from 256. The electrodes are interleaved between the absorbers as in the barrel. In the end-caps it is not possible to compensate for the radial increase of the gap changing the accordion bending angle. The LAr gap increases from 0.9 to 2.8 mm for the outer wheel and from 1.8 to 3.1 mm for the inner wheel. To obtain a uniform response of the detector, an electric field varying continuously with η should be applied. This ideal situation is approximated varying the HV in seven (two) steps in the range 2.5–1.0 kV (2.3–1.8 kV) in the outer wheel (inner wheel).

3.3. *The electrodes and the longitudinal and transverse segmentation*

Information on the longitudinal and lateral development of a shower is essential for identifying electrons and photons against hadrons, for determining the impact position on the calorimeter and for giving the photon direction. In the LAr calorimeters the desired longitudinal and transverse segmentation in pseudorapidity are obtained etching the electrodes in pads. The desired readout granularity in azimuth is reached by ganging together several electrodes at either the back or the front of the calorimeter. Longitudinally the electrodes are divided into three parts named compartment 1 (front/strips), 2 (middle) and 3 (back), each one sampling a part of the shower energy. The depth of each compartment varies with η as shown in Fig. 6. Average values for the front and the middle are 5 and $17 X_0$ and are kept as constant as we could have it in practice. The back varies with η from 3 to $7 X_0$. The end-caps have a similar segmentation. The bulk of the shower energy is measured by the middle compartment, while the first part and the tail are measured by the front and back.

The longitudinal and transverse granularities are sketched in Fig. 7 and detailed in Table 1.

L. Mandelli

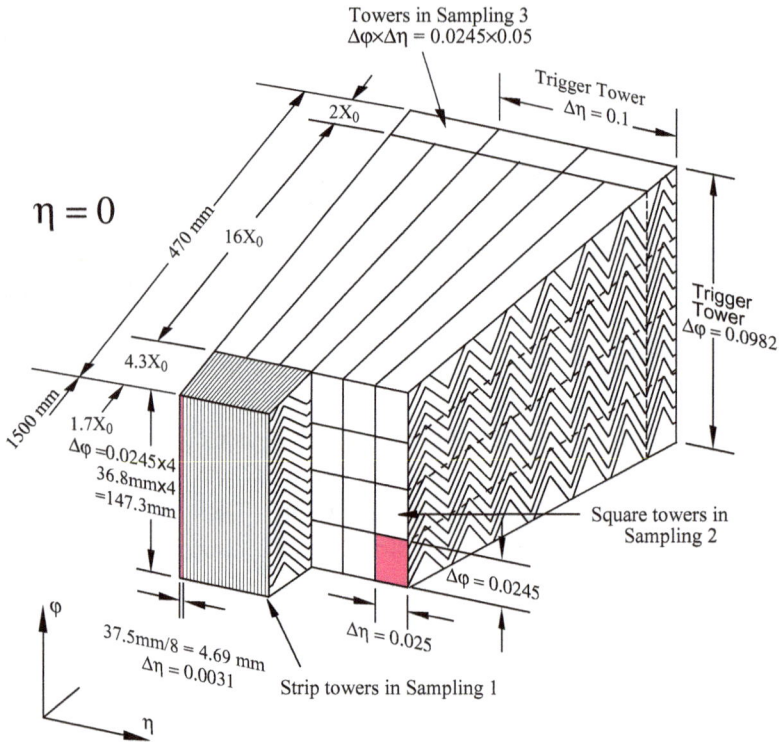

Fig. 7. Sketch of the barrel calorimeter at $\eta = 0$, showing the longitudinal and transverse segmentation.

The transverse segmentation of the three compartments differs considerably and merits an explanation.

The middle, where the bulk of the shower energy is measured, has the granularity expected from the Molière radius ($\Delta\eta \times \Delta\phi = 0.025 \times 0.025$). The back, which measures the tails of the electromagnetic shower, has twice the granularity in η and the same in $\phi(\Delta\eta \times \Delta\phi = 0.05 \times 0.025)$. The front (strips) is typically eight times more segmented in $\eta(\Delta\eta = 0.025/8)$ and four times coarser in $\phi(\Delta\phi = 0.1)$ to limit the number of channels. This can be made without serious deterioration of the performance since in ATLAS, due to the presence of material in front of the calorimeter (see Subsec. 3.4), the combined action of bremsstrahlung radiation and bending by the magnetic field increases the transverse profile in azimuth. This effect influences also photons which materialize in the tracker.

The longitudinal segmentation of the calorimeter with its fine segmentation in the strips gives two benefits: measurement of the shower direction and improved identification of π^0's showers.

Table 1. Longitudinal and transverse segmentation of the calorimeters and the presampler.

	Barrel		End-cap	
	EM calorimeter			
Number of layers and |η| coverage				
Presampler	1	$\|\eta\| < 1.52$	1	$1.5 < \|\eta\| < 1.8$
Calorimeter	3	$\|\eta\| < 1.35$	2	$1.375 < \|\eta\| < 1.5$
	2	$1.35 < \|\eta\| < 1.475$	3	$1.5 < \|\eta\| < 2.5$
			2	$2.5 < \|\eta\| < 3.2$
Granularity $\Delta\eta \times \Delta\phi$ versus $\|\eta\|$				
Presampler	0.025×0.1	$\|\eta\| < 1.52$	0.025×0.1	$1.5 < \|\eta\| < 1.8$
Calorimeter, 1st layer	$0.025/8 \times 0.1$	$\|\eta\| < 1.40$	0.050×0.1	$1.375 < \|\eta\| < 1.425$
	0.025×0.025	$1.40 < \|\eta\| < 1.475$	0.025×0.1	$1.425 < \|\eta\| < 1.5$
			$0.025/8 \times 0.1$	$1.5 < \|\eta\| < 1.8$
			$0.025/6 \times 0.1$	$1.8 < \|\eta\| < 2.0$
			$0.025/4 \times 0.1$	$2.0 < \|\eta\| < 2.4$
			0.025×0.1	$2.4 < \|\eta\| < 2.5$
			0.1×0.1	$2.5 < \|\eta\| < 3.2$
Calorimeter, 2nd layer	0.025×0.025	$\|\eta\| < 1.40$	0.050×0.025	$1.375 < \|\eta\| < 1.425$
	0.075×0.025	$1.40 < \|\eta\| < 1.475$	0.025×0.025	$1.425 < \|\eta\| < 2.5$
			0.1×0.1	$2.5 < \|\eta\| < 3.2$
Calorimeter, 3rd layer	0.050×0.025	$\|\eta\| < 1.35$	0.050×0.025	$1.5 < \|\eta\| < 2.5$
Number of readout channels				
Presampler	7808		1536 (both sides)	
Calorimeter	101,760		62,208 (both sides)	

3.4. *The material upstream of the calorimeter and the need for a presampler*

In ATLAS the cryostat, the solenoid and the tracker with its services are placed in front of the barrel calorimeter. The situation is similar for the end-caps where only the solenoid is absent. Figure 6 shows the number of X_0 upstream of the presampler as a function of $\|\eta\|$. The material thickness varies from about $2X_0$ at $\|\eta\| = 0$ to about $6X_0$ for $\|\eta\| = 1.4\text{–}1.6$, with a peak in the region between barrel and end-caps where cables and services of the tracker exit the detector. Crossing this material, an electron or a photon starts showering, losing energy. As a result of a detailed simulation of the detector, Fig. 10 shows the fraction of energy deposited in front of the calorimeter as a function of $\|\eta\|$. In the region where an electron crosses a substantial number of radiation lengths, a significant fraction of the particle energy is deposited upstream. A measurement of this energy is essential for reaching a good energy resolution. The normal technique, also adopted in ATLAS, is to sample the shower energy as early as possible with a detector. In ATLAS it is an independent detector built only with electrodes which collect the charge deposited in 11 mm (4 mm in the end-cap) of LAr. It is placed just after the cold wall of the cryostat and its readout granularity is $\Delta\eta \times \Delta\phi = 0.025 \times 0.1$. A detailed simulation of electrons and photons in the detector shows that the energy deposited in the material in front is, with large fluctuations, a function of the signal detected in the presampler.

3.5. *The signal readout and the "electromagnetic energy scale"*

Now that we have learnt how the calorimeter is built and its granularity, it is time to learn how the currents induced by the drift of the negative charge carriers are read out. The electrodes are connected by pins to boards (summing boards), where they are ganged together to give the desired granularity in azimuth. Coaxial cables run in front and back of the calorimeter and through feedthrough bring the signals out of the cryostats to warm. The crates with the front-end electronics (FEB) are placed very near to the feedthrough, to minimize cable length.

The block diagram with the main feature of the FEB architecture is shown in Fig. 8. The triangular shape of the current signal (Subsec. 2.1) is preamplified and split in two. Of the two outputs one goes to the analog sum electronics to form the LV1 trigger signal (not discussed here) and the second to the CR–(RC)^2 bipolar shaper. Each shaper has three outputs with gains in the ratio: 1:10:100 to fulfill the dynamic range required by the experiment, as will be discussed below. The amplitude of the shaper outputs is sampled each 25 ns (the LHC bunch crossing frequency) and continuously

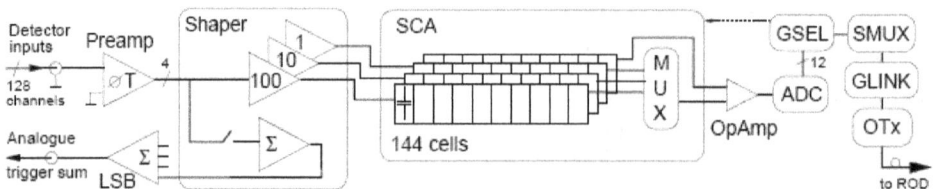

Fig. 8.   Block diagram of the FEB architecture.

stored in three (one for each gain) analog switch capacitor array (SCA) analog pipeline of 144 cells each. In normal operation, for events accepted by the LV1 trigger, the five amplitudes around the peak and one gain are sent to a 12-bit ADC. The optimal gain is hardware-selected (GSEL chip). As will be discussed below, the amplitude maximum is computed from these five amplitudes. The digitized values are sent to the back-end electronics via an optical emitter (OTx).

In Subsec. 2.4 we have computed the μAmp/GeV factor with highly simplified assumptions. In the accordion the sampling fraction is different in different regions and the electric field is not uniform in the regions where the electrodes are bent and particularly in the end-caps. Precise values are obtained from the test beam by comparing the measured currents with the energy deposited in the calorimeter, as predicted by a detailed Monte Carlo of the setup. The present values are shown in Fig. 9.

The current corresponding to an ADC value is found by injecting known currents into the readout. More precisely, calibration boards hosted in the front-end crate generate, through a 16-bit DAC, voltage pulses that when applied to very precise resistors placed on the motherboards inject a reference current into the readout circuit. The current amplitudes will cover the full dynamic range of the experiment, but will differ from the current generated by a particle in the form (exponential versus linear) and the injection point. These differences will be taken into account in computing the pulse amplitude by the optimal filter coefficients (OFCs).

In ATLAS we want to detect very high energy electrons and photons but also the very tiny currents generated by electronic noise (≥ 10 MeV). This

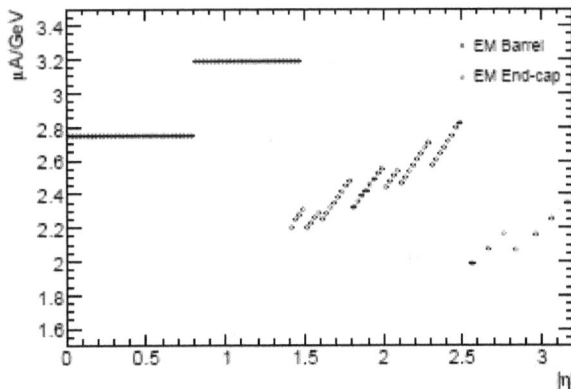

Fig. 9. μA/GeV factor as a function of $|\eta|$.

large dynamic range cannot be reached with a 12-bit ADC. For example, if 1-bit corresponds to 10 MeV, the ADC would already saturate at about 40 GeV (deposited in one calorimeter cell). A dynamic range between 10 MeV and 3 TeV is reached using three different shaper outputs with gain 1, 10 and 100 (normal, medium and high).

3.6. *The back-end electronics; computation of energy and timing of the signal*

The back-end electronics is in a room (USA15) placed at about 70 m from the front-end electronics. For our purposes it is sufficient to mention that its main task is to read out the digital data from the FEB modules and to compute for each cell the energy and the timing of the ionization signal.

The pulse maximum and timing are computed with a sophisticated OFC method which applies a weight to each of the five amplitudes. The weights include the knowledge of the shape of the ionization pulse (function of the capacitance and inductance of a cell and of the transmission line). The coefficients are computed from the pulses generated by the calibration boards but the effects due to the different pulse shape and injection point are taken into account. The coefficients also minimize the contribution of the electronic noise and pileup. Including the factors which for a given gain link the ADC value to the GeV, the energy (E) and time (t) corresponding to a pulse are given by

$$E = \sum_i a_i(s_i - \text{ped}), \quad Et = \sum_i b_i(s_i - \text{ped}), \tag{9}$$

where s_i is the sampled ionization amplitude, a_i, b_i are the mentioned OFCs, and ped is the pedestal value of the channel. The value of the energy computed in (9) is normally referred to as the "energy at the electromagnetic scale." In normal operation the sum is performed over the five samples, but for calibration purposes up to 32 samples can be used. In normal operation the ADC-GeV factor is included in the coefficients. The results, including a pulse quality factor, are transferred to the DAQ. A time resolution of 100 ps has been obtained for electrons with energies ≥ 60 GeV.[12]

Other modules present in the back-end electronics are the trigger timing and control system, which oversees the synchronization of the readout of the FEB, and the L1 receiver, which is an interface with the LV1 trigger processor. It also transforms the cell energy in transverse energy.

4. Computation of the Particle Energy and Direction

4.1. *Corrections applied to the measured energy*

Given the transverse granularity of the calorimeter, the energy of either an electron or a photon is spread over several cells of each compartment. The choice of the transverse dimensions of the cluster is a compromise between a "large" cluster which maximizes the sampled energy and the pileup and electronic noise, both increasing with the number of cells. The cluster sizes normally used in the analysis of the test beam data are $N_\eta \times N_\phi = 3 \times 3$ or 5×5, where N_η and N_ϕ are the number of cells of the middle layer. The cells considered in the presampler and the other calorimeter layers are the ones covering at least the same $\Delta\eta \times \Delta\phi$. In ATLAS, as mentioned, the combined action of the magnetic field and the bremsstrahlung increases the transverse shower dimension in ϕ and the standard cluster dimensions considered are 3×5, 5×5 and 3×7. The choice will depend on the particle type and the analyzed physics channel.

The cluster energy is the sum of the cluster energy of the three compartments. You could think that the work is over, but unfortunately several steps have still to be taken in order to obtain a good value of the energy of the electromagnetic particle. The following corrections have to be applied and are computed by simulating single monoenergetic electrons and photons in ATLAS in the energy range 10–1000 GeV over each cell of the calorimeter. The corrections are functions of a function of η and in some cases the energy deposited in the calorimeter. Symmetry over ϕ is assumed at present.

(1) *The sampling fraction.* It was mentioned above that the sampling fraction value is related to the number of very low energy electrons and photons which are absorbed in the lead and do not ionize the LAr. Since they get more numerous as the shower develops, the sampling fraction varies with the shower depth. The simulation shows a variation of a few percent of the sampling fraction as a function of the shower depth. This effect is parametrized and taken into account.

(2) *Lateral leakage.* The energy deposited outside the transverse dimension of a cluster is also parametrized as a function of the shower depth. As an example, for a 3×7 cluster it varies between 4 and 12% as a function of the pseudorapidity.

(3) *Energy deposited upstream.* As shown in Fig. 10, a significant fraction of the energy of an e/γ is deposited before the presampler and between the presampler and the first compartment of the accordion. In the regions where the upstream material goes up to several X_0 it amounts to several

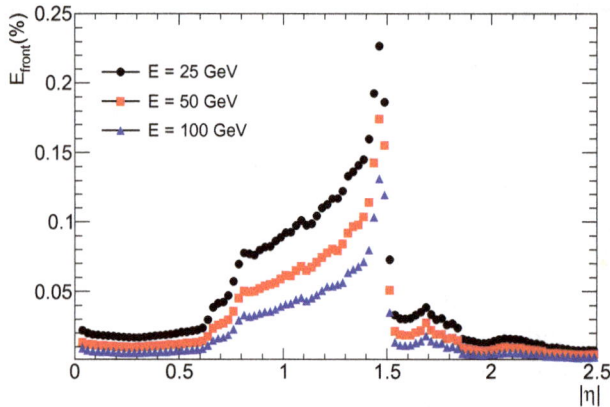

Fig. 10.   Fraction of the electron energy deposited in the material in front of the calorimeter.

percent (Ex.: ~8% for $E = 100\,\text{GeV}$ and $\eta = 1.4$) and has to be computed as accurately as possible to preserve resolution and linearity. A good result is reached by expressing the average energy deposited upstream as a function of the energy deposited in the presampler and in the accordion.

(4) *Longitudinal leakage.* The energy deposited downstream of the third compartment is computed as a function of the shower depth. Normally it is a correction at the % level. Only for showers developing late in the calorimeter or very high energy particles the corrections can reach some percent of the energy measured by the accordion.

(5) *Dependence on the impact position inside the cell.* All the above corrections are averaged over the impact point of a cell's middle compartment. There is residual energy dependence on the impact points in the cell due to the cell granularity (η) and the accordion structure (ϕ). The correction is at the % level.

Figure 11 shows the sampling term obtained for electrons and photons after all the above corrections are applied. One should note that the value of the sampling term is correlated with the amount of material present in front of the presampler. In the $|\eta|$ regions where it amounts to $\sim 2X_0$ the resolution is the nominal accordion resolution (9%), but it raises to 13% (22%) for photons (electrons) at $|\eta| = 1.4$, where the material adds up to $\sim 4X_0$.

Figure 12 shows the ratio between the value of the reconstructed energy and the nominal particle energy (linearity). It is inside the 0.5% over the full

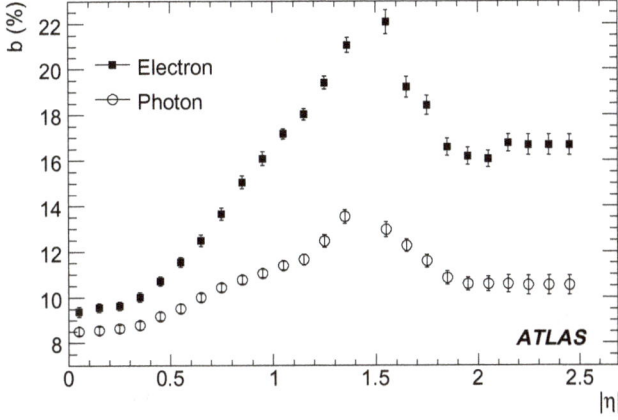

Fig. 11. Sampling term for electrons and photons as a function of $|\eta|$. All corrections were applied.

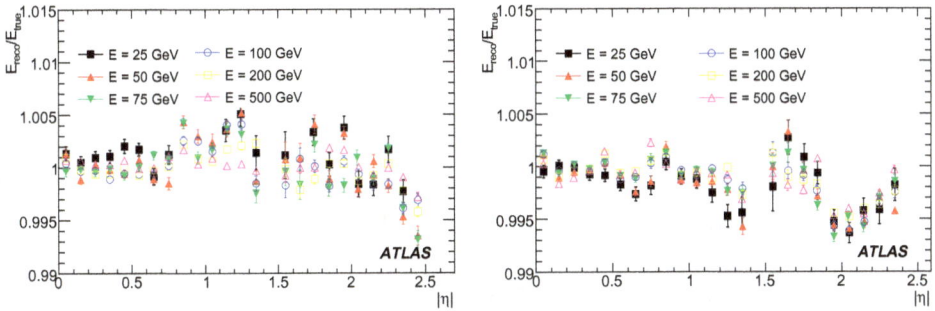

Fig. 12. Linearity for various electrons and photon energies as a function of $|\eta|$.

η range. The above-described calibration method is the default method in the experiment software. Besides computing the particle energy it measures the different terms which contribute to it: $E = E_{\mathrm{upstream}} + E_{\mathrm{acc}} + E_{\mathrm{downstream}}$.

An alternative method frequently used in the analyses of test beam data and also available in the experiment software is the "longitudinal weight" method. The energy of an electromagnetic particle is given by

$$E = A(B + w_{\mathrm{ps}}E_{\mathrm{ps}} + E_1 + E_2 + w_3 E_3), \qquad (10)$$

where E_{ps} and E_i $(i = 1, 3)$ are the energies measured in the presampler and the calorimeter compartments and A, B, w_{ps} and w_3 are weights computed by simulation optimizing the linearity. The weights are a function of the

L. Mandelli

particle type (e/γ), the cluster dimension and the pseudorapidity but not the particle energy. It is worth recalling that in both methods the corrections are at present computed by Monte–Carlo but that very good results are obtained by applying the same algorithms to test beam data.

4.2. *The energy resolution constant term*

In the energy resolution formula (8) the constant term which originates from local nonuniformities and energy reconstruction algorithms has been measured with modules exposed to a test beam. Figure 13 shows the ratio of the reconstructed energy as a function of η and the mean reconstructed energy in the barrel and end-cap modules. A value of $c = 0.5\%$ is measured in the barrel, and $c = 0.7\%$ in the end-cap. We refer to this value as to the "local constant term."

Furthermore, in spite of the great care which was taken in module construction, in calorimeter assembly and calibration procedures, differences due to mechanical distortions, differences in purity and temperature can cause response differences for different parts of the calorimeter. It has been shown[14] that from the measured mass of the Z^0 decaying in two electrons proper corrections can be derived. Assuming an initial gain spread of 2% between $\Delta\eta \times \Delta\phi = 0.2 \times 0.2$ regions, an analysis based on an integral luminosity of $100\,\mathrm{pb}^{-1}$ gives corrections which reduce the gain spread to 0.4%. This value added to the local constant term gives an expected value of $c = 0.7\%$. The error on a $100\,\mathrm{GeV}$ electron will still be dominated by the sampling term.

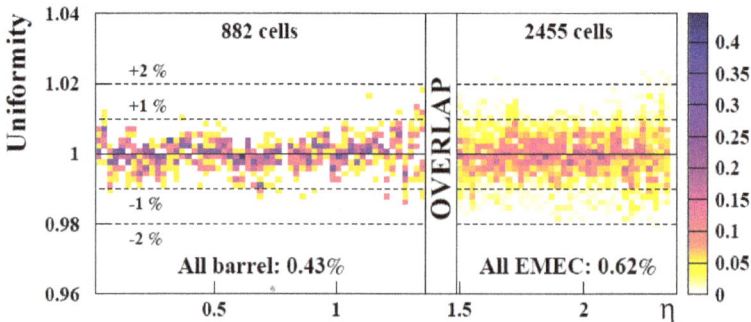

Fig. 13. Ratio of the reconstructed and the mean reconstructed energy over the barrel and end-cap modules as a function of the pseudorapidity. The test beam energy was $\sim 245\,\mathrm{GeV}$ ($\sim 120\,\mathrm{GeV}$) for the barrel (end-cap).

4.3. *Shower position and direction*

The η coordinate of each cluster in the first and second layers is computed as the energy-weighted mean of the cluster cells center. In the first layer, only the three strips around the cluster centers are used. Due to the cell granularity and the transverse shower shape, the η value computed as above is biased (S shape) and is corrected as a function of η, the energy and the particle type. As an example, after correction $\sigma_\eta = 2.5\text{–}3.5 \cdot 10^{-4} \, (6.0 \cdot 10^{-4})$ in the first (second) layer for 100 GeV photons. For instance, at $\eta = 0$ the above resolution in pseudorapidity corresponds to a position resolution along the beam direction of $\approx 400 \, \mu\text{m}$ (1 mm). Given the calorimeter granularity the ϕ coordinate is computed in the second layer only. The accordion waving structure spreads the shower over ϕ and washes out the S shape effect but introduces a small bias corrected as a function of η, the energy and the particle type. As an example, $\sigma_\phi = 0.5\text{–}1.0$ mrad (0.5–1.5 mrad) for 100 GeV photons (electrons) after the correction is applied. The resolution is worse for electrons due to the effect of the bremsstrahlung and the magnetic field.

The η, ϕ values of an electron are given by the values of the track at the vertex. For a photon they are given from the calorimeter values. The ϕ coordinate is simply the value measured in the second layer. Since normally the photon does not originate from the center of ATLAS ($z = 0$), the η values measured by the calorimeter do not correspond to the particle pseudorapidity. The η values measured in the strips and the middle give the coordinate of two points in the plane r–z (r is the distance from the beam axis and z the coordinate along the beam). For the barrel (end-caps) $r(z)$ is the effective depth of the shower. It is determined by Monte Carlo as the value which minimizes the resolution on η. The polar angle of a photon is determined by the polar angle of the straight line given by the two points. When the interaction vertex is known its coordinates will be used in a three-point fit. When a photon converts in the first part of the tracker and the conversion vertex is measured, this point can be used in the fit. The error on the polar angle scales approximately as $1/\sqrt{E}$ and varies between $40/\sqrt{E}$ and $80/\sqrt{E}$ Mrad when only the calorimeter information is used.

5. Rejection Power Against Hadrons and Sensitivity to the $H(\gamma\gamma)$ Signal

The transverse and longitudinal segmentation of the accordion gives a good identification of electromagnetic against hadron showers. The separation is made with algorithms based on the transverse profile of the shower in the front and the middle layer of the electromagnetic and on the leakage in the first compartment of the hadron calorimeter. It can be enhanced, by requiring the absence of tracks above a transverse momentum value in a cone centered on the electromagnetic cluster (isolation criteria). Of particular importance is the fine sampling ($\Delta\eta = 0.025/8$) of the strips, which gives a good rejection toward high energy π^0's which by decaying in two photons can easily fake a single photon. As an example, the distance between two photons coming from the most probable decay of a $P_t = 50\,\text{GeV}$ π^0 is 8.5 mm at $\theta = 90°$. The distance between two strips is 4.7 mm. A detailed Monte Carlo gives a rejection factor against π^0 of about 3 over the whole η range.[8]

More generally, the rejection factor against jets depends on the efficiency chosen to select photons, from the details of the algorithm which is used, from the transverse momentum range, from the parton kind (the quark or gluon which originates the jet) and from the fragmentation model used in Monte Carlo. As an example, a rejection of $8.3 \cdot 10^3$ is obtained for a photon efficiency of about 80% from a sample of two jets and $P_t > 25\,\text{GeV}$ when Pythia Monte Carlo is used. The possibility of detecting the Higgs decay in $\gamma\gamma$ crucially depends on the calorimeter energy and angle resolution, and on the rejection power against hadrons. The background generated by jets faking a photon must be lower than the irreducible $\gamma\gamma$ background due to the QCD two-photon production. For example,[13] assuming $m_{\text{Higgs}} = 120\,\text{GeV}$, the error on the mass is $\sigma_m = 1.4\,\text{GeV}$ (as can easily be computed from the given calorimeter properties). A detailed inclusive and semi-inclusive analysis in which all background reactions are considered shows that a 3.3 s.d. signal can be detected with an integrated luminosity of $10\,\text{fb}^{-1}$.

6. Conclusions

The ATLAS electromagnetic calorimeter, based on the liquid argon technique and the special *accordion* design of the absorbers and electrodes, satisfies all the requirements of a high luminosity experiment at the LHC in terms of radiation-hardness, hermeticity, fast signal response, dynamic range, energy and position resolution. An excellent identification of electrons and photons against hadrons is reached thanks to the three-dimensional fine granularity of the readout.

The detection of a Higgs signal in the two-photon channels, a benchmark reaction for the electromagnetic calorimeter of the experiment, can be achieved at the three-standard-deviation level with an integrated luminosity of $10\,\text{fb}^{-1}$.

Acknowledgments

I am very grateful to my colleagues D. Banfi, L. Carminati, M. Citterio, M. Delmastro, D. Fournier, P. Jenni, M. Livan, L. Serin and I. Wingerter-Seez, for reading the manuscript and giving suggestions.

References

1. W. J. Willis and V. Radeka. Liquid argon ionization chambers as total absorption detectors, *Nucl. Instrum. Methods* **120** (1974) 221–236.
2. V. Radeka and S. Rescia. Speed and noise limits in ionization chamber calorimeters, *Nucl. Instrum. Methods A* **265** (1988) 228–242.
3. I. Abt *et al.* The H1 detector at Hera, *Nucl. Instrum. Methods A* **386** (1997) 310–347.
4. S. Abachi *et al.* The D0 detector, *Nucl. Instrum. Methods A* **338** (1994) 185.
5. C. Grupen. *Particle Detectors* (Cambridge Monographs on Particles Physics, Nuclear Physics and Cosmology).
6. ATLAS Collaboration. The ATLAS experiment at the CERN Large Hadron Collider, *JINST* **3** (2008) S08003.
7. B. Aubert *et al.* Performance of a liquid argon electromagnetic calorimeter with an "accordion" geometry, *Nucl. Instrum. Methods A* **309** (1991) 438–449.
8. ATLAS Technical Proposal CERN/LHCC/94-43 LHCC/p2 (1994).
9. B. Aubert *et al.* Construction, assembly and tests of the ATLAS electromagnetic barrel calorimeter, *Nucl. Instrum. Methods A* **558** (2006) 388.
10. M. Aleksa *et al.* Construction, assembly and tests of the ATLAS electromagnetic end-cap calorimeter, *JINST* **3** P06002.
11. B. Aubert *et al.* Performance of the ATLAS electromagnetic calorimeter barrel module 0, *Nucl. Instrum. Methods A* **500** (2003) 202–231.
12. M. Aharrouche *et al.* Time resolution of the ATLAS barrel liquid argon electromagnetic calorimeter, *Nucl. Instrum. Methods A* **597** (2008) 178–188.
13. R. Wigmans. *Calorimetry* (Oxford University Press).
14. M. Livan, V. Vercesi and R. Wigmans. Scintillating fiber calorimetry. CERN Yellow Report (CERN 95-02 (1995), Genève, Switzerland).
15. Expected performance of the ATLAS Experiment — detector, trigger and physics [arXiv:0901.0512 CERN-Open-2008-020 2008].

Chapter 6

THE CMS ELECTROMAGNETIC CALORIMETER: CRYSTALS AND APD PRODUCTIONS

P. Bloch

Department of Physics, CERN
CH-1211 Genève 23, Switzerland
Philippe.Bloch@cern.ch

After a brief introduction to the CMS crystal calorimeter, we focus on the challenges linked to the lead tungstate crystals and avalanche photodiode production.

1. Introduction

Detection and precise energy measurement of photons and electrons is a key to new physics that is expected at the 100 GeV–TeV scale. The discovery of the postulated Higgs boson is a primary goal at the LHC and $H \to \gamma\gamma$ is the most promising discovery channel if the mass is between 114 and 130 GeV. In this mass range the Higgs decay width is very narrow, but the signal will lie above an irreducible background and so good energy resolution is crucial. A photon energy resolution of 0.5% above 100 GeV has therefore been set as a requirement for the CMS performance.

The CMS experiment has opted for a hermetic homogeneous electromagnetic calorimeter (ECAL)[1] made of lead tungstate (PbWO$_4$) crystals. Scintillating crystal calorimeters offer the best performance for energy resolution, since most of the energy from electrons and photons is deposited within the homogeneous crystal volume of the calorimeter. Several large crystal calorimeters have been successfully operating in high energy experiments such as L3 at LEP, CLEO II at CESR or Babar at PEP2. However, these detectors did not face the challenging requirements of the LHC, particularly in terms of high speed and of the hostile radiation environment. Another important requirement was a highly granular and compact detector: the CMS calorimeters are located inside the solenoid magnet coil and it was mandatory for both cost and technical reasons to strictly limit their dimensions. The choice of lead tungstate, a fast, dense and radiation-hard material, has been dictated by these operating conditions.

Though lead tungstate has already been studied in the 1940s,[2] it was only at the beginning of the 1990s that its scintillation properties were discovered, triggering a renewed interest and proposals to use it for electromagnetic calorimetry at the LHC.[3] At that time, only few-cubic-centimeter samples of a rather yellowish material were available. The progress from this initial stage to the complete production of 75,848 crystals, totaling 91 tons of lead tungstate material, is clearly one of the highlights of the CMS detector construction.

The construction of the CMS ECAL has been an immense challenge in many technical domains. One could cite the light mechanics with minimal gaps between the crystals, the radiation-tolerant low power readout and trigger electronics located inside the detector, transmitting signals through $\sim 10,000$ optical links,[4] or the laser-based light-monitoring system, allowing a follow-up of the crystal transparency with a 0.1% precision.[5] Some of the developments (for example in electronics) were conducted in a common effort between subdetectors and are presented in other chapters of this review. Another genuine development, specific to the CMS ECAL, concerned the photodetectors used to convert the light emitted by the barrel crystals into an electrical signal. The light yield of lead tungstate is rather low (about 5% of CsI or BGO), requiring photodetectors with gain (to achieve a good noise performance) and insensitivity to ionizing particles. The calorimeter has to operate in a 4 T field, ruling out the use of phototubes in the barrel region where the field is transverse to the crystal axis. The CMS ECAL group triggered therefore the development of large area silicon avalanche photodiodes (APDs).

This chapter will detail the innovative developments on crystals and APDs and the corresponding challenge of organizing a large R&D effort and a mass production scale over a limited period of time and on a constrained budget. For completeness, Sec. 2 presents a short description of the calorimeter layout. Section 3 is devoted to the lead tungstate crystals. In Sec. 4 we focus on the development and production of the APDs.

2. Short Description of the CMS ECAL[6]

The ECAL layout is shown in Fig. 1. The ECAL consists of a cylindrical barrel containing 61,200 crystals, closed at each end with end-caps, each containing 7324 crystals. The barrel part of the ECAL covers the pseudorapidity range $|\eta| < 1.479$. It is made up of 36 supermodules (SMs).

The crystals have a tapered shape, slightly varying with position in η. The crystal cross-section corresponds to approximately 0.0174×0.0174 in

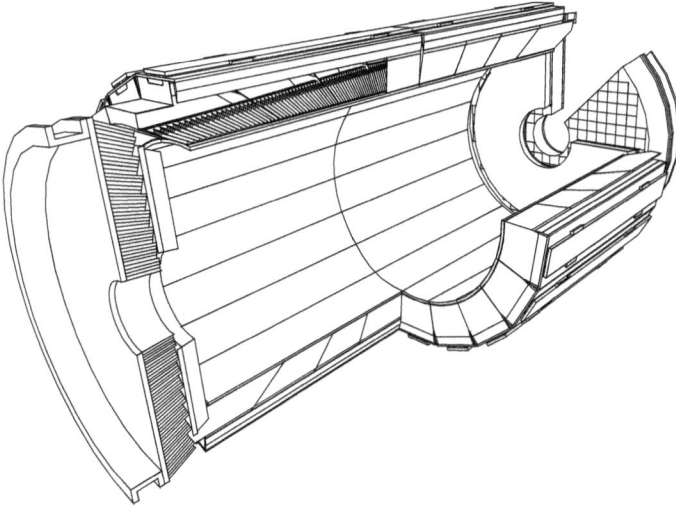

Fig. 1.   CMS ECAL layout.

η–ϕ or $22 \times 22\,\mathrm{mm}^2$ at the front face of the crystal and $26 \times 26\,\mathrm{mm}^2$ at the rear face. The crystal length is $230\,\mathrm{mm}$, corresponding to $25.8\ X_0$. They are mounted in a quasi-projective geometry to avoid cracks aligned with particle trajectories, so that their axes make a small angle ($3°$) with respect to the vector from the nominal interaction vertex, in both the azimuthal and polar angle projections. The barrel crystals are grouped into 5×2 matrices, held in a glass fiber alveolar submodule, of which 40 or 50 are then mounted into a module. The modules are held by an aluminum grid, which supports their weight from the rear. Four modules (of different types, according to the position in η) are assembled together in an SM, which thus contains 1700 crystals.

The end-caps cover the rapidity range $1.479 < |\eta| < 3.0$. Each end-cap consists of identically shaped crystals grouped in mechanical units of 5×5 crystals (supercrystals, SCs) consisting of a carbon-fiber alveolar structure. The crystals and SCs are arranged in a rectangular x–y grid, with the crystals pointing at a focus $1300\,\mathrm{mm}$ beyond the interaction point, so that the off-pointing angle varies with η. They have a rear face cross-section of $30 \times 30\,\mathrm{mm}^2$, a front face cross-section of $28.62 \times 28.62\,\mathrm{mm}^2$ and a length of $220\,\mathrm{mm}$ ($24.7\ X_0$), slightly shorter than for the barrel because of the presence of a $3\ X_0$ preshower in front of them. The SCs are mounted on rigid back plates to form four half end-caps or "dees," each with 3662 crystals.

APDs are used as photodetectors in the barrel and vacuum phototriodes (VPTs) in the end-caps. In the first case, the orientation of the magnetic field rules out the use of phototubes and imposes the use of solid state devices. On the other hand, the radiation level prohibited using APDs for the end-caps while VPTs suffer an acceptable gain loss when oriented quasi-parallel to the magnetic field.

Scintillation light detected by the photodetectors is read out by the front-end electronics, located in the detector behind the crystals. The signal is first shaped and then amplified in parallel by three amplifiers with nominal gains of 1, 6 and 12. Multiple gains are necessary in order to preserve the excellent ECAL precision over a dynamic range larger than that provided by the 12-bit ADCs used. This functionality is built into the multigain pre-amplifier (MGPA),[7] an ASIC developed in 0.25 μm technology. The shaping is done by a CR–RC network with a shaping time of ~ 40 ns. The three analog output signals of the MGPA are digitized in parallel by a multichannel, 40 MHz, 12-bit ADC, the AD41240,[8] also developed in 0.25 μm technology. An integrated logic selects the highest nonsaturated signal as output and reports the 12 bits of the corresponding ADC together with two bits coding the ADC number. The data consist then of a series of consecutive digitizations, corresponding to a sequence of samplings of the signal at 40 MHz. It is envisaged that a time frame of 10 consecutive samplings will be read out in LHC operation.

The electronic readout chain follows a modular structure whose basic elements are matrices of 5×5 crystals corresponding to a trigger tower in the barrel or an SC in the end-cap. The digitized data are locally stored in the front-end electronics board and the trigger primitives, which are elementary quantities such as the energy sum in a trigger tower, are generated and transmitted to the trigger electronics located in the service cavern by an 800 Mbit/s optical link system. On receipt of a level-1 trigger, with a latency of $\sim 3 \mu$s, the data are transmitted to the off-detector electronics by a similar optical link.

A preshower detector with fine granularity is deployed in front of the end-cap crystals to improve the π^0/photon discrimination: indeed, in this region the distance between the decay photons of a typical 50 GeV $p_t\pi^0$, which constitutes the main source of reducible background for the low mass Higgs search, is only about 6 mm. The preshower consists of a lead absorber, two radiation lengths deep, followed by a plane of silicon strip detectors with 1.9 mm pitch, followed by one radiation length of lead, followed by a second plane of silicon with the readout strips orthogonal to the first. The preshower detector is operated at $-10°$C, to keep the effects of radiation damage below an acceptable level. The device covers the rapidity interval $1.65 < |\eta| < 2.61$.

3. Lead Tungstate Crystals

3.1. *Physical properties of lead tungstate*

The main physical and optical properties of lead tungstate are given in Table 1, in comparison with other crystals widely used in high energy calorimetry.

On the positive side, one notes:

- The high density and the small Moliere radius, allowing the building of a compact and granular calorimeter;
- The fast response, with 85% of the light collected in 100 ns and the absence of slow component or phosphorescence, which matches well the requirements for the LHC, where particles bunches cross every 25 ns.

On the more negative side, one finds:

- The low light yield, requiring an optimization of the light collection. The light yield plays a strong role in both the stochastic contribution to the energy resolution and the relative contribution due to the electronics noise; it also limits the timing resolution, particularly in the low energy domain.
- The strong temperature dependence of the light yield, about $2\%/°C$ at room temperature, requiring a very good temperature stabilization (typically $0.1°C$) in order to guaranty the 0.5% energy resolution at high energy without the need for time-dependent temperature corrections.

One could add that lead tungstate is relatively brittle, requiring careful design of the cutting and polishing equipment to obtain a high yield in the mechanical processing.

The crystallographic structure of synthetic lead tungstate synthetic crystal has been determined by x-ray diffraction and identified as sheelite-type with tetragonal symmetry. The emission spectrum is shown in Fig. 2; it peaks at 425 nm. A detailed discussion on the luminescence centers and of the scintillation mechanism can be found in Ref. 9.

3.2. *Crystal production*

The melting temperature of lead tungstate is $1123°C$. The production crystals were grown by two different suppliers using different growth techniques:

- At the Bogoroditsk Technical Chemical Plant (BTCP, Russia), the crystals were grown with the Czochralski method. Figure 3 shows a picture of a typical growth oven. The melt is contained in a large open platinum

160

P. Bloch

Table 1. Physical properties of the lead tungstate in comparison with other commonly used crystals: density (ρ), radiation length X_0, Moliere radius, light yield (LY) and its temperature dependence, scintillation decay time τ_{SC}, peak emission wavelength (λ_{em}) and index of refraction.

	ρ (g/cm^3)	X_0 (cm)	Moliere radius (cm)	LY (ph/MeV)	LY temperature dependence (%/°K @ RT)	τ_{sc} [ns (%)]	λ_{em} (nm)	Index of refraction
PbWO$_4$	8.28	0.89	2.2	200	−1.98	5 (73%) 14 (23%) 110 (4%)	420	420 nm: 2.36/2.24 600 nm: 2.24/2.17
BGO	7.13	1.12	2.33	8200	−1.6	300	480	2.15
CsI (Tl)	4.53	2.43	3.5	52,800	+0.3	1050	560	1.80

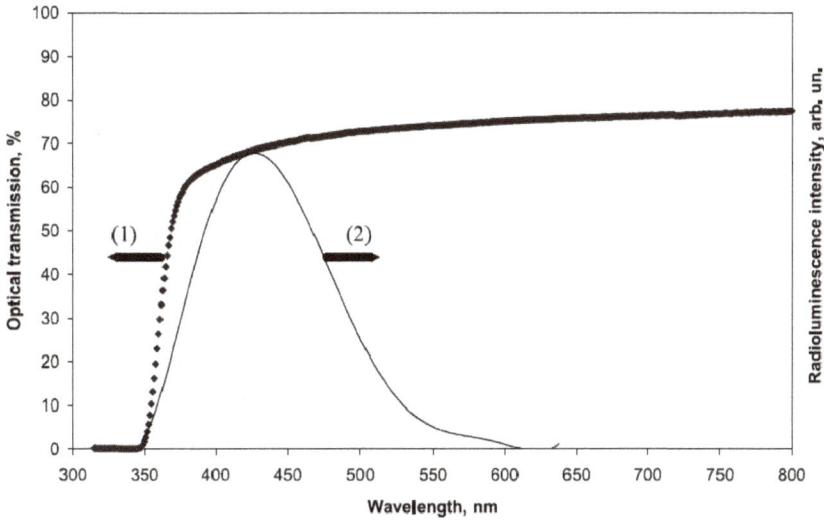

Fig. 2. Room temperature longitudinal transmission (1) and radioluminescence (2) spectra of PbWO$_4$ (from Ref. 9).

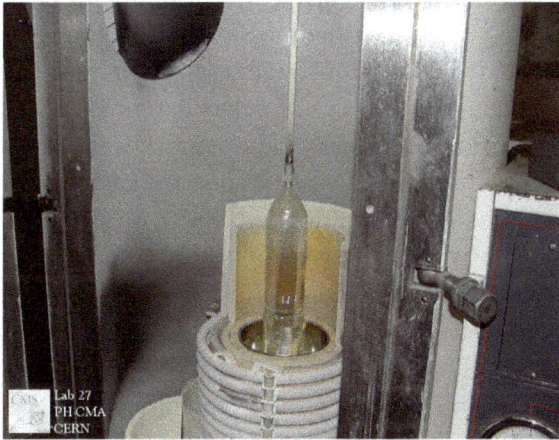

Fig. 3. Picture of a BTCP oven, at the end of a CMS crystal growing cycle (*courtesy of BTCP*).

crucible. The monocrystal grows at the interface between the liquid and the solid part, which is pulled up by a rotating platinum rod. Each oven is used to grow one crystal at a time. The cycling time for an oven to get an ingot suitable for a 23-cm-long CMS crystal is typically 72 h, with a growth rate of 6–8 mm/h. At the maximum rate of the production, 158 such ovens were used in parallel, allowing the BTCP to deliver up to 1200 qualified crystals per month.

- At the Shanghai Institute for Ceramics (SICCAS, China), the crystals were grown using the modified Bridgman method. The component powder is included in a sealed platinum crucible of slightly larger and longer shape than the final crystal. The crucible is pulled down very slowly through a thermal gradient, the solid monocrystal being cooled from the one end where the seed crystal is located. The time to obtain an ingot suitable for a 23-cm-long CMS crystal is 15–20 days. To increase the production rate, each oven contains up to 28 crucibles, allowing the SICCAS to deliver 200–300 qualified crystals per month.

Altogether, after a four-year R&D period mostly devoted to radiation hardness improvement and industrialization of the process, the production of the 76,000 crystals took about ten years, albeit about 55% of them were grown during the last three years.

3.3. *Mechanical processing and light collection uniformity*

After annealing, the ingots are mechanically processed to give them the right shape and surface quality, using first a diamond cutting tool, then a lapping tool (yielding a surface roughness of 300–500 nm), and finally optical polishing equipment (to provide a roughness better than 20 nm).

One of the challenges with crystal calorimeters is to obtain a good longitudinal light collection uniformity. The longitudinal development of electromagnetic showers fluctuates by about one radiation length (X_0) and the absence of longitudinal segmentation prohibits measuring the longitudinal profile shower by shower. As a consequence, a nonuniformity in the region of the shower maximum impacts directly the constant term of the energy resolution. The tapered geometry of the crystals produces a genuine nonuniformity in the longitudinal light collection: the light is more focused toward the photodetector when emitted at the front part of the crystal. This focusing effect is only partially compensated for by the light absorption within the crystal. In previous crystal calorimeters, the light collection uniformity was tuned crystal by crystal either by a differential wrapping or by painting. These techniques were found cumbersome and not flexible enough for the production of 76,000 pieces, and an alternative technique was developed. For the barrel crystals, it was found[10] that by depolishing in a controlled way one of the lateral faces, one could tune the nonuniformity and bring it down within a specified upper limit of 0.35%/X_0 (see Fig. 4). Thanks to the reflective coating inside the alveolae containing the barrel crystals, the light loss is small. The good uniformity of the production process allowed the definition of a unique roughness parameter for each crystal geometry, and depolishing

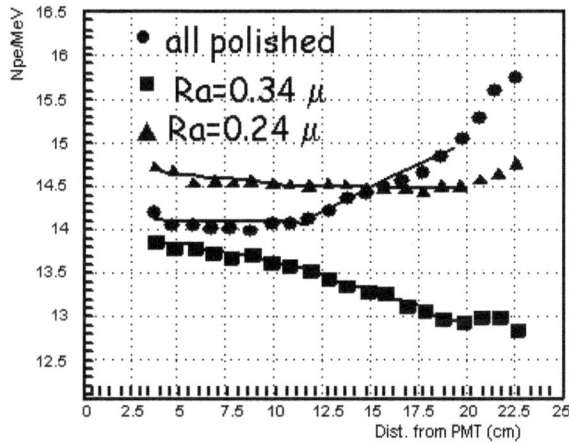

Fig. 4. Light yield as a function of distance from the photomultiplier for a fully polished crystal and after depolishing one face with two different roughnesses, $0.34\,\mu$m and $0.24\,\mu$m.

was applied during the mechanical processing directly at the supplier's. Less than 1% of the crystals required further treatment by CMS.

End-cap crystals could not be depolished, as the light loss was prohibitive in their black carbonfiber, uncoated container. However, due to the very small tapering of the end-cap geometry, the nonuniformity was acceptable without treatment.

3.4. Production quality control

A key point in such a large scale production is the quality control. It was clear right from the beginning that both automation and an excellent database system were a necessity. The quality control and assembly work was shared between two "regional centers" located at CERN and at INFN (Roma, Italy). Robotic equipment was developed at both centers to provide automatic control of dimensions, of optical transmission, light yield and uniformity.[11,12] A picture of one of the two control devices located at CERN is shown in Fig. 5. Each device was able to qualify 60 crystals in about 8 h. Similar equipment was installed at the producer's sites, to allow prescreening before shipping. All measurements were stored in the CRISTAL database,[13] specifically developed to collect the CMS ECAL construction data and guide the operators during assembly. Table 2 summarizes the optical and mechanical specifications for accepting a crystal.[14] Figure 6 shows the distribution of some of these parameters for the production samples. Radiation hardness qualification is discussed in the next section.

P. Bloch

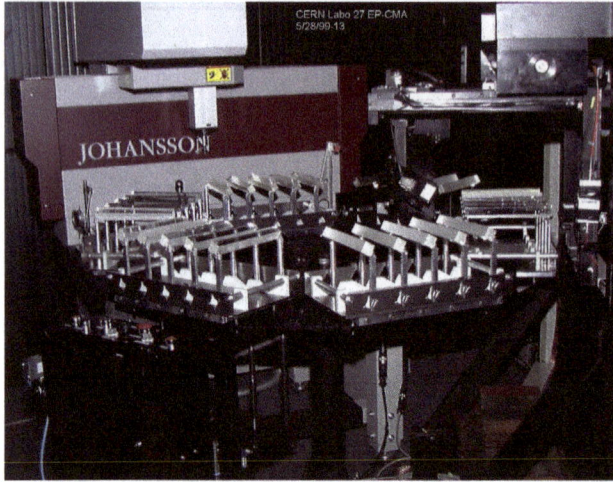

Fig. 5.   Automatic control equipment ACCOCE at CERN.

Table 2.   Optical and mechanical specifications for the CMS crystals.

Criteria	Specification	Comment
Longitudinal transmission at 360 nm	$T > 25\%$	Radiation hardness
Longitudinal transmission at 420 nm	$T > 55\%$	At emission wavelength
Longitudinal transmission at 620 nm	$T > 65\%$	Absence of core defects
Wavelength spread along crystal axis of transversal transmission at 50%	$\Delta\lambda < 3\,\text{nm}$	Crystal uniformity and radiation hardness
Light yield with a ^{60}Co radioactive source	LY > 8 p.e.	Measured with a PM
Decay time	90% of the light emitted in $1\,\mu s$ collected in 100 ns	Absence of slow component
Light yield uniformity	$|\text{FNUF}| < 0.35\%/X_0$	See Subsec. 2.2
Dimensions	$[+0, -100]\mu m$	Minimize intercrystal cracks
Face planarity	$<20\,\mu m$	Minimize crystal-to-crystal stress transfer

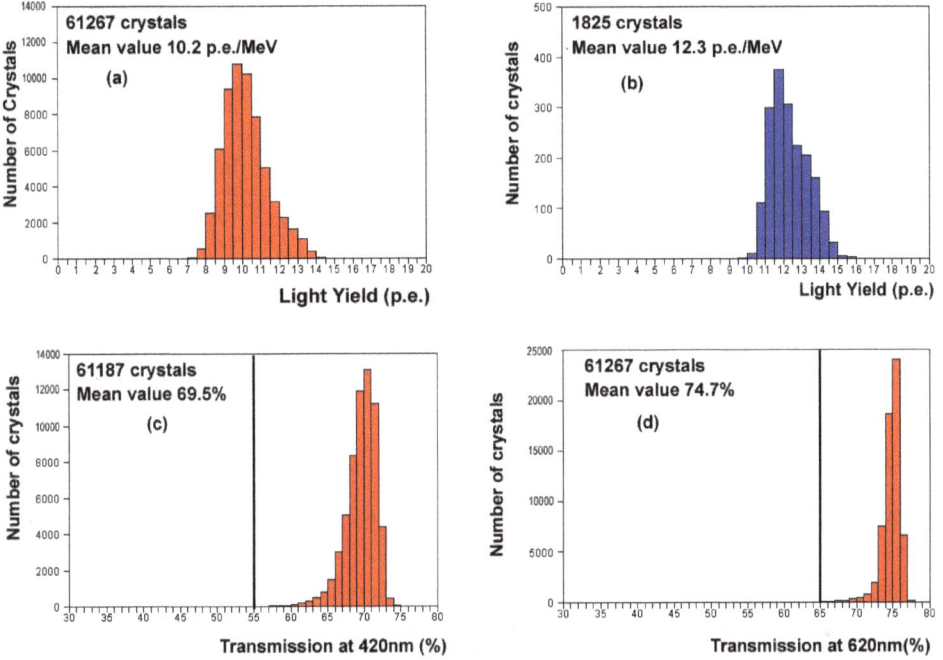

Fig. 6. Distribution on the whole barrel production[15] of (a) light yield of BTCP barrel crystals; (b) light yield of SICCAS barrel crystals; (c) transmission at 420 nm (BTCP barrel crystals); (d) transmission at 620 nm (BTCP barrel crystals).

3.5. *Radiation hardness and its control*

Another important challenge was to ensure the radiation hardness of the crystals and to control it during the whole production.

Operating at a peak luminosity of 10^{34} cm^{-2}s^{-1}, the LHC will produce a very harsh radiation environment for detectors. The ECAL will be exposed to fast hadrons, mostly pions, which, in interactions with the ECAL itself, produce secondary hadrons, and build up a flux of low energy neutrons, with energies typically below 10 MeV. In addition, electromagnetic showers inside the crystals provide a significant dose. The radiation environment can be characterized by three quantities: the absorbed dose, the density of inelastic hadronic interactions (stars[16]), and the neutron fluence below 20 MeV.

Table 3 gives, as a function of pseudorapidity, these three radiation quantities for an integrated luminosity of 500 fb^{-1}. Another important quantity is the dose rate, which varies from 0.17 to 0.25 Gy/h at the shower maximum and at nominal LHC peak luminosity in the barrel to 5 Gy/h at $\eta = 2.5$ in the end-caps.

Table 3. Radiation exposure of the ECAL crystals at some selected pseudorapidity values. All values are given for $500\,\text{fb}^{-1}$ integrated luminosity.

Pseudorapidity (η)		0.1	1.4	1.7	2.1	2.5	2.9
Dose (kGy)	ave	0.83	1.32	2.34	9.73	32.6	115
	peak	2.34	3.51	5.72	22.2	66.6	166
Star density ($10^{11}\,\text{cm}^{-3}$)	ave	1.03	2.05	3.10	13.1	37.1	51.0
	peak	1.64	2.75	3.56	14.8	40.8	60.7
n-fluence ($E < 20\,\text{MeV}$)	ave	3.86	4.67	11.7	47.2	115	96.8
($10^{13}\,\text{cm}^{-2}$)	peak	4.56	5.27	15.9	60.5	165	107

It has been shown that under the LHC irradiation conditions mentioned in Table 3 the scintillation mechanism of lead tungstate is not affected.[17] However, when lead tungstate crystals are exposed to ionizing radiation, pre-existing point structure defects may act as traps for electrons or holes. The resulting charged defects have discrete energy levels and optical transitions can be induced, absorbing part of the scintillation light during its transport to the photodetector. This is the mechanism of radiation-induced color centers, which are the main source of damage in lead tungstate at the LHC. Up to five types of color centers have been identified in lead tungstate, with corresponding absorption bands at 350–400, 470, 520, 620 and 715 nm. The induced absorption spectrum and relative intensity of these five bands strongly depend on the nature and density of pre-existing structural defects, which themselves depend on the crystal growth conditions. Undoped crystal grown from stoichiometric raw material has an absorption spectrum with two dominating broad bands peaked near 380 and 620 nm. Much effort has been spent to optimize the quality of the raw materials and of the growth parameters in such a way as to reduce the density of point structure defects at the origin of color centers under irradiation. Although significant, the improvement was not good enough to meet the severe radiation hardness specifications for the CMS ECAL crystals. It was therefore decided to use a doping strategy to charge-compensate for the remaining defects in the structure and to suppress electric carriers trapping on them.

After some R&D, the doping of lead tungstate crystals by specified impurities like La, Y or Nb at different stages of the growth process has been used for the production of CMS crystals to improve their radiation hardness. A very significant suppression of the electron/hole trapping processes is already observed for a doping concentration of the order of 100 ppm if the crystal stoichiometry is well controlled.[9,18,19]

A parameter which describes well the damage of a crystal's optical properties induced by radiation exposure is the radiation-induced absorption coefficient (μ), defined as

$$\mu = \frac{1}{L} \cdot \ln \frac{T_{\text{init}}}{T_{\text{irr}}},$$

where T_{init} and T_{irr} are respectively the values of the crystal transmission measured before and after irradiation at the peak emission wavelength and L is the length of the crystal. For a given wavelength, the induced absorption coefficient is directly proportional to the total density of all color centers absorbing at this wavelength:

$$\mu \propto D = \sum_i D_i.$$

The situation is, however, complicated by the fact that the radiation-induced absorption in the crystal results from the balance between the creation of color centers and their destruction due to annealing at the detector operation temperature. At equilibrium under continuous irradiation at a fixed dose rate R, the radiation-induced absorption coefficient is given by

$$\mu = \sum_i \frac{\mu_i^{\text{sat}} \cdot R \cdot b_i}{a_i + R \cdot b_i},$$

where μ_i^{sat} is the induced absorption coefficient at saturation due to defects of type i, b_i is the damage constant (related to the cross-section of free carrier capture) and a_i the recovery rate, again of defects of type i.

As a consequence, two radiation specifications were defined.[14] The purpose of the first specification is to prevent the total damage exceeding a certain level when the radiation damage in the crystal is fully saturated throughout its volume. This level corresponds to an induced absorption of $1.5\,\text{m}^{-1}$ at the peak emission wavelength. This maximum value prevents the change of front nonuniformity induced by irradiation becoming large enough to have an impact on the energy resolution of the calorimeter. The second specification is related to the operation of the calorimeter at typical LHC (barrel) dose rates and the need to monitor it precisely. The light loss in these conditions should not exceed 6%.

For the very large sample of crystals produced at the BTCP, the radiation hardness could not be checked individually for each crystal. Studies made during the R&D period allowed one to find correlations between radiation hardness behavior and initial optical parameters that can be measured easily for each crystal.[14] It was found that the presence of an initial absorption in the band edge region (350–360 nm) is harmful to the radiation hardness.[20] Based on this correlation, certification limits were defined for transmission spectra at short wavelengths, which were systematically measured for each crystal using the automatic control machines. Crystals having one of the optical certification parameters close to or above acceptance limits were subjected to systematic irradiation tests. All other crystals were randomly irradiated either at a high dose rate (350 Gy/h) to induce damage saturation or under conditions comparable to those in the LHC (0.15 Gy/h front irradiation with a ^{60}Co source). Test irradiations for a sizeable fraction of the crystals were also performed directly by the producers. Figure 7 shows the two specifications' parameters for randomly chosen BTCP crystals.

For the SICCAS crystals, no correlation between radiation hardness and initial optical parameters could be found. All crystals were therefore irradiated by the producer for 24 h at a dose rate of 30 Gy/h. Sample irradiations were performed by CMS to crosscheck the producer's results and to measure the damage at a low dose rate. Figure 8 shows the correlation between the SICCAS and CMS measurements, and the results of low dose rate irradiations.

3.6. Conclusions

The CMS ECAL required an unprecedented production of more than 10 cubic meters of radiation-hard lead tungstate crystals. Thanks to a large effort of the groups involved in the project, of the suppliers and of other academic partners, industrial methods for production and quality control were developed and successfully used. This achievement has not only allowed

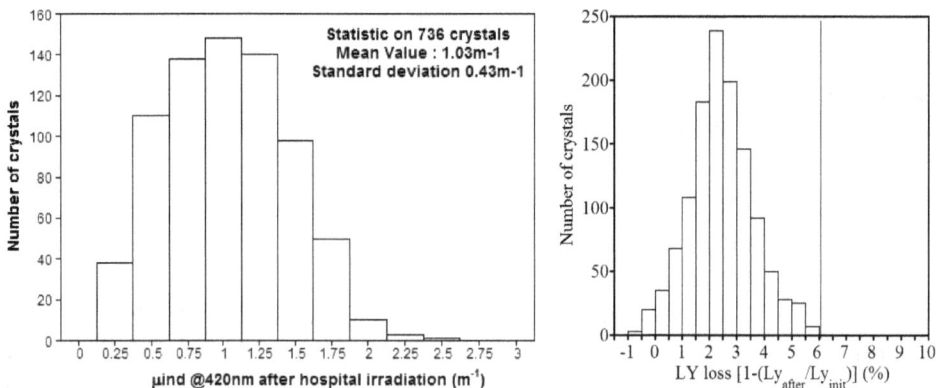

Fig. 7. *Left*: Induced absorption coefficient after high dose rate (350 Gy/h) irradiation. *Right*: Distribution of relative LY loss after 1.5 Gy at a rate of 0.15 Gy/h for BTCP crystals randomly selected from among those having optical parameters within ECAL specification. (From Ref. 21.)

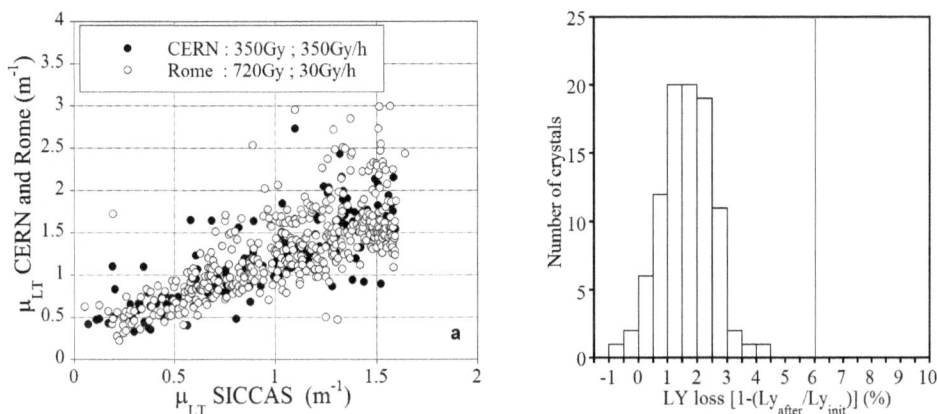

Fig. 8. *Left*: Correlation between the induced absorption coefficient measured at the SICCAS and by CMS. *Right*: Distribution of relative LY loss after 1.5 Gy at a rate of 0.15 Gy/h for SICCAS crystals randomly selected from among those having optical parameters within ECAL specification. (From Ref. 21.)

the installation of both the barrel and end-cap ECAL in CMS in time for the first LHC beam, but has also paved the way for future projects using a large amount of lead tungstate crystals.[22]

4. Production and Tests of 130,000 Avalanche Photodiodes

4.1. *Introduction*

In the CMS barrel electromagnetic calorimeter, the photosensors needed to transform the light produced in the crystals into an electronics signal have to fulfill very strong requirements. These include operation in a 4 T field, radiation hardness, high speed and, of course, compatibility with the demanding ECAL energy resolution requirements. The last implies a large area, good sensitivity to the light emitted by the crystal, stability and insensitivity to voltage and temperature fluctuations, as well as low capacitance, series resistance, noise and dark current. Furthermore, the low light yield of the crystals brings in two more important requirements: gain and insensitivity to ionizing particles traversing the diode. These last two points ruled out the use of standard PIN diodes as in previous crystal calorimeters. In fact beam tests carried out in 1993 with Si PIN photodiodes showed that the nuclear counter effect was unacceptably large due to the low light yield: this effect was largely eliminated when an APD was used.[26,27] A final requirement is a very high reliability, since the mechanical construction of the ECAL excludes a later replacement of broken or bad APDs. A reliability of 99.9% for ten years of operation at the LHC was set as a target.

Unlike PIN diodes, APDs contain a region with a high electric field so as to provide internal photocurrent gain by impact ionization. At the time when lead tungstate was chosen as the active medium for the CMS ECAL, no APD with a large-enough area was available on the market (not to talk about the radiation hardness). After ten years' R&D by producers in close contact with members of the CMS ECAL, requiring the development and production of about 100 different prototypes of APDs, the model S8141, produced by Hamamatsu, was selected.[23] The structure of this APD is shown in Fig. 9. Incident photons are converted in a very thin layer at the surface just behind the p^{++} contact layer with a thickness of about 100 nm and their electrons are amplified in a high field p–n junction. The n^- layer decreases the capacitance and the sensitivity of the gain to the applied bias. The V-shaped grooves reduce the surface current. Electrons from ionizing particles traversing the diode are amplified only if produced in the conversion layer of $5\,\mu$m thickness. Table 4 summarizes the properties of the S8141 APD.

The gain is a steep function of the applied bias voltage (Fig. 10). In CMS, the APDs are operated at a gain (M) of 50. The breakdown voltage (V_B) is defined to be that at which the dark current reaches $100\,\mu$A. This occurs if the applied bias voltage is about 40 V higher than the required bias voltage (V_R) for a gain of 50.

As a compromise between light collection efficiency, capacitance (linked to the noise for the front-end electronics) and cost, it was decided to use two APDs per crystal, connected in parallel to a single preamplifier, requiring a total of 130,000 devices to be produced. The serial production was launched

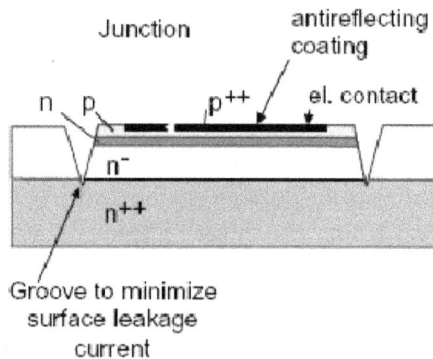

Fig. 9.   APD structure.

Table 4. APD parameters at gain 50 and 25°C.

Active area	$5 \times 5 \, \text{mm}^2$
Operating voltage	$\sim 380 \, \text{V}$
Maximum gain	> 1000
Capacitance	$80 \, \text{pF}$
Serial resistance	$3 \, \Omega$
Dark current	$< 50 \, \text{nA}$
Quantum efficiency at 430 nm	75%
$1/M \times dM/dV$ at gain $M = 50$	3.1%
$1/M \times dM/dT$ at gain $M = 50$	-2.4%
Excess noise factor at gain $M = 50$	2

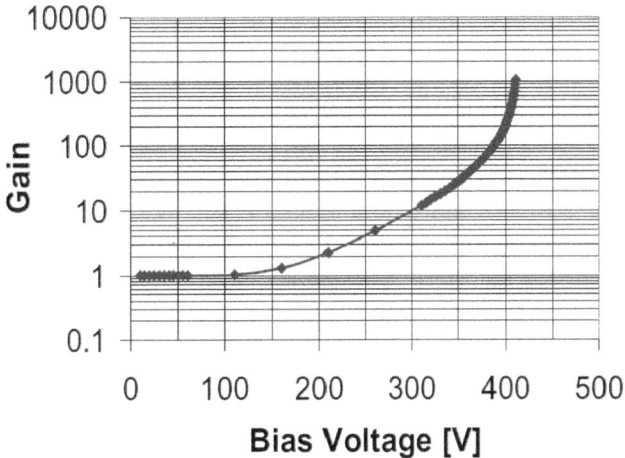

Fig. 10. Gain dependence on the applied bias voltage.

during winter 2001 and took about four years. One of the most difficult issues during the production was the quality control linked to the radiation hardness. This is the subject of the next section.

4.2. *APD radiation hardness and production screening*[24]

Irradiation of APDs with hadrons could cause displacements of atoms in the silicon lattice, resulting in a change of the doping profile in the bulk. This would induce a high dark current. Extensive tests showed that the operational parameters (gain, quantum efficiency) of the APD did not change due to irradiation with hadrons.[23] Irradiation with gammas causes modifications in the surface region like loss of quantum efficiency due to a clouding of the epoxy or high surface currents due to SiO_2 breakup. Surface effects were tested by exposing APDs to gammas emitted by a ^{60}Co source. It turned out that V_B–V_R is a valuable parameter for judging if an APD is radiation-hard. For initially developed samples, 10–15% had a strongly reduced breakdown voltage. After an intense R&D program, Hamamatsu Photonics modified the design, especially in the region of the V-shaped grooves, and serial production could be started.

However, it turned out that still about 5% of the APDs did not survive the irradiation. These problems occurred again only on the surface and were caused by defects in the masks or by dust particles. There is no way for the producer to control this down to the per mill level. A screening method had to be developed to reject unreliable APDs before they are built into the detector.

The screening was done in two steps. At the PSI institute (Switzerland), all APDs were irradiated with a ^{60}Co source to 5 kGy at a dose rate of 2.5 kGy/h. After a relaxation time of one day, the breakdown voltage and the dark current as a function of gain were remeasured to detect APDs that had been damaged by irradiation. A few days later, at CERN, the noise power was measured at frequencies up to 1 MHz, at gains from 1 to 300. Then APDs were annealed under bias in an oven at 80°C for four weeks. After this step, the breakdown and the dark current were remeasured. Based on these four measurements, faulty APDs could be rejected. This treatment does not change the APD parameters. Only the dark current rises. It is typically less than 10 nA before irradiation, about 300 nA one day after irradiation and about 50 nA after annealing.

APDs were rejected if, after irradiation or annealing, V_B had changed by more than 5 V, or if the dark current (I_d) or the noise was anomalously large (see Figs. 11 and 12). The cuts were applied relative to the mean for the wafer, due to wafer-to-wafer variations in I_d and the noise, and to accommodate measurement offsets in V_B. APDs were also rejected if the ratio I_d/M rises between $M = 50$ and 400. If I_d is due to surface currents, it will rise with bias voltage ohmically, and thus I_d/M should fall steadily with M. A rise in I_d/M well below the normal breakdown point could come from current at a local defect being amplified. Figure 12 shows I_d/M vs. M after irradiation for APDs from one wafer where most of them are well behaved but three are rejected due to rises in I_d/M.

In order to tune the screening procedure, a number of APDs were screened twice. The idea of these double screenings is that if the screening is effective, all weak APDs are found in the first screening and no new

dV$_B$ after Co irradiation

APD number

I$_d$ after Co irradiation

APD number

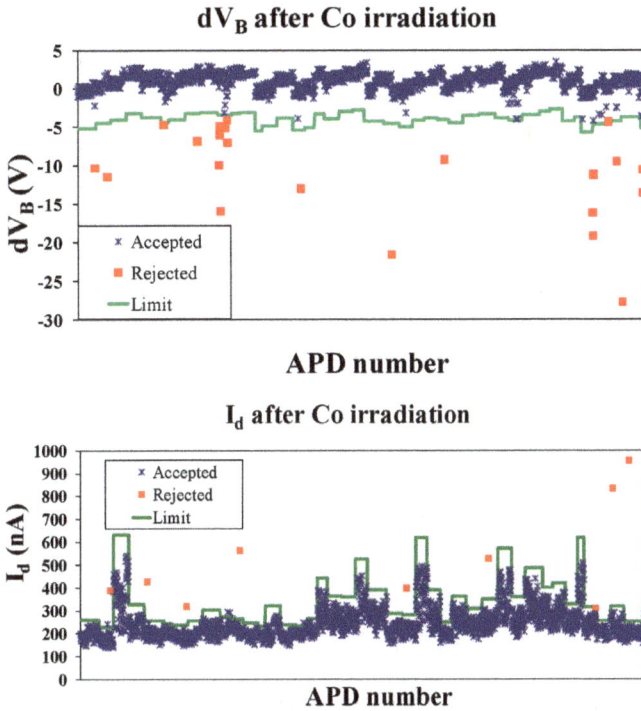

Fig. 11.   Change in breakdown voltage (top) and induced dark current (bottom) for 3000 APDs (from Ref. 24). The lines mark the rejection cuts, set for each wafer.

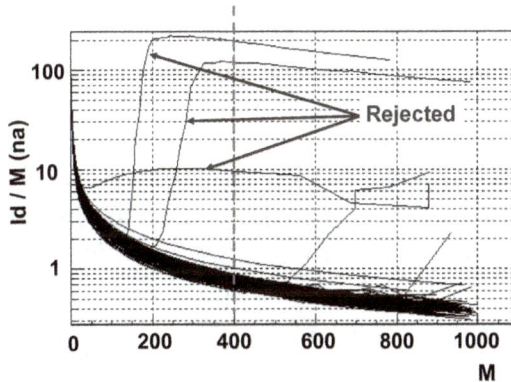

Fig. 12. I_d/M vs. M for APDs from one wafer (from Ref. 24). For good APDs, I_d/M does not rise between $M = 50$ and 400.

ones will be found in the second screening. If the screening is efficient, a large fraction of APDs found to be weak in the first screening will again be weak in the second screening. For 834 APDs which passed the first screening, only 1 failed the second screening, implying a reliability around the required

99.9% level. The behavior of this APD suggests that it might become noisy in CMS, but not die. On the other hand, for 221 APDs which failed the first screening, 102 (46%) also failed the second screening.

4.3. *Performance in situ*

Out of the serial production, 125,800 APDs have been assembled in the form of capsules containing 2 APDs each and glued to the back face of the crystals. As 50 capsules are connected to only one HV bias channel, APDs reaching gain 50 at the same bias within ±1 V have been grouped together. The APDs were retested after capsule assembly and after gluing. No APD failure was observed. The CMS barrel ECAL has been operated almost continuously from August till the end of October 2008 (often with an APD increased gain, $M = 200$, to improve the signal-to-noise value for cosmic ray muons), and again no APD failure was reported.

5. Conclusion

In this short report, we have presented two of the most innovative developments for the construction of the CMS electromagnetic calorimeter. In 2004 a fully equipped barrel supermodule was tested in the CERN H4 beam. The energy resolution measured with electron beams having momenta between 20 and 250 GeV/c confirmed the expectations described above. Since the electron shower energy contained in a finite crystal matrix depends on the particle impact position with respect to the matrix boundaries, the intrinsic performance of the calorimeter was studied by using events where the electron was limited to a $4 \times 4\,\text{mm}^2$ region around the point of maximum containment (central impact). The energy is reconstucted by summing the energy in a 3×3 crystal matrix. A typical energy resolution was found to be [25]:

$$\left(\frac{\sigma}{E}\right)^2 = \left(\frac{2.8\%}{\sqrt{E}}\right)^2 + \left(\frac{.125}{E}\right)^2 + (0.030\%).$$

where E is in GeV.

In mid-August 2008, the complete calorimeter had been installed and fully commissioned and was ready for the detection of the first LHC beam interactions. Data recorded and with cosmic rays *in situ* have confirmed the excellent quality of the detector, which is meeting all the specifications set at the start of the project.

Acknowledgments

As already mentioned in the introduction, the biased choice of the few topics covered in this chapter does not do justice to the very large and innovative effort which was necessary to design, assemble and integrate the CMS electromagnetic calorimeter. The success of this long project has only been possible thanks to the professionalism, the enthusiasm and the dedication of hundreds of physicists, engineers, technicians and students making up the CMS ECAL community.

References

1. CMS ECAL Technical Design Report CERN/LHCC 97–33 (1997).
2. F. A. Kroger. *Some Aspects of the Luminescence of Solids* (Elsevier, Amsterdam, New York, 1948).
3. P. Lecoq (ed.) *Proc. Crystal 2000 International Workshop* (Editions Frontieres, Gif-sur-Yvette, 1993). See presentation of V. Katchanov, I. Nagornaya and M. Kobayashi.
4. *The Front-End Electronics System for the CMS Electromagnetic Calorimeter — Proc. 11th Workshop on Electronics for LHC and Future Experiments* (2005), p. 221.
5. M. Anfreville *et al.* Laser monitoring system for the CMS lead tungstate crystal calorimeter. CMS Note 2007/038.
6. A more extensive description can be found in: S. Chatrchyan *et al.* (CMS Collaboration), The CMS experiment at the CERN LHC, JINST **3** (2008) S08004.
7. M. Raymond, J. Crooks, M. French and G. Hall. The MGPA electromagnetic calorimeter readout chip for CMS, *IEEE Trans. Nucl. Sci.* **52** (2005) 3.
8. G. Minderico *et al.* A CMOS low power, quad channel, 12-bit, 40 MS/s pipelined ADC for applications in particle physics calorimetry, *Proc. 9th Workshop on Electronics for LHC and Future Experiments* (2003).
9. A. Annenkov *et al.* Lead tungstate scintillation material, *Nucl. Instrum. Methods Phys. Res. A* **490** (2002) 30.
10. E. Auffray *et al.* Development of a uniformisation procedure for the $PbWO_4$ crystals of the CMS electromagnetic calorimeter. CMS Note 2001/004.
11. E. Auffray *et al.* Performance of ACCOS, an automatic crystal quality control system for the PWO crystals of the CMS calorimeter, *Nucl. Instrum. Methods Phys. Res. A* **456** (2001) 325.
12. S. Baccaro *et al.* An automatic device for the quality control of large-scale crystal's production, *Nucl. Instrum. Methods Phys. Res. A* **459** (2001) 278.
13. A. Bazan *et al.* The use of production management techniques in the construction of large scale physics detectors, in *Proc. 45th IEEE Nuclear Science Symposium, IEEE Trans. Nucl. Sci.* **46–3** (1999) 392.

14. E. Auffray *et al.* Specifications for lead tungstate crystals preproduction. CMS Note 1998/038.
15. E. Auffray on behalf of the CMS ECAL group. Overview of the 63000 Barrel crystals production for CMS-ECAL, *IEEE Trans. Nucl. Sci.* **55** (2008) 1314.
16. M. Huhtinen *et al.* High-energy proton induced damage in PbWO$_4$ calorimeter crystals, *Nucl. Instrum. Methods Phys. Res. A* **545** (2005) 63.
17. A. N. Annenkov *et al.* Systematic study of the short-term instability of PbWO$_4$ scintillator parameters under irradiation, *Radiat. Meas.* **29**(1) (1998) 27.
18. M. Nikl *et al.* Radiation-induced formation of color centers in PbWO$_4$ single crystals, *J. Appl. Phys.* **82**(11) (1997) 5758.
19. E. Auffray *et al.* Improvement of several properties of lead tungstate crystals with different doping ion, *Nucl. Instrum. Methods Phys. Res. A* **402**(1) (1998) 75.
20. E. Auffray, I. Dafinei, F. Gautheron, O. Lafond-Puyet, P. Lecoq and M. Schneegans. Scintillation characteristics and radiation hardness of PWO scintillators to be used at the CMS electromagnetic calorimeter at CERN, in *SCINT'95: Proc. Int. Conf. Inorganic Scintillators and Their Applications* (Delft, The Netherlands; Aug. 28–Sep. 1, 1995), p. 257.
21. P. Adzic *et al.* Radiation hardness qualification of PbWO$_4$ scintillation crystals for the CMS Electromagnetic Calorimeter. To be published.
22. A detector for antiprotons physics. GSI Scientific Report 2001.
23. K. Deiters *et al. Nucl. Instrum. Methods Phys. Res. A* **453** (2001) 223.
24. Z. Antunovic *et al.* Radiation hard avalanche photodiodes for the CMS detector, *Nucl. Instrum. Methods Phys. Res. A* **537** (2005) 379.
25. P. Adzic *et al. Energy resolution of the barrel of the CMS electromagnetic calorimeter, JINST* **2** (2007) P04004.
26. *CMS Status Report and Milestones, Report to the LHCC*, Nov. 1993 and *CMS Technical Proposal*, CERN LHCC 94-38, Dec. 1994.
27. E. Lorentz *et al.*, Max Planck Institute, Munich, MPI-PHE-93-23 (1993).

Chapter 7

ATLAS ELECTRONICS: AN OVERVIEW

Philippe Farthouat

Physics Department, CERN
CH-1211 Genève 23, Switzerland
philippe.farthouat@cern.ch

The readout electronics development of ATLAS has been driven by the large number of channels of the different subdetectors and by particular environment constraints. This chapter gives an overview of the developments made for the front-end electronics, the front-end links, the back-end electronics and the power distribution. It also shows how common issues across the experiment have been handled.

1. Introduction

The readout electronics of ATLAS has been extensively described in a number of publications.[1-8] This chapter will concentrate on the reasons and parameters that drove the readout architecture design, as well as on the common problems that were encountered and the way they were tackled.

In Sec. 2, the main parameters, including the interface to the trigger and data acquisition (DAQ) system, will be reviewed. Section 3 will detail the front-end electronics architecture and some of the main components. Section 4 will treat trigger and off-detector electronics, and Sec. 5 will cover common issues (radiation tolerance and electromagnetic compatibility) and the way they were handled.

2. Main Parameters

The ATLAS experiment, like the other LHC experiments, is made up of large subdetectors, each having an important number of readout channels. The beam characteristics, the trigger and DAQ functional model, the harsh environment and the large number of channels were the main factors that drove the definition and design of the readout electronics.

P. Farthouat

2.1. *Number of channels*

The overall ATLAS detector is shown in Fig. 1, while Table 1 lists the different ATLAS subdetectors and their number of readout channels.

Such a large number of channels could not be handled without integrating a large fraction of the readout electronics inside the detector

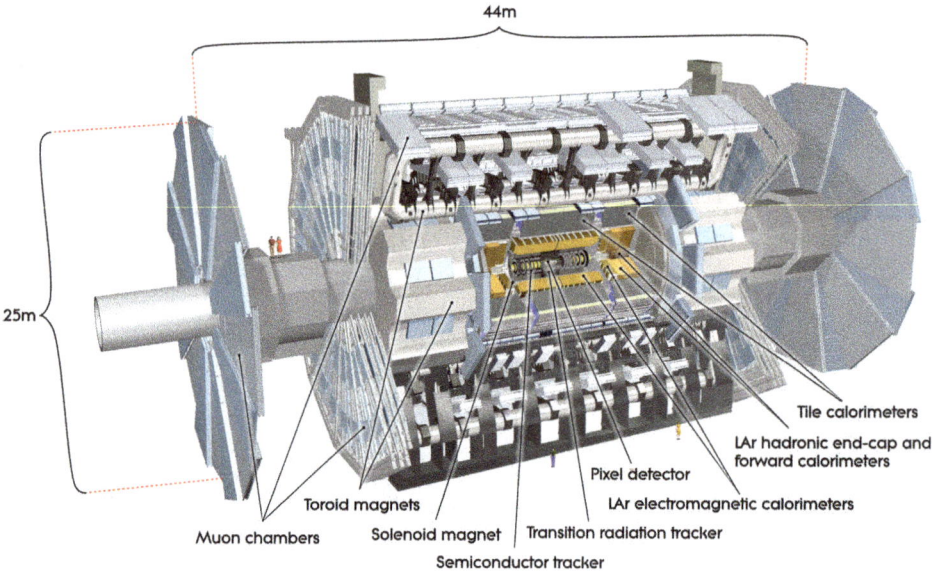

Fig. 1. ATLAS detector. The dimensions of the detector are 25 m in height and 46 m in length. The overall weight of the detector is approximately 7000 tonnes.

Table 1. List of the ATLAS subdetectors and their number of readout channels.

Subdetector	Number of readout channels
Pixel detector	$80 \cdot 10^6$
Silicon strip detector (SCT)	$6.2 \cdot 10^6$
Transition radiation tracker (TRT)	$3.5 \cdot 10^5$
Liquid argon electromagnetic calorimeter	$1.74 \cdot 10^5$
Liquid argon hadronic end cap (HEC) calorimeter	$5.63 \cdot 10^3$
Forward calorimeter	$3.52 \cdot 10^3$
Tile calorimeter	$9.85 \cdot 10^3$
Muon monitored drift tube (MDT) chambers	$3.54 \cdot 10^5$
Muon thin gap chambers (TGCs)	$3.22 \cdot 10^5$
Muon resistive plate chambers (RPCs)	$3.74 \cdot 10^5$
Muon cathode strip chambers (CSCs)	$3.1 \cdot 10^4$

(front-end electronics), and performing a large amount of multiplexing before transmission to the off-detector counting rooms (back-end electronics). As a consequence the front-end electronics is subject to very harsh environmental conditions and needs to be implemented in integrated circuits due to the high channel density.

2.2. *Interface to the trigger and data acquisition system*

The beam packets cross in the center of ATLAS at a 40 MHz frequency; the data acquisition system (DAQ) not being able to work at that rate, a trigger system is needed. This system has three distinct levels: L1, L2, and the event filter. Each trigger level refines the decisions made at the previous level and, where necessary, applies additional selection criteria. The first level trigger uses a limited amount of the total detector information to make a decision in less than $2.5\,\mu$s, reducing the rate to about 75 kHz, which is manageable by the DAQ.

The readout electronics has to cope with this architecture, of which a schematic view is presented in Fig. 2.

The front-end electronics includes different functional components:

(i) The front-end analog or analog–digital processing.
(ii) The L1 buffer, in which information (analog or digital) is stored and is retained for a time long enough to accommodate the L1 trigger latency. This element had to be designed to accommodate an L1 trigger latency of at least $2.5\,\mu$s.
(iii) The derandomizing buffer, in which the data corresponding to an L1 trigger accept (L1A) are stored before being sent to the following level. This element is necessary for accommodating the maximum instantaneous L1 rate without losing a significant amount of data.
(iv) Dedicated links or buses for transmitting the front-end data stream to the back-end electronics. Their speed, in conjunction with the derandomizing buffer size, defines the amount of data lost because of the random nature of the L1 trigger. This amount should not exceed 1% with a 75 kHz L1 rate.
(v) Dedicated timing, trigger and control (TTC) links or buses for transmitting the necessary signals (bunch crossing, reset signals, L1A, etc.) to the front-end electronics.

The back-end electronics is the functional element of the readout electronics system where one can reach a higher level of data concentration and multiplexing by gathering information from several front-end links. Elementary

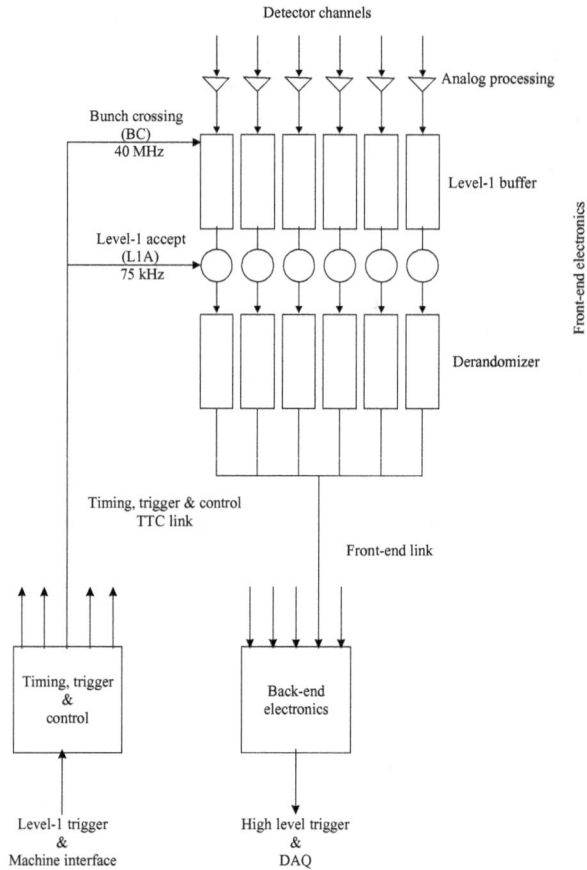

Fig. 2. Architecture of the readout electronics. The information (analog or digital) from each detector channel is stored in the level-1 buffer at each bunch crossing and is retained for a time long enough to accommodate the L1 trigger latency (2.5 μs). The data corresponding to an L1 trigger accept (L1A) are stored in a derandomizing buffer before being sent to the back-end electronics through dedicated front-end links, one link being shared by several channels.

digitized signals are formatted as raw data prior to being transferred to the DAQ.

2.3. *Environmental constraints*

There are four main environmental constraints that dramatically influence the design of the readout electronics:

(i) The radiation level in the experimental cavern;
(ii) The magnetic field in the experimental cavern;

(iii) The very limited access to the detector;

(iv) The location of the services caverns (counting rooms) housing the back-end electronics.

2.3.1. *Radiation*

There is a large radiation background when the LHC operates at design luminosity and the electronics is sensitive to radiation. For instance, threshold voltages in CMOS technologies are sensitive to the total ionizing dose (TID), ß of bipolar transistors are sensitive to displacement damages caused by nonionizing energy loss (NIEL), and digital electronics is sensitive to single event upsets (SEUs) caused by energetic particles. An exhaustive list of effects can be found in Ref. 9.

Table 2 shows the maximum radiation levels at the location of the front-end electronics of the different subdetectors.

These are simulated radiation levels and, to be safe, the electronics must be validated for higher levels, as described in Sec. 5.

2.3.2. *Magnetic field*

The front-end electronics of the inner detector will be subjected to a magnetic field as high as 2 teslas (central solenoid), while the front-end electronics of the muon spectrometer will see as much as 1 tesla (toroid). Although the influence on the electronics itself is very little, the presence of such a field

Table 2. Maximal simulated radiation levels seen by the front-end readout electronics of the different ATLAS subdetectors. Yearly equivalent 1 MeV neutron fluence (NIEL), yearly total ionizing dose (TID) and yearly fluence of hadrons able to generate single event upsets (SEUs) are given.

Subdetector	NIEL $(1\,\text{MeVn} \cdot \text{cm}^{-2} \cdot \text{year}^{-1})$	TID $(\text{Gy} \cdot \text{year}^{-1})$	SEU $(>20\,\text{MeV hadron} \cdot \text{cm}^{-2} \cdot \text{year}^{-1})$
Pixel B-layer	$1.62 \cdot 10^{14}$	$1.1 \cdot 10^5$	$2.30 \cdot 10^{14}$
Pixel	$7.58 \cdot 10^{13}$	$4.98 \cdot 10^4$	$8.98 \cdot 10^{13}$
SCT	$1.59 \cdot 10^{13}$	$7.83 \cdot 10^3$	$1.22 \cdot 10^{13}$
TRT	$7.23 \cdot 10^{12}$	$1.56 \cdot 10^3$	$2.86 \cdot 10^{12}$
Lar	$1.85 \cdot 10^{11}$	4.77	$3.78 \cdot 10^{10}$
Tile	$2.74 \cdot 10^{10}$	$3.78 \cdot 10^{-1}$	$6.74 \cdot 10^9$
CSC	$8.34 \cdot 10^{11}$	23.0	$1.59 \cdot 10^{11}$
RPC	$5.71 \cdot 10^9$	$1.15 \cdot 10^{-1}$	$1.04 \cdot 10^9$
TGC	$6.53 \cdot 10^9$	$1.65 \cdot 10^{-1}$	$1.75 \cdot 10^9$
MDT	$4.80 \cdot 10^{10}$	$7.41 \cdot 10^{-1}$	$8.25 \cdot 10^9$

imposes constraints on the selection of components (nonmagnetic material,
difficulties in using ferrites) and has a very strong influence on the power
supplies to be used in the ATLAS cavern.

2.3.3. *Access*

The access to the front-end electronics is very limited and in some cases
(inner detector) almost impossible. This has some implications for the way
the electronics system is designed in taking into account reliability issues,
avoidance of single points of failure and of loss of large parts of the detector
in case of failure of one component.

2.3.4. *Services caverns*

Although a lot of electronics is embedded in the detector, a large part of
the readout system is located off-detector in the ATLAS counting rooms.
Two counting rooms are available (US 15 and USA 15), as seen in Fig. 3, to
house the back-end electronics. Data are transmitted from/to the front-end

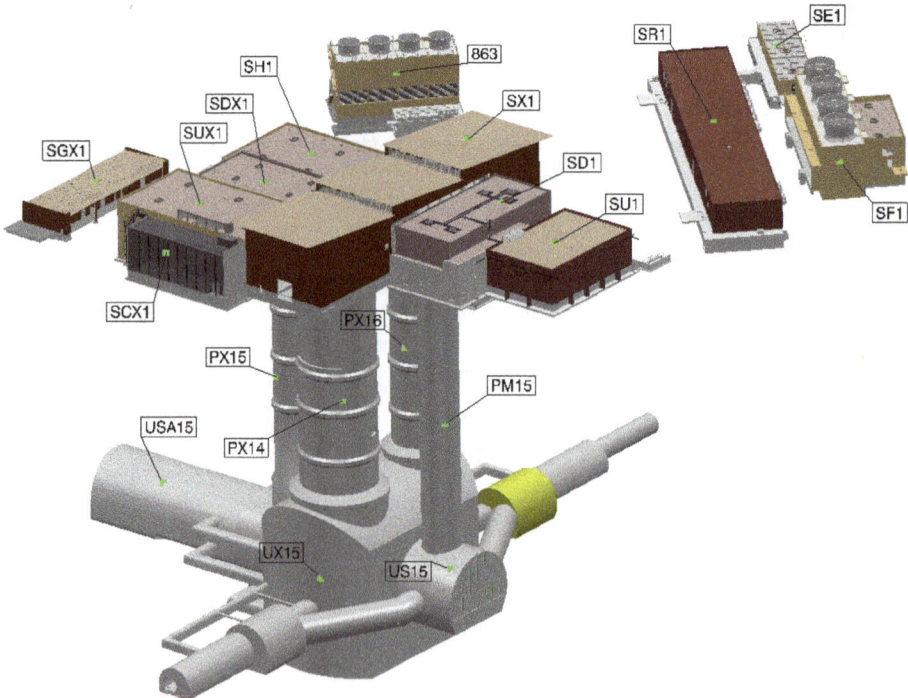

Fig. 3.   View of the ATLAS cavern (UX 15) and of the two counting rooms (US 15
and USA 15).

electronics to/from the back-end electronics through so-called front-end links. The technology to be used for these links is defined by the type of data to be transmitted (analog or digital), the speed at which it has to be transmitted and the length of the links, which ranges from 50 to 150 m.

3. Front-End Electronics

As seen in Fig. 2, the front-end electronics consists of an analog processing, a level-1 buffer (analog or digital), an analog-to-digital conversion (before or after the level-1 buffer) and a front-end link. In the case of ATLAS, all the detectors but the liquid argon calorimeter and the CSC muon chambers perform the analog-to-digital conversion before the level-1 buffer. In the case of the liquid argon calorimeter and the CSC muon chambers, the level-1 buffer is an analog memory and the analog-to-digital conversion is done only for those data corresponding to a level-1 trigger, before the data are transmitted to the back-end electronics. No analog information is sent from the on-detector front-end electronics to the off-detector electronics (with the exception of the analog calorimeter trigger towers used to form the level-1 trigger; see Sec. 4). As a consequence, all the front-end links are transmitting digital information. Due to the necessary length and speed of these links, they all use optical technology.

3.1. *Front-end ASICs*

Specific developments had to be made for the different subdetector front-end electronics. This led to the design of a large number of ASICs but, due to the radiation hardness requirements, they were fabricated using a somewhat limited number of technologies. At the time the designs of the inner detector front-end electronics started, only one radiation-hard technology meeting the requirements was available (DMILL). Later, the use of deep submicron (0.25 μm) CMOS technology together with special layout techniques was introduced.[10] These two technologies have been used for most of the front-end ASICs installed in places subject to high radiation levels, i.e. up to the electromagnetic calorimeter. Support for these technologies was organized centrally across experiments for design issues (access to libraries and design kits) and also for purchasing and prototyping (frame contract availability and multiproject wafer organization). At the periphery of the detector (muon spectrometer) it has been possible to deploy other technologies; this has allowed designers from different institutes to benefit from previous developments or existing development infrastructures. Tables 3–6 give a list of all the ASICs that have been installed in ATLAS. Two radiation-hard technologies have been used in the inner part of the detector (DMILL and $0.25\,\mu$m CMOS). At the periphery of the detector, where the radiation level is lower, other technologies were used. In total, 39 ASICs using 11 different technologies have been designed and produced. Note that some of them are common to different experiments (e.g. the TTCrx and the GOL). The large number of ASICs designed for the liquid argon calorimeter is due to the fact that its complex front-end electronics system (in addition to the readout function, it includes the formation of tower trigger signals and high performance calibration features) was first designed with components off the shelf (COTS), including complex FPGAs, which appeared to be not radiation-tolerant enough. It was deemed safer and faster to design replacement in radiation-hard technologies rather than seeking for radiation-hard-enough COTS. The large number of different technologies used could also be questioned and one could wonder why a single CMOS technology had not been used. This is mainly due to the fact that the development of the different ASICs had not been done at the same time and that for a number of collaborators the cost of changing technology was deemed too high.

Table 3. List of the ASICs used in the ATLAS tracking detectors. The quantity of ASICs given is the quantity installed in the detector; more were produced to include spares and/or the minimum amount of wafers to be processed.

Tracking detectors	ASIC	Technology	Functionality	Quantity and remarks
Pixel	FEI	$0.25\,\mu$ CMOS	Front-end	~28,000
	Module controller	$0.25\,\mu$ CMOS	Control	~1800
	VDC	$0.25\,\mu$ CMOS	VCSEL driver	~500
	DORIC	$0.25\,\mu$ CMOS	Timing and control receiver	~400
Silicon strips	ABCD	DMILL	Front-end	~50,000 BiCMOS design
	DORIC	$0.35\,\mu$ BiCMOS	Timing and control	~4100
	VDC	$0.35\,\mu$ BiCMOS	VCSEL driver	~8200
TRT	ASDBLR	DMILL	Amplifier–shaper–discriminator	~38,000 Bipolar design
	DTMROC	$0.25\,\mu$ CMOS	Digitizer	~19,000

Table 4. List of the ASICs used in the ATLAS calorimeter. The quantity of ASICs given is the quantity installed in the detector.

	ASIC	Technology	Functionality	Quantity and remarks
LAr calorimeter	HAMAC-SCA	DMILL	Analog memory	~52,000
	BiMUX	DMILL	Multiplexer	~8000 Bipolar design
	OpAmp	DMILL	Operational amplifier	~17,000 Bipolar design
	DAC	DMILL	16-bit DAC	~130
	SPAC slave	DMILL	Control interface	~2000
	Configuration	DMILL	Control	~1600
	SMUX	DMILL	Digital multiplexer	~1600
	Calibration logic	DMILL	Glue logic	~800
	SCA controller	$0.25\,\mu$ CMOS	Control	~3200
	Gain selector	$0.25\,\mu$ CMOS	Control	~13,000
	Clock fan-out	$0.25\,\mu$ CMOS	Clock driver	~11,000
	DCU	$0.25\,\mu$ CMOS	Control	~3200e CMS design
	HEC	GaAs	Front-end amplifier	~700 Inside liquid argon
Tile calorimeter	TileDMU	$0.35\,\mu$ CMOS	Pipeline	~256 Gate array

3.2. *Front-end and TTC links*

The front-end links transmit the data from the on-detector front-end electronics to the off-detector back-end electronics. As already mentioned, they only transport digital data and they all use optical technologies.

The speed of these links is an important parameter and is dictated by the amount of data to be transferred each time a level-1 accept signal has been issued. Table 7 summarizes the characteristics of the different links.

There are two classes of links: one running at relatively low speed (40–80 Mbits/s) and one running in the Gbits/s range. The first class is

Table 5.   List of the ASICs used in the ATLAS muon spectrometer. The quantity of ASICs given is the quantity installed in the detector.

	ASIC	Technology	Functionality	Quantity and remarks
MDT	ASD	$0.5\,\mu$ CMOS	Amplifier–shaper	~ 5000
	AMT	$0.3\,\mu$ CMOS	Time to digital conv.	$\sim 15{,}000$
CSC	ASM1	$0.5\,\mu$ CMOS	Preamplifier	~ 1300
	ASM2	$0.5\,\mu$ CMOS	Multiplexer	~ 1300
	Clock driver	$0.5\,\mu$ CMOS	Clock driver	~ 200
	HAMAC-SCA	DMILL	Analog memory	~ 2600 Common with LAr
RPC	ASD	GaAs	Amplifier–shaper	$\sim 47{,}000$
	CMA	$0.18\,\mu$ CMOS	Coincidence matrix	~ 3300
TGC	ASD	Bipolar	Amplifier–shaper	$\sim 81{,}000$
	HpT	$0.35\,\mu$ CMOS	Trigger	~ 800
	PP	$0.35\,\mu$ CMOS	Trigger	$\sim 15{,}000$
	SLB	$0.35\,\mu$ CMOS	Trigger	~ 3000
	JRC	$0.35\,\mu$ CMOS	JTAG controller	~ 1400

Table 6.   List of common items.

	ASIC	Technology	Functionality	Quantity and remarks
Common items	TTCrx	DMILL	Timing and trigger	Used by all experiments
	PHOS4	DMILL	Programmable delay	GOL used in TRT ($\sim$800)
	Delay25	$0.25\,\mu$ CMOS	Programmable delay	and MDT ($\sim$1200)
	GOL	$0.25\,\mu$ CMOS	Gbit serializer	

used by the silicon pixels and strips detectors,[11] for which it was impossible to multiplex a large number of channels in the front end (there is basically one or two fibers per detector module). The second class is used when it was possible to read out more channels on a single link; that implies the use of high speed serializers but has the advantage of reducing the amount of fibers to be installed.

Table 7. Characteristics of the front-end links used in ATLAS. They all transmit digital information over optical fibers. The speed is either in the range 40–80 Mbits/s or in the Gbits/s range. Special packages have been developed for the pixel, silicon strips and liquid argon calorimeter to meet mechanical requirements. Two types of serializers have been used: the commercially available GLink and the radiation-tolerant GOL designed at CERN.

Subdetector	Link speed (Mbits/s)	Quantity	Remarks
Pixel	40–80	∼ 250	8-way 850 nm VCSEL arrays in custom package
Silicon strips	40	∼ 8200	850 nm VCSEL in custom package
TRT	1600	∼ 400	Uses GOL serializer and a commercial 850 nm VCSEL
LAr calorimeter	1600	∼ 1600	Uses GLink serializer and a 850 nm VCSEL in a custom package
Tile calorimeter	1600	512	Uses GLink serializer and a redundant commercial 850 nm VCSEL
MDT	1600	∼ 1200	Uses GOL serializer and a commercial 850 nm VCSEL
CSC	1600	∼ 400	Uses GLink serializer and a commercial 850 nm VCSEL
RPC	1600	512	Uses GLink serializer and a commercial 850 nm VCSEL
TGC	1600	∼ 200	Uses GLink serializer and a commercial 850 nm VCSEL

Two main serializers have been used. One is the commercial GLink chipset[12] and the other one is the GOL, a radiation-hard device developed at CERN in $0.25\,\mu$m CMOS technology.[13] The latter could have been used everywhere; however, at the time when final prototypes of some front-end systems were being validated and decisions for production being taken, the GOL was not yet available.

Only multimode fibers and a 850 nm wavelength have been used.

The TTC links transmit to the front-end electronics the LHC beam-crossing clock, the L1A signal and some controls and timing signals, either globally defined for ATLAS (such as reset signals used for clearing event counters) or specific to a given subdetector (such as sending a test pulse).

A standard optical TTC network had been developed and used by all experiments.[14] It has been used in every ATLAS subsystem in the electronics counting rooms housing the back-end electronics and up to the front-end

Table 8.   Characteristics of the TTC links.

Subdetector	Link speed (Mbits/s)	Quantity	Remarks
Pixel	40	~250	8-way 850 nm optical links
Silicon strips	40	~4100	850 nm optical links
TRT	40	~400	Electrical links
CSC	1600	~200	850 nm optical links using GLink
All others	80	~2600	Standard optical TTC links

electronics for all subsystems but the inner detector and the muon CSC. For the latter, the TTC functions have been included in the front-end electronics and special links, either optical or electrical, were developed. Table 8 summarizes the characteristics of the different TTC links used.

3.3. Front-end modules

The front-end ASICs are mounted on front-end modules. These can be hybrids housing naked dies (pixel and silicon strip detectors), small printed circuit boards (TRT and muon spectrometers) or large printed circuit boards (calorimeters). Their form factors, the number of readout channels they contain and their number are very much detector-dependent. In both inner detector and muon spectrometer there are a large number of small front-end modules, each of them dealing with a small fraction of the total number of channels (for example, a silicon strip front-end hybrid manages 768 channels, representing 0.01% of the full silicon strip detector), while the calorimeters use a reduced number of large front-end modules, each of them managing a larger fraction of the detector (for example, a liquid argon calorimeter front-end module houses 0.07% of the full detector). Figure 4 shows pictures of some of the front-end modules.

3.4. Power in the front-end electronics systems

The power dissipated by the front-end electronics requires special attention, for two reasons:

(i) The power is dissipated in confined places and the total amount of heat which can be dumped in the experimental cavern is strictly limited;
(ii) Bringing power inside the detector might require thick cables.

Table 9 summarizes the power consumption of the front-end electronics per subsystem and also the amount of current needed.

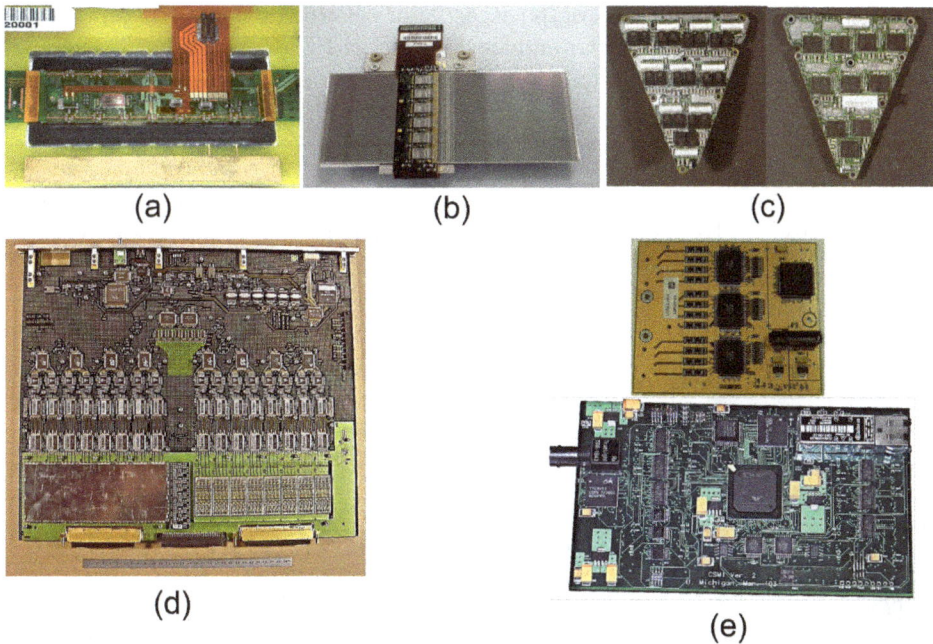

Fig. 4. View of some front-end modules. (a) Top view of a pixel module, showing the hybrid housing the module control chip. (b) View of a silicon strip barrel module, showing the readout hybrid housing six naked ABCD chips. The size of a hybrid is about $10\times2\,cm^2$. (c) View of a barrel TRT front-end module housing on one side 16 ASDBLR chips and on the other side 8 DTMROC chips. There are 12 different types of such 14-layer printed circuit boards. (d) View of a liquid argon calorimeter front-end board housing the front-end electronics of 128 channels. A total of 1600 such $40\times40\,cm^2$ boards are installed in readout crates in the detector. (e) View of a precision muon chamber (MDT) front-end module (top) and of a chamber service module (CSM; bottom). Each of the 1200 chambers needs one CSM reading out up to 18 front-end modules.

The distribution of the power has been done in different ways, due to different constraints and the availability of components.

For the inner tracker, a large number of distributed low power front-end modules must be powered. The lack of sufficiently radiation-tolerant voltage regulators or DC-to-DC converters to be used in the inner detector volume led to the direct and individual powering of the front-end modules. This is done either directly from multichannel power supplies[15] located in the counting rooms (silicon strips), or through some custom boards housing radiation-tolerant voltage regulators[16] located inside the muon spectrometer volume and powered with bulk low voltage power supplies.[17] These supplies are housed either in one counting room or in the experimental cavern (pixel

Table 9. Power consumption of the on-detector electronics. This includes the trigger elements for the calorimeters and for the RPC and TGC chambers. The amount of current is the current consumed by the on-detector electronics systems on their low voltage supply lines.

Subdetector	Power/channel (mW)	Total power in the on-detector electronics (kW)	Total current in the on-detector electronics (kA)
Pixel	0.100	8	3.4
Silicon strips	3.6	23	6.0
TRT	60	22	6.5
Inner detector total		53	15.9
LAr calorimeter	700	140	27
Tile calorimeter	4400	44	7.7
Calorimeter total		184	37.7
MDT	71	25	9.1
CSC	310	9	1.4
RPC	88	33	9.9
TGC	93	30	9.0
Muon spectrometer total		107	29.4
Total power dissipated by the on-detector electronics			334 kW
Total current needed by the on-detector electronics			80 kA

detector and TRT). The cost to be paid is the large quantity of cables to be installed in the detector and a large power loss on these cables as the full 16 kA must flow through them (the overall efficiency in terms of power is of the order of 30%).

The calorimeter front-end electronics is housed in front-end crates (liquid argon calorimeter) or in the detector drawers (tile calorimeter). Hence a relatively low number of high power units have to be powered. In both cases this has been done using DC-to-DC converters[18,19] located close to the front-end (583 kW units for the liquid argon calorimeter and 256 200 W units for the tile calorimeter). These converters are powered with high voltage and low current lines (280 V for the liquid argon calorimeter and 200 V for the tile calorimeter).

The muon spectrometer also uses a large number of relatively low power front-end modules. A commercial multichannel power supply system has been used.[20] It is based on radiation- and magnetic-field-tolerant DC-to-DC converters located in the experimental cavern and getting their primary power from AC-to-DC converters located in the counting rooms.

Figure 5 shows pictures of some power elements.

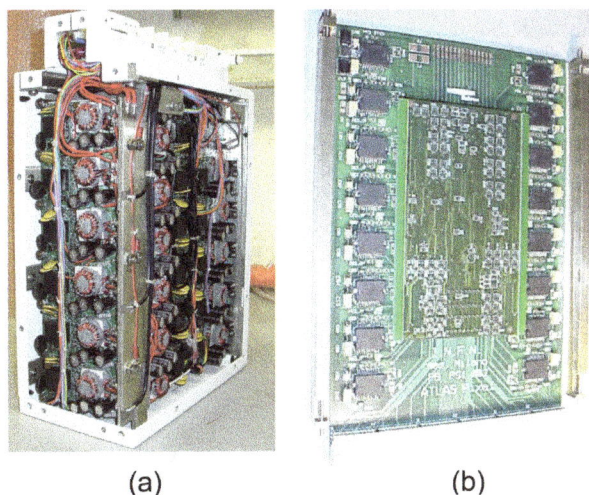

(a)   (b)

Fig. 5. Pictures of some power elements. (a) A 3 kW DC-to-DC converter used to power one liquid argon calorimeter front-end crate. (b) A power distribution board used to power the pixel front-end electronics. This board uses radiation-tolerant voltage regulators to generate low voltage power supplies for five front-end modules from a single bulk supply.

4. Trigger and Off-Detector Electronics

4.1. *L1 trigger electronics*

The ATLAS level-1 trigger[21] is a fixed-latency, pipelined digital system using custom electronics. The trigger formation is based on information from the calorimeter and muon trigger processors. The trigger information consists of multiplicities for candidate electrons/photons, taus/hadrons, jets and muons, and of flags for total scalar transverse energy (E_T), total missing transverse energy and total scalar jet transverse energy. An overview of the ATLAS level-1 trigger is shown in Fig. 6. The level-1 trigger also supplies information on so-called regions of interest (RoIs) to a higher level trigger. For exclusive objects these RoIs are the geographical coordinates of the detector regions where they were found. This information is used by a higher level trigger to seed its selection process.

The calorimeter[22] and muon trigger[23,24] processors provide trigger information to the central trigger logic[25] (CTL). The CTL contains the central trigger processor (CTP), which forms the level-1 accept (L1A) decision and fans it out to the timing, trigger and control (TTC) partitions of the experiment.

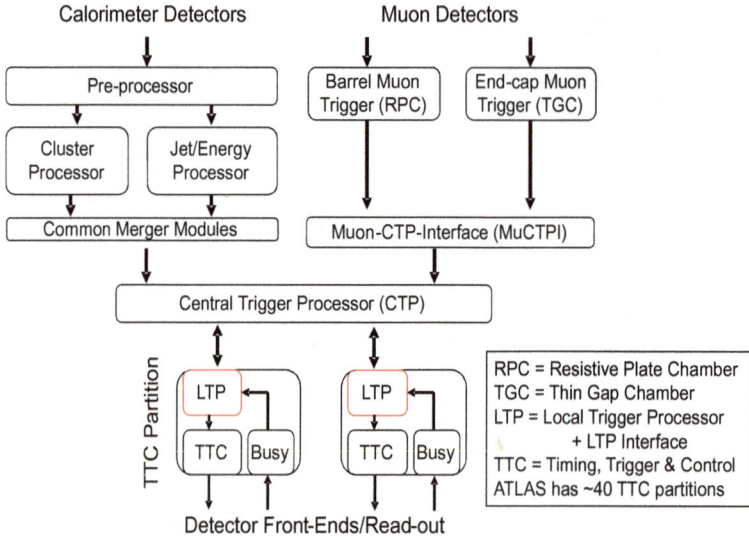

Fig. 6. Overview of the ATLAS level-1 trigger.

Two important parameters for the L1 trigger were defined: the 75 kHz maximum L1A rate and the maximum 2.5 μs latency of the system. These parameters had to be defined very early in the ATLAS life, as they define some key parameters of the front-end electronics, such as the size of the L1 buffers and the bandwidth of the front-end links. The low latency value was adopted to limit the size of the front-end L1 buffers, which was a tight constraint before the introduction of deep submicron CMOS technologies or for those front-end systems using analog L1 buffers (such as the LAr calorimeter).

Table 10 gives the latency for each part of the level-1 trigger system, in time unit (ns) and in number of bunch crossing periods (BC).

Table 10. Latency for each part of the level-1 trigger system.

Level-1 trigger system component	Latency (ns)	Latency (BC)
Detector-specific MUON trigger	1293	52
Calorimeter trigger	1465	59
CTL	250	10
TTC electronics	100	4
TTC cables and fibers (80 m)	400	16
Total	2215	89

The level-1 calorimeter trigger system looks at jets, isolated e/γ, missing ET and total ET, and provides multiplicity values for different energy thresholds to the CTP. About 8000 analog sums are formed on the detector and transmitted to the off-detector electronics, where they are digitized and processed. The choice for having the digitization of the trigger tower signals done off-detector was driven by power considerations (at that time there were no low power 40 MHz ADCs). The block diagram of the off-detector electronics is shown in Fig. 7.

The level-1 muon system is based on information provided by special trigger chambers [resistive plate chambers (RPCs) in the barrel region and thin gap chambers (TGCs) in the end-cap]. These chambers are logically divided into 224 sectors and in each of them the on-detector sector logic looks for muon candidates at different transverse momentum (P_T). Each sector logic transmits to the off-detector so-called muon-to-CTP-interface (MUCTPI) 32-bit words every 25 ns. The MUCTPI computes the total multiplicity per P_T and extracts the RoI. As for the calorimeter part, the main system issues are in very fast data movement. For instance, about 300 Gbits/s are transmitted to the MUCTPI. Although high speed serial links would have been easier to implement, they would have induced an extra latency incompatible

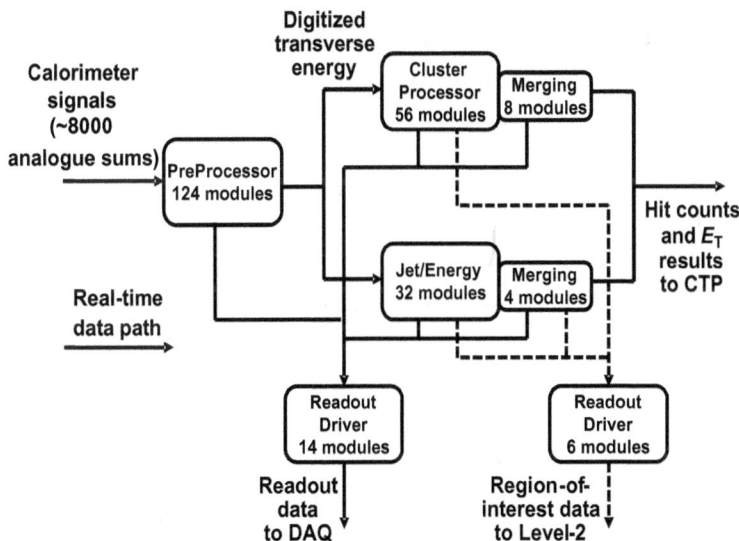

Fig. 7.   Level-1 calorimeter off-detector system. About 250 9U VME modules housed in 16 crates are used. High speed LVDS (between the preprocessors and the processors) and optical (between the processors and the RODs) links, together with dedicated backplanes, are needed to process the data.

Fig. 8. Photograph of the MUCTPI in the ATLAS underground counting room. Eight MIOCT modules (on the right side) still have to receive their input cables.

with the requirements. As a consequence, somewhat bulky parallel electrical links have been used, as seen in Fig. 8.

4.2. Off-detector readout electronics

As shown in Fig. 2, the back-end electronics receives data from the front-end electronics through the front-end links. The main back-end module is called the readout driver (ROD), and it has been decided very early that this module would be detector-specific, with well-defined interfaces to the data acquisition system and to the trigger and timing distribution system. It was also decided that the RODs should be VME64x modules and that a common VME64x single board computer[26] would be used to control the crates housing the RODs. This allowed the development of a common ROD crate data acquisition framework (ROD crate DAQ). The ROD processes the data received from the front-end electronics, prepares an event fragment

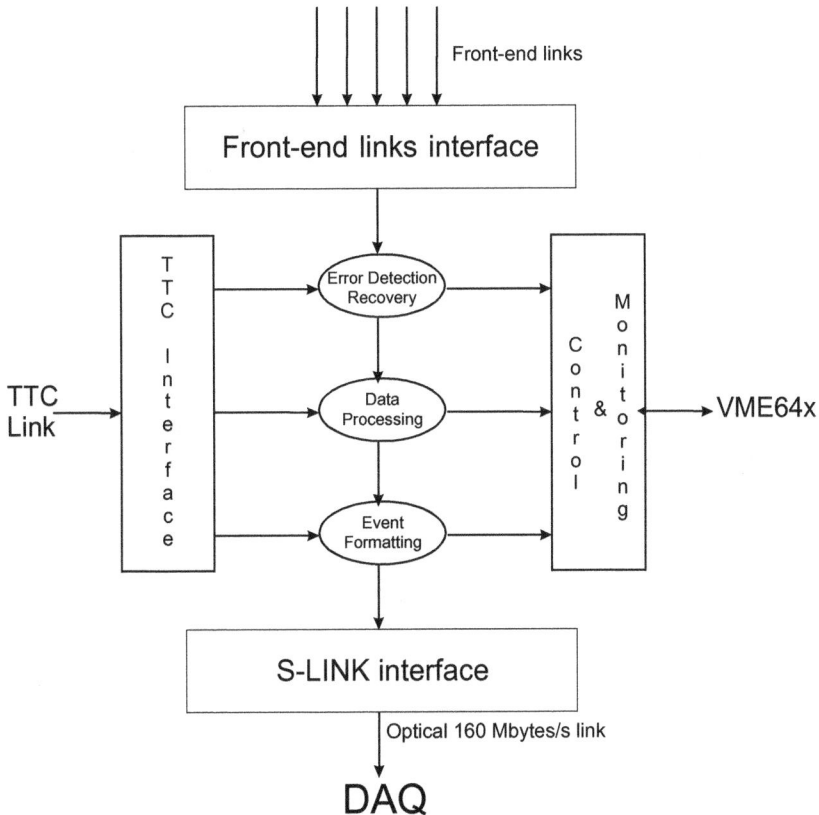

Fig. 9. Block diagram of an ROD. The ROD gathers data from several front-end links, processes the data (digital filtering, zero suppression, data compression, etc.), prepares subevent fragments following an ATLAS defined data format and ships these fragments to the data acquisition system through an ATLAS standard link (S-link). The ROD receives timing and trigger information from the TTC system, checks the consistency of data received from the front-end and, when necessary, flags and corrects errors and throttles the L1 trigger if buffers are overfilling.

following an ATLAS data format and ships them to the data acquisition system through a 160 Mbytes/s ATLAS standard optical link (S-link[27]). In addition the ROD has monitoring and control functionalities. A generic block diagram of a ROD is shown in Fig. 9.

The data processing performed by the ROD is different from one system to the other. It goes from data compression (e.g. Huffman coding in the TRT RODs) to digital filtering and feature extraction (e.g. in the calorimeter RODs), and is done either with large FPGAs or with digital signal processors. Pictures of RODs are given in Fig. 10.

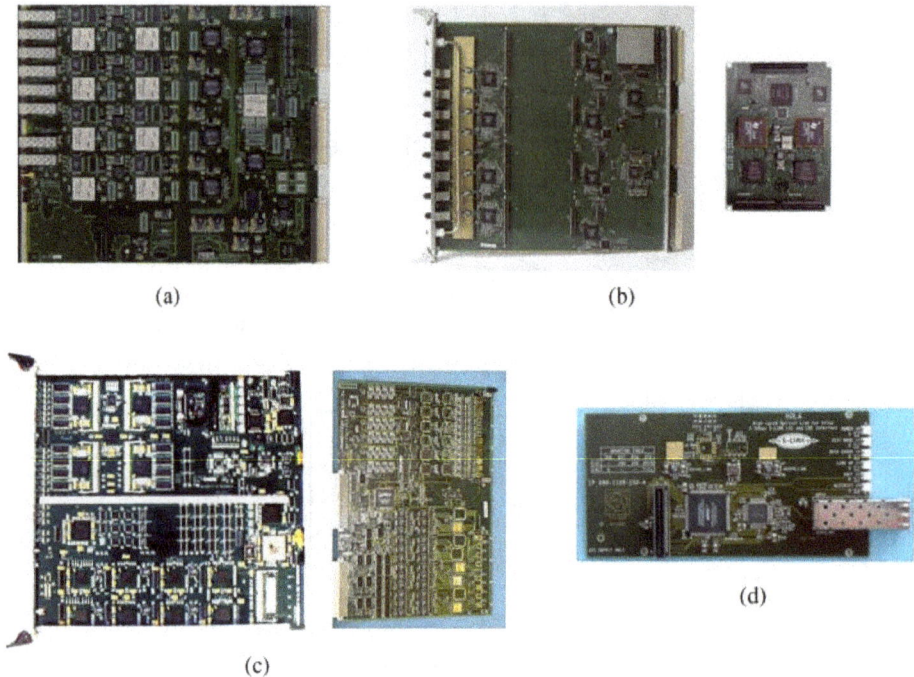

Fig. 10. View of a few RODs. (a) ROD of the muon MDT. It receives data from up to eight chambers through gigabit optical links. FPGAs compress the data and build a partial event, which is then pushed to the DAQ through an S-link. Monitoring is performed on board with digital signal processors. (b) ROD motherboard and DSP mezzanine of the calorimeters. It receives data from 1024 channels through 8 gigabit optical links. Data are processed (digital filtering) by the DSPs before being shipped to the DAQ. (c) Silicon strips and pixel ROD motherboard (left) and back of crate (BOC) board of the SCT. 96 optical links running at 40 Mbits/s are connected to the BOC (48 at 80 Mbits/s in the case of the pixel detector); data are then handled by FPGAs on the motherboard. DSPs are used for monitoring and calibration purposes. (d) S-link mezzanine card used by all systems to transmit data to the DAQ at 160 Mbytes/s.

A view of the liquid argon calorimeter RODs as installed in the experiment is given in Fig. 11. Table 11 summarizes the main characteristics of the different RODs.

5. Common Issues

A number of issues are common to all subdetectors and have been tackled and resolved globally. Out of them, four will be presented: the development of a common "slow control" module for ATLAS, the qualification of electronics against radiation and the electromagnetic compatibility (EMC) issues.

Fig. 11. Liquid argon calorimeter RODs in USA 15. 192 RODs housed in 16 VME64 × 9U crates readout 1524 128-channel front-end boards.

5.1. *The embedded local monitor board*

The ATLAS detector control system (DCS)[28] needs to monitor and control a large number of devices distributed in the experiment. The DCS architecture is organized in front-end (FE) equipment and a back-end (BE) system. In most of the cases, the connection of the BE system to the FE is using the industrial fieldbus CAN.[29] A general purpose I/O and processing unit called the embedded local monitor board (ELMB)[30] was developed, which can operate in the high magnetic field and under ionizing radiation in the cavern. It can either be embedded in the electronics of detector elements or directly read out sensors in a stand-alone manner.

The $50 \times 67\,\mathrm{mm}^2$ ELMB provides 64 differential analog inputs, an analog multiplexer and an analog-to-digital converter (ADC) with a resolution of 16 bits, as well as 32 digital I/O lines. In addition, a 3-wire serial peripheral interface (SPI) can be used to control additional components. Finally, in-system-programming (ISP) and universal synchronous asynchronous receiver

Table 11. Characteristics of the ATLAS subdetector RODs.

Subdetector	Input	Readout function	Output	Monitoring and calibration	Number of RODs
Silicon strips and pixel	48–96 × 40–80 Mb/s links	Event building in FPGAs	1 S-link 160 MB/s	Calibration and monitoring histograms done by DSPs	182
TRT	8 × 1.6 Gb/s links 640 MB/s	Data compression and event building in FPGAs	1 S-link 160 MB/s	Sampling monitoring	96
Calorimeters	8 × 1.6 Gb/s links Up to 1120 MB/s	Optimal filtering, feature extraction and event building with DSPs	4 S-links 640 MB/s	Calibration and monitoring task done by DSPs	224
Muon MDT	5–8 × 1.6 Gb/s links 425–680 MB/s	Zero suppression and event building in FPGAs	1 S-link 160 MB/s	Spying, histogramming with DSPs	192
Muon CSC	10 × 1.6 Gb/s links 1.536 GB/s	Data processing, feature extraction, zero suppression and event building in FPGAs and DSPs	1 S-link 160 MB/s	Sampling monitoring	16
Muon TGC	11 × 800 Mb/s links 100 MB/s	Event building in FPGAs	1 S-link <20 MB/s used	Monitoring data transmission through an Ethernet interface	26
Muon RPC	LVDS links 2 Gb/s	Event building in FPGAs	1 S-link 160 MB/s	Sampling monitoring	32
L1 calorimeter	18 × 800 Mb/s links 1.8 GB/s	Data compression and event building in FPGAs	4 S-links 640 MB/s	Sampling monitoring	20

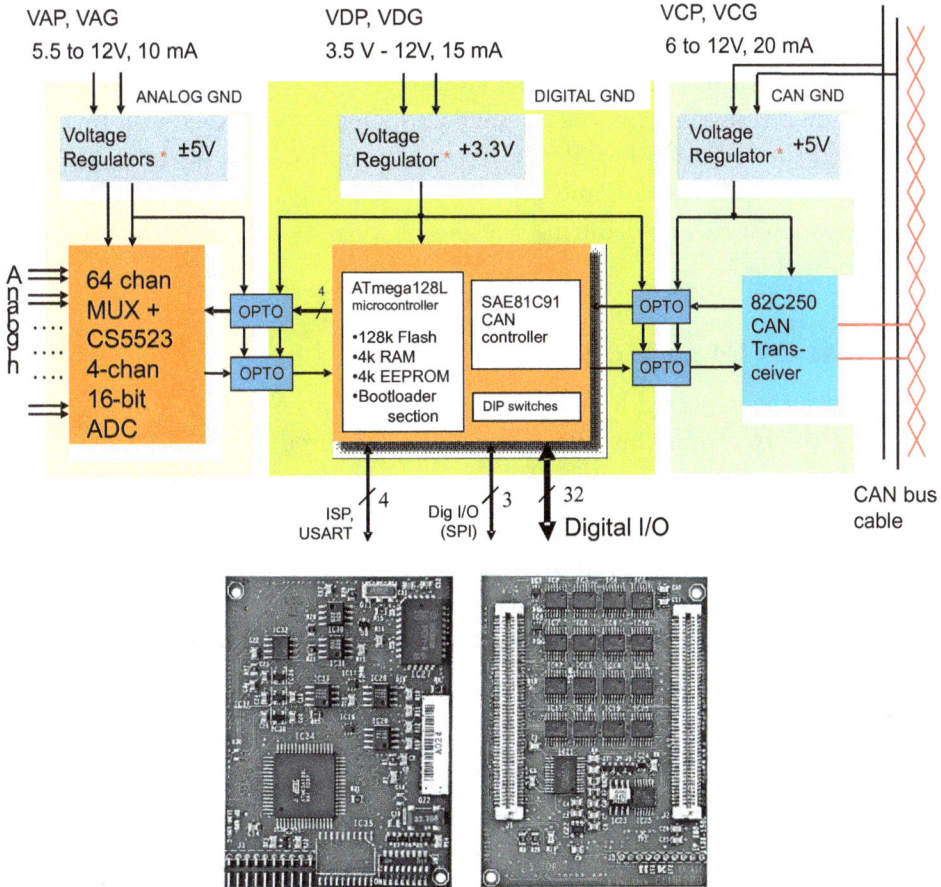

Fig. 12. ELMB block diagram and picture. About 10,000 devices have been produced and are used all over ATLAS.

transmitter (USART) serial interfaces are available. A block diagram of the ELMB and a picture of it are shown in Fig. 12.

5.2. *The ATLAS policy on radiation-tolerant electronics*

All electronics devices located in the experimental cavern are subject to radiation. The high level of radiation expected in the inner tracker volume imposed the use of radiation-hard electronics, while in the outer layers the use of components off the shelf (COTS) could be envisaged. However, this requires a qualification process to make sure that the selected electronics and the produced systems will work according to specification in this environment.

It appeared rapidly that a unified qualification was necessary across the experiment and for that purpose a policy on radiation-tolerant electronics was defined.[31] This policy defined standard test procedures for total ionizing doses, nonionizing energy loss damage and single event effects. In addition, it defined the radiation hardness criteria in every location of the detector as well as COTS preselection and procurements. Lastly, the radiation hardness of systems was reviewed during the design process and prior to launching the production.

5.3. *The ATLAS EMC policy*

To achieve the expected level of performance of the ATLAS experiment, the EMC of the electronic installations must be ensured. Three topics must be covered:

- Compliance with electrical safety rules applicable at CERN;
- Immunity from conducted and radiated emissions present in the experimental area;
- Measurement of conducted and radiated emissions of each installation in the experimental area.

An ATLAS policy was put in place,[32] covering all these aspects and defining some limits, such as the allowed common mode emission of electronics

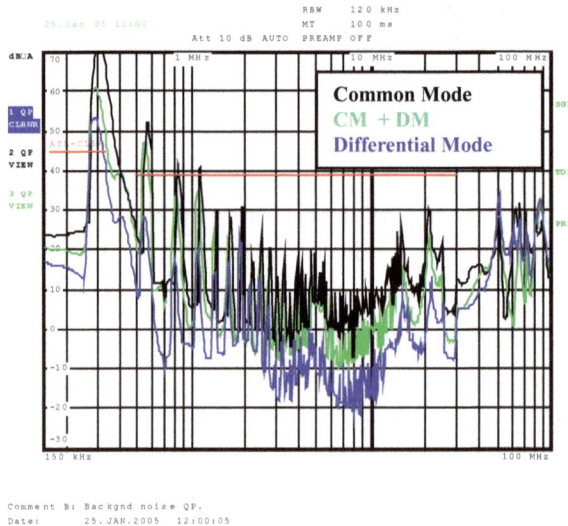

Fig. 13.   Example of EMC measurement showing the excess of common mode and differential mode current of a power supply in the frequency range 200 kHz–1 MHz.

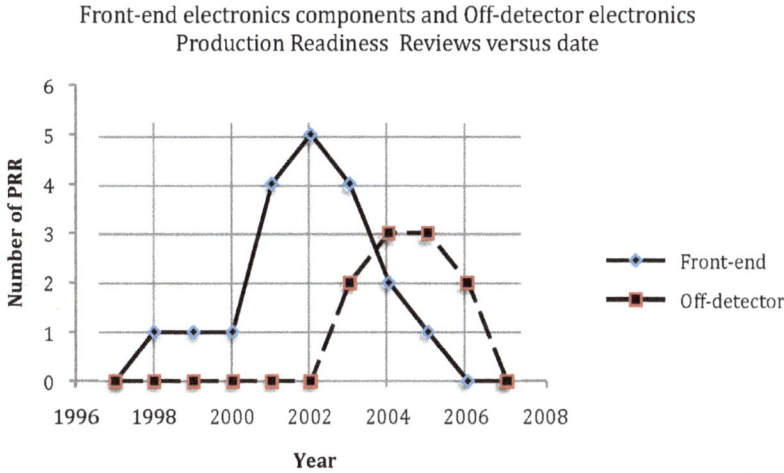

Fig. 14. Number of production readiness reviews per year for the front-end electronics components (plain line) and the off-detector electronics (dashed line).

systems or the susceptibility of front-end electronics to noise on the power lines. A lot of measurements were done during the detector assembly (see Fig. 13) and commissioning both on the surface and in the experiment, and this systematic approach has allowed us to find and fix difficult problems.

6. Development Schedule

The schedule of ATLAS has been changed several times, due to delays in the LHC construction, and some decisions were taken with the assumption of an earlier startup and the necessity to start production of the electronics. That has been the case, for instance, for the front-end components of the silicon strips. As formal production readiness reviews (PRRs) were organized before launching the production of any electronics system, it is of interest to look at the dates of reviews of the front-end components and of the off-detector electronics (see Fig. 14). The early PRRs are related to front-end ASICs for the silicon strips detector and the liquid argon calorimeter. The delay between the front-end production and the off-detector electronics is in a range of 2–3 years and the last PRR happened about a year prior to the expected startup of the LHC.

7. Conclusions

A very quick overview of the ATLAS electronics system and of the developments made have been given. The very harsh environment has led to the development of a large number of radiation-hard or radiation-tolerant ASICs using different technologies. The nonrecurring engineering cost of very deep submicron technologies (nowadays) would not allow us to do the same for a new project. The power distribution in the front-end electronics has been a difficult challenge and it is a field where dramatic improvements will be needed in future experiments or upgrades. The latency of the L1 trigger, a key parameter for the front-end electronics design, was defined very early in the design phase of the experiment and had to be kept as low as possible. This constraint on latency should disappear for future experiments or upgrades, thanks to the use of very small feature size technologies for their front-end electronics.

The readout electronics of ATLAS has now been used for more than a year for cosmic data taking, and for some subsystems it has already been running for several years. The next milestones are the confirmation of the good behavior of the front-end systems in the presence of radiation and the reliability of all the electronics components.

References

1. G. Aad. ATLAS pixel detector electronics and sensors, *JINST* **3** (2008) P07007.
2. F. Campabadal *et al.* Design and performance of the ABCD3TA ASIC for readout of silicon strip detectors in the ATLAS semiconductor tracker, *Nucl. Instrum. Methods A* **552** (2005) 292.
3. E. Abat. The ATLAS TRT electronics, *JINST* **3** (2008) P06007.
4. N. J. Buchanan. ATLAS liquid argon front end electronics, *JINST* **3** (2008) P09003.
5. A. Bazan. ATLAS liquid argon calorimeter back end electronics, *JINST* **2** (2007) P06002.
6. K. Anderson *et al.* Design of the front-end analog electronics for the ATLAS tile calorimeter, *Nucl. Instrum. Methods A* **551** (2005) 469.
7. S. Berglund. The ATLAS tile calorimeter digitizer, *JINST* **3** (2008) P01004.
8. Y. Arai. ATLAS muon drift tube electronics, *JINST* **3** (2008) P09001.
9. F. Faccio. Radiation issues in the next generation of high energy physics experiments, *Int. J. High Speed Electro. Syst.* **14**(2) (2004) 379.
10. G. Anelli *et al.* Radiation tolerant VLSI circuits in standard deep submicron CMOS technologies for the LHC experiments: practical design aspects, *IEEE Trans Nucl. Sci.* **46**(6) 1690 (1999).
11. A. Abdesselam. The optical links of the ATLAS SemiConductor Tracker, *JINST* **2** (2007) P09003.
12. Detailed information about the GLINK chipset and protocol is posted at http://literature.agilent.com/litweb/pdf/5966-1183E.pdf.

13. P. Moreira *et al.* A radiation tolerant gigabit serializer for LHC data transmission, in *Proc. 7th Workshop on Electronics for LHC Experiments* (Stockholm, Sweden, 2001).
14. S. Baron. The TTC system, http://ttc.web.cern.ch/TTC/intro.html.
15. P. Phillips. The ATLAS SCT power supply system, in *Proc. TWEPP* (Prague, 2007).
16. STm, LHC4913 and LH7913 low drop voltage regulator.
17. W-IE-NE-R, Plein & Baus GmbH, Mullersbaum 20, D-51399 Burscheid, Germany, www.wiener-d.com.
18. J. Kierstead. PRR of the liquid argon low voltage power supplies. Available at http://indico.cern.ch/conferenceDisplay.py?confId=a04497.
19. I. Hruska. Radiation-tolerant custom-made low voltage power supply system for ATLAS TileCal detector, in *Proc. TWEPP* (Prague, 2007).
20. C. A. E. N. Easy 3000. Documentation available at http://www.caen.it.
21. ATLAS Collaboration. First-level trigger Technical Design Report. CERN/LHCC/98-14 (June 1998).
22. R. Achenbach *et al.* The ATLAS level-1 calorimeter trigger, *JINST* **3** (2008) P03001.
23. The level-1 trigger muon barrel system of the ATLAS experiment at CERN, *JINST* **4** (2009) P04010.
24. The ATLAS level-1 end-cap muon trigger, http://atlas.web.cern.ch/Atlas/project/TGC/www/doc/MuonEndcap_rev01.pdf.
25. S. Ask *et al.* The ATLAS central level-1 trigger logic and TTC system, *JINST* **3** (2008) P08002.
26. Datasheet of the VP110 VME CPU by Concurrent Technologies. See http://www.gocct.com/sheets/vp11001x.htm.
27. O. Boyle *et al.* The S-link interface specification, http://cern.ch/s-link.
28. A. Barriuso Poy. The detector control system of the ATLAS experiment, *JINST* **3** (2008) P05006.
29. CAN in automation, CAN, http://www.can-cia.org.
30. B. Hallgren. The embedded local monitor board (ELMB) in the LHC front-end I/O control system, in *Proc. 7th Workshop on Electronics for LHC Experiments* (Stockholm, Sweden, 2001).
31. M. Dentan. Overview of the ATLAS policy on radiation tolerant electronics, in *Proc. 6th Workshop on Electronics for LHC Experiments* (Krakow, Poland, 2000).
32. G. Blanchot. Overview of the ATLAS electromagnetic compatibility policy, in *Proc. 10th Workshop on Electronics for LHC Experiments* (Boston, USA, 2004).

Chapter 8

INNOVATIONS IN THE CMS TRACKER ELECTRONICS

G. Hall

Blackett Laboratory, Imperial College
London, London SW7 2AZ, UK
g.hall@imperial.ac.uk

The CMS silicon microstrip tracker readout system is unprecedented in size, with over nine million channels. It is an analogue readout system, implemented using CMOS ASICs and linear, semiconductor laser transmitters which send pulse height data off-detector for digitisation and the first level of data processing. The basic components which define the architecture originated in R&D projects in the early 1990s and were crucial in allowing this system to be realised. The availability of several key technologies was critical to achieving the design preferences, and the components and technologies were utilised also to build the control and monitoring system and, later, to implement critical elements of other CMS sub-detector systems. The background to the technology choices and early development of the system is described and an attempt is made to draw some lessons which could be relevant for the future.

1. Introduction

The electronic readout, timing and control system of the CMS tracker is based upon three major elements: CMOS integrated circuits, fibre-optic transmission with customised optical transmitters and receivers in miniaturised packages, and large off-detector digital electronic boards relying on large field programmable gate arrays (FPGAs). Almost none, if any, of the components eventually deployed in the system could have been procured when the system was originally conceived and its realization was possible only because of technology evolution during the lifetime of the project. This was in part a consequence of the high level of radiation tolerance required for all particle tracking systems at the LHC; being in close proximity to the collision point they are exposed to the highest fluxes of particles emerging from the proton collisions. However, this was not the only factor: the requirements to be met were extremely challenging in terms of numbers of channels, noise performance, power to be dissipated, material budget constraints and cost, to name a sample.

Based on discoveries and progress in particle physics during the preceding couple of decades, at the outset it was very clear that the LHC (or an equivalent accelerator) would define the future and therefore it was vital that detector systems should be devised. Since no one knew how to design working experiments for the extreme conditions of the LHC, the particle physics community was granted a significant period of imaginative R&D lasting about a decade, in which many ideas were developed and many others eventually fell by the wayside. With hindsight it still seems impressive that there was enough belief that these unprecedented problems could be overcome.

It is possible to give an account of the CMS tracker system by describing the requirements and summarising the specifications and performance achieved but it is probably illuminating, and perhaps more interesting for the reader with less specialised knowledge, to give a first hand semi-historical account of how this system evolved and the factors which determined the overall design. The system includes several notable innovations, most of which were considered challenging, or even controversial, at times during the development period. Many of the features, and in several instances actual components, have been successfully incorporated into other CMS sub-detectors, which has been beneficial for standardisation within CMS and in minimising the overall development costs and hopefully long term maintenance. There are important lessons to be learned from this experience and it may be valuable to review the memories of the tracker system evolution in view of large and ambitious future projects, which include possible upgrades to cope with higher luminosity operation of the LHC.

2. Requirements for an LHC Tracker

In the early 1990s, the importance of tracking compared to calorimeters and muon systems still needed defending, as the physics justification was often considered less clear-cut, especially for large scale detectors to provide comprehensive track reconstruction. Arguments, such as the importance for electron identification of energy–momentum matching and improvement of muon momentum measurements, as well as the ability to distinguish multiple events in congested high luminosity conditions, have proved valid. From today's perspective, the capabilities of the tracker at lower momentum have far exceeded original targets, and techniques such as particle flow have been developed for CMS and look to be increasingly important in future.

To quote from one of the early status reports submitted to the CERN Detector R&D Committee (DRDC),[1] "Although the main physics goals of an LHC experiment are clear, the role and specifications for an inner tracking system are less well defined.... Cost constraints will play an important part in final decisions of overall detector design."

The challenges of developing a tracking detector for LHC operation were well identified; it is easy to forget that not only were they unparallelled but there was no obvious certainty that they could be met. The first and most apparent was radiation tolerance, where the $\sim$10 Mrad (100 kGy) and $\sim 10^{14}$ hadrons $\cdot$ cm^{-2} requirement[a] was compared to the interior of nuclear reactors. The optimal sensor material for LHC tracking was not evident and there were doubts whether silicon would be adequate; gallium arsenide seemed promising and diamond and silicon carbide were also discussed. However, a silicon microstrip detector with several million channels and spatial resolution $\sim 20\,\mu$m was thought to be possible.

From the electronics perspective the initial focus was on the front end chip; the significant challenges were the on-chip data storage required for the level 1 trigger latency, the likely power and the extraordinarily large data volume and processing rate. The target front end electronics specifications are listed in Table 1. It seemed plausible that AC-coupled double-sided

Table 1. Target specifications for front-end electronics defined by the RD20 project to read out 300 μm silicon sensors at LHC; adapted from Ref. 1.

Minimum ionising particles deposit a most probable signal of $\sim$25,000 electrons.

Equivalent noise charge of 2000 electrons from all sources throughout the lifetime of the detector, to include the increase due to radiation-induced leakage currents
Precise timing, with the ability to localise the signal origin to single beam crossing
Minimum pile-up
Dead-time-free
Analogue data read out
Data available for level 2 trigger
Immunity to extrinsic noise
Minimum power. In practice 1–2 mW/channel is believed to be the minimum feasible.
Reliable, robust, calibratable
Easy to implement in radiation-hard form

[a]It is conventional to use Mrad as a unit for the ionizing dose; 1 Gray = 100 rad. Hadron fluences often refer to 1-MeV-equivalent neutrons and must be corrected for energy spectra and particle type.

sensors might be used, which also raised the unappetising prospect of constructing very large systems with half of the electronics held at the sensor bias voltage, which might change with irradiation to maintain charge collection efficiency. There was little or no experience with building DC-coupled sensors and matching front-end electronics.

Little was known about multi-Mrad radiation hardness of integrated circuit electronics in the particle physics community other than from textbooks[2-4] which focused on environments where such concerns had been encountered; these were principally in the nuclear power industry, and space and military applications. A more recent overview can be found in Ref. 5. Various meetings introduced potential protagonists to one another; the LHC efforts benefited from earlier R&D investment for the US Superconducting Super Collider (SSC), because manufacturers who were interested in SSC and LHC business were mainly American and were already conscious of the approaching end of the Cold War era and its expected impact on military spending. Space had been identified as an expanding market but collider experiments seemed to offer them a much larger possible demand. This introduced another parameter into the equations: cost.

It quickly became clear that manufacturers interested in the SSC and LHC integrated circuit market were used to customers with large budgets. Radiation tolerance in space was achieved in part by shielding and in part by selecting small numbers of components and "qualifying" them, or exposing them to radiation levels in the expected range, adding a suitable safety margin, and ensuring that satisfactory operation was maintained. Once this had been done, other components from the same production batch were considered safe to use, without necessarily imposing significant requirements on the internal design. However, this was practical only for small numbers of components and meant that space-qualified parts were far behind the commercial state of the art when launched. For defence applications, significant government investments underpinned technologies in use, and restrictions on use and information transfer were widespread, and cause for concern, especially because the LHC tracker radiation tolerance requirements were well above even conventional military specifications.

The primary circuit damage mechanisms were adequately understood, namely atomic displacement effects in bipolar circuits and oxide charging in MOS electronics.[2-5] However, analogue performance did not seem to be the most critical factor for commercial products and it was questionable whether MOS devices would be usable. The techniques employed to achieve hardening were obviously commercially sensitive. This generated a need for studies at the transistor level, measuring the noise spectral density pre- and post-irradiation of components with dimensions typical of amplifier input devices. The number of technology variants was not small and there appeared to be prospects for some non-standard processes, e.g. silicon on insulator, to achieve improved performance.

Bipolar circuits offered attractive possibilities; noise in bipolar amplifiers is predictable using only the DC currents drawn, although the current-gain parameter (β) must be measured to confirm that its degradation under high levels of irradiation is tolerable. In addition, bipolar circuits offer their best analogue performance with pulse shaping times in the range required for the LHC, of a few tens of ns, with capacitive loads typical of microstrip sensors. However, CMOS technologies were better-established and were already dominating the market, as well as being more accessible via multi-project wafer (MPW) submissions both commercially and through academic training consortia in Europe. CMOS was also intrinsically immune to bulk (displacement) damage, since the currents are carried in extremely shallow surface layers under the gate oxide; this was easily verified and would simplify qualification studies by removing a need for large scale studies with neutrons or protons. In contrast, there was some evidence that bipolar circuits might also be affected by the surface charge and dose rate, adding to the possible evaluations required. The uncertainty for CMS transistors was the actual noise level achieved in accessible processes, often higher than the theoretical minimum, and whether this would degrade during irradiation, which required studies of each process.

Overall, the early 1990s was a period of some confusion with widely differing opinions about the technologies with the best prospects, the likely costs, and long term access to the foundries where contracts might be placed, as well as sensitivities on national security grounds.

3. Front-End Readout

The CMS microstrip tracker readout ASIC originated in the RD20 R&D project,[6] which was set up to investigate a wide range of issues involved in constructing an LHC tracker, including sensors, electronics and mechanics and cooling. From an electronic point of view, most of the proponents shared the view that analogue readout was to be preferred if feasible, based on experience in LEP experiments and improvements in spatial resolution from charge sharing in microstrip systems.

Although bipolar circuits were not rejected, it was felt important to explore the possibility of using CMOS because of existing expertise and access to processes at modest cost in shared MPW submissions. Another important argument was that LHC circuits required quite a lot of logic and memory functionality and this would only be practical using CMOS. It was not easy to see how high density bipolar front-end amplifiers, typically with 128 channels on 40–50 μm pitch, would be easily connected to nearby CMOS logic without a large number of wire bonds, raising noise issues and questions of supply voltage consistency. While not ruled out, the practical issues of building modules and chip handling appeared severe.

However, a very important and persuasive argument was that, for a given power, CMOS could not match the noise performance of bipolar amplifiers where an ∼1 mW analogue power/channel looked feasible, especially for pulse shaping required for the 15 ns LHC machine clock then envisaged.

3.1. *Analogue versus binary*

Demonstration of pipeline memory circuits was considered to be a crucial point at the outset of LHC R&D. LEP experiments had stored only one or two samples at most and there was no experience of relatively long term storage of a large number of analogue samples with high speed reading and writing, and power constraints. A binary system would avoid this problem, at the price of sacrificing signal amplitude information. However, several prototype designs were developed by different projects and were able rapidly to confront concerns about capacitor leakage, chip area, control logic and noise generated by the digital circuitry, although it was only when pipelines were fully integrated into full size chips that the noise contribution was optimized. Since analogue samples were not significantly larger than the sensor input signals, it proved to need care to ensure that the amplifier writing to the memory had enough gain[b] and storage capacitors were dimensioned so that the stored charge ($Q = CV$) did not suffer from added noise from the amplifier used to read samples from the pipeline. Capacitor leakage turned out not to be a serious issue for the LHC trigger latency of a few μs.

The possible access to analogue samples encouraged RD20 to examine means of processing consecutive pipeline samples on-chip to overcome the difficulties in developing a fast-enough CMOS amplifier. The early efforts focused on the leading edge of the pulse, and whether precise timing information could be extracted with amplitude. It was well known that zero crossing techniques would add noise, as well as circuit complexity.

---

[b]The chips developed by RD20 wrote voltages into the pipeline, hence the APV nomenclature, which stood for Analogue Pipeline, Voltage mode.

Starting from simulations trying to solve a matrix problem of inverting the measured pulse shape from a few samples to obtain the fast input signal, the deconvolution method was devised.[7,8] The signal processing is implemented in a relatively simple analogue circuit following the pipeline by forming a weighted sum of three consecutive samples. Because it does not process every event, but only those which pass the level 1 trigger, the circuit runs with a clock much slower than the beam crossing rate, which saves power. It is dead-time-free, because the time to process each event is less than the average interval between level 1 triggers and sufficient buffer depth is provided to allow for fluctuations in first level trigger arrival times.[c]

Although deconvolution appeared a novel idea to particle physicists, it turned out to be textbook material in *digital* signal processing. However, the noise implications of this type of *analogue* processing had not been studied previously and required further calculations and experimental measurements to confirm the behaviour. As expected,[7,8] the method effectively shortened the pulse shape and led to an increase in series noise (associated with the input transistor and sensor capacitance) and reduction of parallel noise (associated with leakage currents).

The power gains associated with deconvolution did not fully overcome the possible weaknesses in using a CMOS amplifier, compared to bipolar — effectively, a fast leading edge pulse is needed — but they were compelling enough to convince many that this was a viable solution for analogue readout. Nevertheless, it was not universally accepted that analogue readout was needed; commercial electronics trends certainly favoured digital system implementation. However, commercial systems were not necessarily considered a good guide to implementing a low noise, highly sensitive detector with an enormous number of channels operating in an environment of which there was no experience and where radiation-induced degradation might be inevitable. The subject remained controversial; some of the arguments are given in Table 2.

3.2. *Front-end chip evolution*

Over the period from 1992 to 1997, a series of circuits was implemented incorporating more of the final features, and evaluated in laboratory and

---

[c]Actually extra logic would be required to ensure that the three samples could be used even for closely spaced triggers, so a rule requiring a minimum gap of two clock intervals between triggers is imposed. This gives a 0.5% dead time at a 100 kHz trigger rate.

Table 2.   Possible pros and cons of analogue readout.

Pros	Cons
Improved spatial resolution from charge sharing information	Potential extra complexity, especially compared to binary
Greater and more flexible common mode noise rejection	Possible need for on-chip ADC
Greater diagnostic capability	Larger data volume
Greater system monitoring capability for long term operation	Likely larger chip size, compared to binary
Access to analogue data offers robustness against unexpected system noise	Transmission of analogue data may incur a noise penalty
Absence of on-chip thresholds to manage	

beam tests. A 32-channel APV3 was constructed in 1994, followed by a 128-channel APV5 in 1995, and the APV6 in January 1997. All were implemented using a Harris (USA) 1.2 μm radiation-hard process, which was formally qualified to a standard 300 krad level but which tests had demonstrated to be adequate to much higher doses.

The APV6, which was evaluated during the first half of 1997,[9] looked promising for CMS production. A good relationship had been established with the manufacturer and, while no contracts were in place, the production schedule and cost were beginning to look acceptable. The yield of well-functioning circuits was the largest uncertainty, since it was known to be design-dependent, and the company advised that several large production runs would be needed to draw conclusions.

The APV6 was also rather easy to study; for the first time internal bias current and voltage settings had been implemented by programming internal registers in the chip using an industry standard serial interface protocol. This was extremely beneficial for production testing, as well as operation in the laboratory, since a range of settings could be explored simply by reprogramming the registers. It was possible to make firm estimates of the time needed to test wafers and, thus, the large scale testing requirements, including considering whether this could be outsourced.

During 1996, Harris had moved the production line back to a facility on their main site — for straightforward economic motives, as production demand was limited. One consequence was that, to maintain the radiation hardness qualification for commercial parts, transistors from the process had to be characterised for performance following irradiation, as the production line parameters were validated and tuned. The results looked promising and there seemed little reason to doubt that the behaviour would be similar to that observed in 1995 before the foundry move. To surprise on all sides, this turned out not be the case, as was discovered during the summer of 1997 after irradiations with Co^{60} gammas.[10] The APV6 performance degraded after a few Mrad and it was not feasible for the company to fine-tune the process for LHC applications.

During the APV series development, a parallel chip development (APVM) had begun for MSGCs (microstrip gas chambers) using the DMILL technology,[11] which had been developed in France with LHC applications in mind and which was to be taken over by a commercial manufacturer. The CMS tracker design in 1997 still foresaw the outer part of the detector being constructed from gas detectors, which it was believed should be less expensive than silicon microstrips. There was therefore a requirement for about ten million channels of MSGCs, compared with only three million silicon channels.[12] MSGC signal amplitudes were similar in magnitude but with a different time structure, with large fluctuations during the signal pulse. The deconvolution method intended for a fast δ-like current impulse was not ideally adapted for gas detector signals but it seemed that a comparable method of processing using multiple samples could be adopted, following performance studies with prototype detectors. It seemed that the DMILL process, which had not yet been production-qualified, could be the fallback technology and an effort began to convert the APV6.

However, at the same time as the radiation sensitivity of the Harris process was identified, work in CERN had begun to establish the prospects of commercial deep sub-micron processes, in the range below $0.5 \, \mu m$ minimum feature size. Although there had been hints that such processes could offer substantial levels of hardness,[13,14] data from mostly commercial parts under irradiation had led to an inconsistent picture. The picture in 1997 was not nearly as clear-cut as some later accounts[6] suggested; experts in the field had referred to variations between production lots,[d] between wafers in the same lot, and even between chips on the same wafer, and this was not well understood.[15-17] The work from CERN[18] established a few important facts:

- The fundamental radiation hardness of commercial sub-micron CMOS processes could be in the multi-Mrad range required for tracking detectors, as predicted by studies on oxides;

---

[d]A "lot" is a batch of production wafers typically 24 wafers in size.

- By using enclosed transistor designs, which avoided current leakage paths between the source and the drain of the transistor, the less predictable currents associated with much thicker oxides on other parts of the circuits could be eliminated.

The results were very encouraging, but more information was needed on analogue performance if this was to provide an alternative to radiation-qualified processes. Studies soon confirmed excellent noise behaviour at the single transistor level. Tolerance to radiation-induced latch-up was shown to be good and single event upset (SEU) immunity could be enhanced by circuit architecture.[19-21]

It was decided to attempt a translation of the APV6 into an IBM 0.25 μm process in a single step. This offered some very significant potential advantages, with apparently few risks (Table 3) except for the challenges of working in a new technology in which there was limited experience and constructing a large, complex chip in a single iteration.

The translation went remarkably quickly, with a team from the UK and CERN collaborating closely, and the first design iteration produced an excellent chip in autumn 1999.[22,23] It was aided by the fact that the process turned out to be well characterised, and transistor models reliable. This was one example of the benefit of using a large volume commercial process whose customers rely on predictable features, including prices, high yields and fast processing.

Table 3. Benefits and risks of adopting the 0.25 μm process, as seen in 1998.

Benefits	Risks
Modern commercial process with high yield and low cost per unit wafer area, so substantial cost saving	New, more complex process with limited design experience
Significant reduction in both analogue and digital power consumption	Development time might exceed expectation
Higher density circuit for the same functionality, so reduced size chip, even with longer pipeline memory	Unexpected radiation effects might emerge
Well-characterised process with lower noise with reduced power	Increased sensitivity to single event upsets
Faster manufacturing time and larger wafers, so reduced wafer handling and faster acceptance testing	Relatively high non-recurrent engineering (mask) costs

3.3. *Production*

Although the first version of the APV25, produced inexpensively in a shared MPW run, was almost perfect and only minor tuning in a second iteration was necessary before the chip was considered ready for large scale manufacture, it was some time before large scale manufacture began, during which there was a lengthy postscript to the development.

Unexpected variations in yield between early wafer lots were observed. Initially it seemed that something had gone wrong in manufacturing but after investigation, wafer replacement and resumption over a one-year period it became clear that the yield was not stable. Despite large amounts of data and additional processing of test structures to test hypotheses, no clear link could be demonstrated between designs and failure rates, and no weak point in processing could be identified. However, we were informed that problems had not been observed in deliveries to commercial foundry customers.

Regular intercontinental conference calls took place between designers, the evaluation team, foundry specialists and failure analysis experts from within IBM. A few points in some chips where failures could be isolated were identified. IBM used sophisticated microscopy techniques to section chips and examine their internal structure, which rapidly narrowed down the hypotheses.

The problem was related to intensive use of one metal layer in HEP designs. The large area of this metal influenced the duration of a chemical-metal polishing (CMP) step responsible for wafer planarisation and, consequently, minor variations in metal thickness could give rise to small gaps, e.g. $\sim 0.1\,\mu$m, between vias and metal layers which they were intended to connect. In a series of wafer lots CMP timing was varied, to optimise for the APV25 and other designs, following which yields of $\sim 80\%$ or more were regularly observed during manufacture of over 150,000 APV25 chips.[24,25]

It would have been impossible for HEP users alone to have replicated the IBM failure analysis and diagnosed the fundamental problem. It had also been believed that a high volume process (typically $\sim 40,000$ wafers per month) would not be tuned for small volumes. Yet yield variations and unexpected process changes have been a feature of many past experiences with smaller scale ASIC production, usually without any final convincing explanation. A close working relationship between foundry vendor and user is clearly highly desirable.

3.4. *Performance*

The APV25 achieved an equivalent noise charge of 270e + 38e/pF in peak mode and 430e + 61e/pF in deconvolution mode[22] for a power consumption of 2.3 mW/channel; about half of the power is used by the analogue circuits. In practice about 0.4 mW/channel extra is required on the hybrid to bias an inverter stage, which allowed the chip to operate with both sensor signal current polarities. The peak mode pulse closely approximates an ideal CR-RC pulse shape with a 50 ns time constant, as required for the deconvolution filter. Consequently the deconvoluted pulse shape is close to ideal, with a Gaussian-like shape and a peak amplitude after 25 ns and the whole pulse contained in 50 ns. The output is very linear to about 12 fC, deviating by 10–15% at signals of 30 fC.

There was a substantial radiation testing programme and many chips were irradiated to 10 Mrad at typical dose rates of 0.6 Mrad/h. The effects are minimal. Several chips have been irradiated to much higher levels, and even at over 100 Mrad there is little degradation of performance, with minor tuning of biases to optimize the pulse shape. SEU studies were also carried out, concluding that the impact on CMS will be very small.[26]

4. Optical Links

Given the large number of sensor channels to be read from tracking systems, it was attractive to examine fibre-optic links as a means of transmitting data through the detector to the underground counting room about 100 m distant. This would avoid certain significant challenges, such as the power to drive electrical signals with insignificant attenuation and, perhaps more important, long return current paths, with scope for ground loops and potential noise impact. A CERN DRDC project, RD23,[27] realised that tracking systems requiring a moderate dynamic range with modest deviations from linearity were suited to analogue optical transmission. The proposal also identified a need for fast, parallel signal transmission and processing for the first level trigger, for example using transition radiation detectors for electron identification, which was expected to be a more important objective.

4.1. *Early developments*

There was much less experience of optical technology in particle physics experiments than with electronic integrated circuits, previously mainly confined to a few fibre systems using LEDs in short distance laboratory set-ups. The RD23 project allowed interactions with industry to narrow down alternatives. While in 2009 fibre-optic data transmission is now commercially well established, most of the backbone infrastructure was installed only during the 1990s, especially local telecommunications links, which are probably closer to LHC applications than very long distance, trans-oceanic, data transmission. However, there are some important differences to note at the outset: commercial applications place important emphasis on reliability, robustness and ease of installation, for obvious reasons, including the cost of highly qualified personnel. But access to repair should also be possible, even if failure rates should be acceptably low.

The LHC tracking applications required low mass, low power and low cost, so that transmitters and receivers could be installed inside a lightweight structure. The link speed for a development should not be too high, to be compatible with ASICs operating in the multi-MHz range, which did not seem to be consistent with a very high level of multiplexing. It became clear that commercial components were still developing to meet market needs and that potential transmitters were mainly semiconductor lasers. LEDs could be used for very short distance links but would be less efficient in coupling light to fibres which would be a drawback for longer distances. The LHC radiation environment again posed special challenges, which were not encountered in any commercial systems, and LEDs did not seem to be sufficiently radiation-hard.

Lithium niobate modulators using Mach–Zehnder (MZ) modulation were briefly considered. For these links a continuous wave (CW) laser could be situated external to the detector, solving a big part of the radiation tolerance problem, and light sent to a waveguide divided into two optical paths on the surface of a lithium niobate crystal. One of the optical paths could be modulated with an electric field, proportional to the signal, to change the relative phase of the two light paths and allow them to interfere destructively. The lithium niobate was expected to be radiation-hard but required a rather high voltage of 10–20 V to switch. The optical path length necessary for creating a large enough phase shift was rather large, $\sim$10 cm, and the modulators were bulky and relatively high-mass. Issues requiring maintenance of light polarization in the fibres were also identified. It did not look easy to fit MZ modulators into a tracking system, so, although the idea had attractive features, the weak points seemed too important. However, these investigations began to demonstrate that the optical power budgets, losses and noise issues would be compatible with tracking requirements.

An alternative modulator technology looked more promising and was prototyped by a vendor for RD23 up to 1996. It used passive reflective modulator technology,[28] with external CW lasers providing the optical power, as with the MZ concept. The reflective modulators, acting as transmitters, were III–V semiconductor multi-quantum-well (MQW) electro-absorptive structures, tuned around a 1.55 μm wavelength. They translated the front-end chip output into an optical signal by modulating incident optical power from the external lasers. Power consumption within the tracker volume would have been limited to the very low photocurrent in the modulator. Sample modulators were irradiated to high levels using neutrons and gammas with no significant changes, so radiation hardness did not appear to be a major concern.

However, the system was quite complex and component development was a slow and expensive process, which would have required significant investment to reach large scale production. CMS would have been the unique customer of a single manufacturer. Since only small series of prototypes had been studied, important technical questions remained outstanding, such as dynamic range, noise sources intrinsic to the reflective system, tuning of laser wavelength and modulator bias voltage, possibly required at the module level. Modulators were challenging to manufacture and it became evident that the system cost was likely to exceed the CMS budgetary target with a high technical risk, including uncertain reliability, since no comparisons could be made with similar systems or components. Therefore in 1996 a recommendation was made to CMS to develop a system based on directly modulated lasers as transmitters.[29]

4.2. *Semiconductor lasers*

The most promising alternatives to modulators were actively modulated laser diodes. A number of semiconductor lasers from several manufacturers were successfully evaluated early on,[30] and extremely long mean time to failure rates were quoted. A commercial trend towards reliable, low power and, possibly, low cost solid state lasers, with many applications for data transmission, seemed to be emerging. Since these early studies, a substantial qualification programme has been undertaken by the CERN CMS optical link team.[31]

The semiconductor laser is based on a p–n diode in a direct band gap material in which forward bias creates the population inversion necessary for laser light emission. The operating wavelength depends on the band gap, which can be tailored by utilising compound semiconductors, e.g. InGaAs, to match the wavelength windows for low attenuation transmission in optical fibres. The other essential condition for laser operation is an optical cavity to generate oscillations and stimulate emission. It is constructed in a Fabry–Perot laser by cleaving the wafer crystal to create optical facets which act as partially reflective mirrors. Other types of edge emitting laser exist. Vertical cavity lasers (VCSELs), as the name suggests, are constructed to emit transverse to the wafer surface.

A minimum current, the threshold current, is required before laser action commences, since a certain fraction of photons are lost from the cavity via external emission and internal absorption. However, threshold values as low as a few mA and forward voltage drops of 1–2 V are achievable with output power of many mW. Above threshold the light output power is often highly linear with current, which made lasers attractive for an analogue system.

A major concern was radiation damage. Traps created in the active volume of the laser are likely to act as non-radiative recombination centres which reduce the optical gain. However, semiconductor lasers are usually constructed using III–V materials, which are known to be relatively radiation-hard. Lasers minimise the active volume using heterostructure quantum well designs to increase the laser gain by optimising the overlap of carriers with the optical field in the cavity. The lateral dimensions of the laser are also designed to be small, to minimise electrical power requirements and maximise efficiency.[32] Since the magnitude of radiation damage effects is related to volume, the drive for lower threshold lasers seemed to coincide with requirements for radiation-tolerant transmitters. Irradiation tests were carried out with neutrons and gammas, with excellent results,[31] including studies of receivers and fibres in addition to transmitters.

4.3. *Optical system issues*

By 1996 the system design which was implemented for CMS was reasonably fully evolved (Fig. 1). Despite a wish to undertake a complete system design and confront all major issues, the architecture originated in the principal components seen to be crucial to the implementation, namely the front-end readout chip and the optical link (of course matching requirements defined by the sensors). Both involved new technologies, and failure in either to meet the requirements would have necessitated a major rethink.

The choice of analogue readout, which had been strongly supported by most of the CMS sensor community, drove the design, which was then largely motivated by the desire for a "simple" system. Thus, originally, each front-end chip was attached to a transmitter sending its data synchronously to the off-detector data acquisition. The ability to place this part of the system outside the experimental cavern was seen to be very advantageous in minimizing

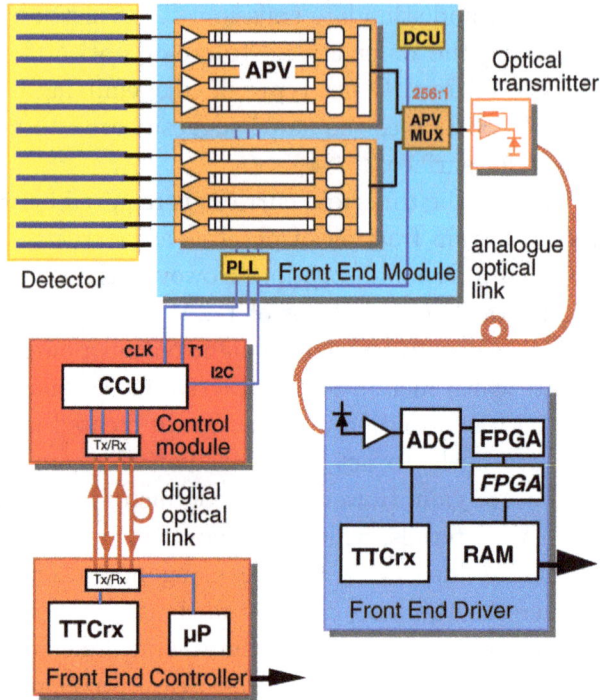

Fig. 1. A schematic of the CMS tracker readout system as envisaged in 1996.

the use of custom radiation-hard electronics, especially in view of resources available. A second important factor was that the tracker teams understood quickly that the detector would be completely inaccessible once assembled, even before installation, so it had to be constructed to a very high standard and operate reliably for a very long time. This was unprecedented; interventions on a detector during each winter shutdown were considered normal before the LHC. Placing ADCs in the counting room reduced the number of custom components to be developed for the radiation zone and gained performance as well as allowing access and avoiding potential problems of magnetic field operation and space constraints. Some of the arguments are listed in Table 4, but in many cases they were impossible to evaluate fully until the system had been built.

Although cost did not permit the use of an optical channel for each readout chip, 2:1 multiplexing allowed retention of the synchronous analogue concept and made better use of the available optical bandwidth, with each fibre effectively delivering almost $400\,\mathrm{Mbit\cdot s^{-1}}$.[33] The access to analogue information has proved to be as beneficial as originally foreseen, especially

Table 4. Additional pros and cons of analogue readout using optical analogue data transmission.

Pros	Cons
No radiation-hard on-chip ADC, saving design effort	Cost of analogue links
Smaller front-end chip than with ADC	Linear optical transmitter required
Larger analogue ADC range possible than with on-chip ADC	Increased sensitivity to connections in optical system
Lower power than alternatives	Greater sensitivity to noise after
Synchronous system	front-end

during commissioning and performance evaluation studies, as well as in the early phases of LHC data taking, so far limited to cosmic data.

4.4. *Optical link cost and performance*

Low cost devices are essential for the large scale of the CMS tracker. This is usually achieved in semiconductor manufacture by high volume production with large numbers of devices on each wafer. Automatic testing before dicing then maximises yield at the packaging stage. This is not so easy if cleaving is required before operation, so one reason for the relatively high cost of lasers is the testing requirement. The structure of VCSELs, which emit transverse to the wafer surface and seemed to simplify testing by manufacturers and offer potentially lower cost, was a reason that VCSELs seemed so promising to some users.

However, the total cost of an optical system involves a number of factors which this article cannot explore in detail. For the transmitter, in addition to the cost of the laser, packaging to accurately couple the emitter to the core of an optical fibre is an important cost driver. One factor which influences the packaging is the choice of fibre, which can be single-mode (SM), as widely used for long distance communication links, or multi-mode (MM), as used in many short distance links. SM fibre is less expensive but connectors require greater mechanical precision since the core fibre diameter is $\sim 8\,\mu$m, compared to $\sim 50\,\mu$m for MM, and so connector costs represent a larger contribution in SM systems. SM systems operate at longer wavelengths, typically $1.3\,\mu$m or $1.55\,\mu$m, with different implications for power, cost and eye safety and radiation hardness. Thus many factors enter into the choice between systems, including the availability and performance of

Table 5.   Estimated typical noise contributions for microstrip detectors using the APV25 in deconvolution mode.[34] Various sensor lengths are used at different places in the tracker.

	Noise contribution (electrons)	
	Inner sensors ($\sim$300 μm thick)	Outer sensors ($\sim$500 μm thick)
APV25	$400 + 60 \times 15\,\mathrm{pF} = 1300$	$400 + 60 \times 20\,\mathrm{pF} = 1600$
Metal strip resistance (200 Ω)	500	500
Leakage current of 1 μA	350	350
Optical link	600	600
Total	1560	1810

all the elements needed to construct a complete link, and evaluation should not focus simply on a single component.

So far the record of performance and reliability has been very good.[35] An analogue link adds noise to the front end but typically only 5–10% of the total, when combined in quadrature with other contributions (Table 5).

The issues which emerged during installation and operation were largely anticipated, even if sometimes difficult to avoid: individual fibres are strong but need careful handling, especially where space is tight and damage might be expected. Cleanliness at connectors, which are very precise, is essential. The laser current thresholds have a well-known dependence on temperature which generally results in baseline shifts which can be easily handled but are sometimes inconvenient, for example if temperature changes occur relatively quickly. In most cases, this was trivial to handle because of the temperature-controlled environment imposed on the tracker for other reasons.

5. Overall System Design

The early years of the readout system development were spent primarily on the radiation-hard front-end chip and optical links, including a series of ASICs needed for clock and trigger regeneration and opto-electronic functions, which already stretched the small teams contributing to the system design. It was envisaged early on that the same optical components could be deployed to transmit system clocks, control and monitoring data to and from the interior of the tracker, but it was only in 1996 that effort became available to define and develop control features, which began by trying to understand the amount of traffic which would be required and the protocols needed for data transfer. While safety-sensitive control operations, such as power on and off, were to be hardwired in conventional "slow control" mode, it was necessary to configure the front-end chips and laser driver gains by a faster, radiation-hard, communication interface, which would then also allow one to monitor the internal state (temperature, leakage currents, local voltages, etc.) via other ASICs on the tracker.

A particularly important requirement was to synchronise all the front ends, which was far more demanding for the LHC than in the past, given the large number of channels, deep pipeline storage, time of flight and pile-up from particles looping in the magnetic field. Monitoring hits in the tracker and comparing them with the beam structure was expected to provide some indication but not be sufficient to configure or control such a large system. The flexibility of the control system and access to the APV25 sampled pulse shape data provided a solution to this.[36]

Another special requirement was to ensure that buffers on the APV25s could not overflow by sending too many triggers in a short interval. Even though this is a rare situation with an L1 trigger rate of 100 kHz, it is also inevitable, but the APV25 cannot warn of impending overflow in time to be useful because of the $\sim 0.5\,\mu$s transit time from the counting room to the interior of the tracker. Counting triggers sent to the system is not a reliable or efficient solution because of the long delay in receiving data, made up of the transit time through CMS and the time needed to process APV data frames lasting $7\,\mu$s each. This problem was solved by emulating the APV25 control logic outside the experiment. In principle, a single chip could replicate the state of every APV25, but a more powerful method was to emulate the pipeline logic in an FPGA on a module communicating directly with the trigger throttling system.[37]

5.1. *Control and monitoring system*

The tracker control system consists of three main elements:

- A VME Front End Controller (FEC) card to manage the communication network and interfaces to the CMS control system;
- A network based on a simple token-ring architecture to communicate between control room and embedded electronics with long sections implemented using optical fibres and short ones on low mass copper cables;

Read-out and control architecture

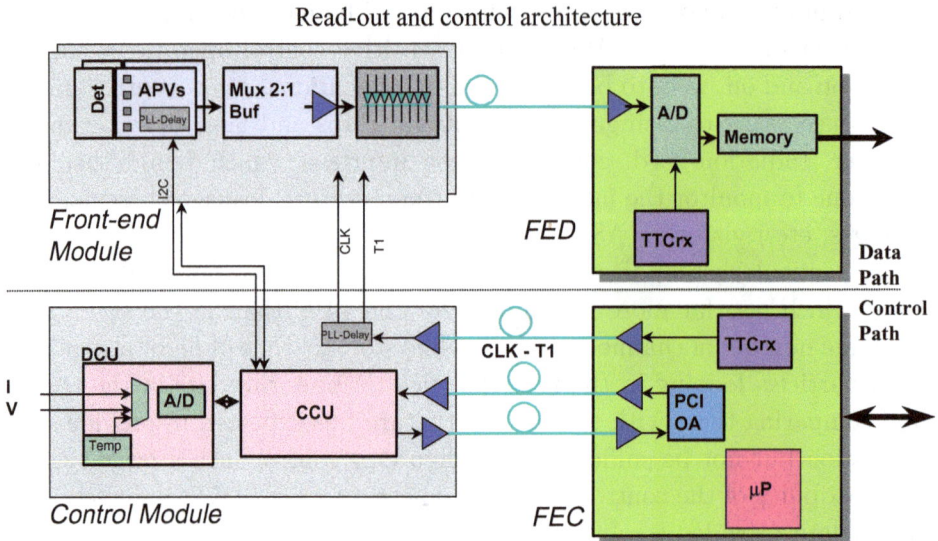

Fig. 2. An alternative view of the CMS tracker electronic system, emphasizing the control features.

- An embedded Communication and Control Unit (CCU) to link the communication network and front end or any monitoring ASIC. The CCU is complemented by a Phase-Locked Loop (PLL) ASIC on every module for recovery and local distribution of clock and trigger signals.

The communication architecture[38] is based on a ring to connect the FEC to CCUs and the CCUs between themselves with a specific protocol transporting data packets (messages) from the FEC to the channel controllers. Then control modules talk to front-end chips via channels with a second protocol. Each ring controls many front-end channels and high reliability is essential. A malfunctioning control element would be unacceptable, so a redundancy scheme based on doubling interconnection lines and bypassing of CCUs is included.

5.2. Off-detector electronics

The microstrip tracker has more than nine million detector channels and will generate at least half the CMS recorded data volume. The front-end driver digitises the data, finds hit clusters and performs zero suppression before transmitting to the central DAQ. At projected CMS trigger rates, the total input data rate on each FED will be approximately $3 \text{ GB} \cdot \text{s}^{-1}$. After zero suppression this reduces to roughly $50 \text{ MB} \cdot \text{s}^{-1}$, depending on strip occupancy, which will vary from 0.5% to 3% in high luminosity LHC running. This was evident from the outset and was not a minor issue.

In total, 440 FEDs are needed to read out the 73,000 APV25s and had to meet tight budgetary constraints while maintaining data processing flexibility. It is likely that cluster finding will evolve during CMS operation and common mode and other noise conditions were not certain until the tracker had been built. The CMS cost estimates were closely scrutinised in the phase where CMS was seeking approval and the idea that more than about 32 optical channels could be integrated onto a single 9U board (390 mm by 400 mm) was considered aggressive enough to require demonstration. The final FED layout (Fig. 3) has managed to achieve 96 optical channels per board.[39]

Interestingly, although the cost of a board was considered a critical point in making early cost estimations, it was very hard to establish the real cost of many items, such as for an LEP project. Records of costs and how they were broken down were rather difficult to obtain.

Despite the demonstration of the integration density, the cost target remained a challenge because the design required an ADC (10 bits, 40 MHz), then considered the most demanding component for cost, power and integration density reasons. However, digital processing was originally foreseen to require ASIC developments but it was also predicted that this could

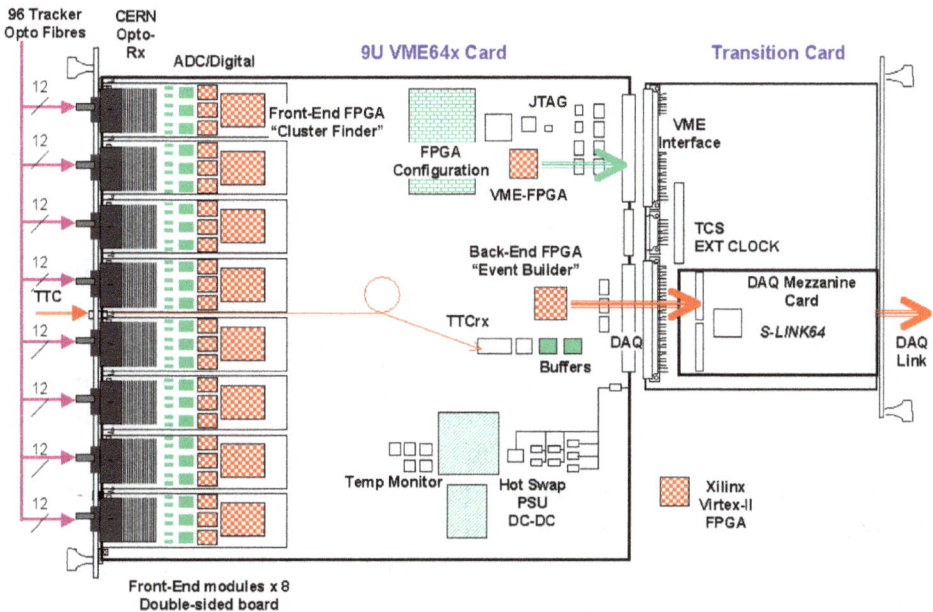

Fig. 3.   The layout of the final version of the front-end driver (FED).

be implemented in a more flexible form using field-programmable gate arrays (FPGAs) provided that the schedule permitted, allowing large-enough devices to reach the market at an affordable price. Only during 2000 did this begin to seem a realistic prospect, which would have been a severe problem if the official LHC schedule at that time had been maintained, although this would have been true for almost every other part of CMS. In fact devices with the size and functionality became available, at a high price, for prototyping the first versions of the final boards, and the cost fell remarkably quickly, as predicted by experts who had followed the evolution of earlier FPGA series.[40] By the time the full scale manufacture began in 2005, FPGAs, which had been state-of-the-art components at the layout stage, were well below the top of the line, whose prices had probably reached their minimum, being superseded by more advanced parts.

Another important choice was to aim for a monolithic board, rather than build a series of daughter boards to mount on a carrier. The decision arose from a number of factors: cost (as the mother–daughter board solution would have been more expensive), reliability of connectors, density of boards in the system. However, daughter boards potentially offer easier maintenance. To date, the reliability of the boards has vindicated the choice.

The production of the 500 boards needed was still a concern, as there were many reports of difficulties with high quality production of large, multilayer PCBs and disparate conclusions about the best metallisation finish, which had been identified by some users as the origin of problems in assembling boards with high density ball grid array packages. The FED is double-sided to accommodate approximately 6000 components, with a high component density, especially in the sensitive analogue front-end region, with 14 layers and almost 25,000 tracks.

In fact, manufacture went remarkably well, but this should not be attributed to chance or good fortune. A very close working relationship was developed with the vendor, with frequent on-site visits and an exchange of expert views on details of the design before it was constructed to optimize the layout from the manufacturing perspective. A simple-to-use test set-up was developed, and it was installed in the factory so that boards could be tested before delivery and any faults not picked up by the automated inspection process could be rectified quickly. The final yield was remarkable: 99.8%, with relatively little rework required following delivery and usually of a minor nature, such as occasional dry joints not easily identified by close visual or x-ray inspection. At such levels, the quality of the components is a significant factor and it was beneficial to allow the vendor to be responsible for procurement of parts. The project clearly benefited from its overall scale, since orders of a few boards clearly cannot merit such detailed attention.

Operational experience has been equally good, with occasional problems which usually seem to be traceable to components malfunctioning under certain conditions. However, one small feature of the design has proved to be of major benefit. Because of concern about the valuable, and irreplaceable, opto-electronic components, the designers suggested incorporating on-board temperature sensing to shut down the board in case of over-temperature. The board air flow was studied carefully in lab tests to try to optimize it since the opto-receivers are mounted vertically in a line, so that those at the top are cooled less efficiently than those below.

During both commissioning and early operation this minor feature proved its value many times over, although not for the reasons originally anticipated. Rack and crate cooling proved to be a major headache: in achieving the quality of the plumbing, ensuring adequate water circulation, managing the detector control and safety systems, among other issues. Cooling systems have failed several times, usually at inconvenient times of the day or night when no one is at hand. Even worse, other CMS boards have occasionally overheated, for example because of a poor ground connection, for which rack smoke and temperature sensors are the final protection. Users taking data often did not appreciate an FED switching itself off during early commissioning, but it was always found to be a symptom of a deeper problem with the environment, and so was of great importance in preventing damage.

5.3. *CMS choices*

During development of a completely novel system on a large scale, such as the CMS tracker readout, difficult issues are raised which cannot easily be decided on technical grounds, at least without prior knowledge of the future. One of the most important examples was the commitment to the type of optical system which had been chosen. Although the very earliest cost projections from industry later proved to be optimistic, realistic appraisal of the final costs was achieved within a few years and remained reliable for the lifetime of the project. The participants in the CMS tracker project made a deliberate choice to work closely with industry, which required the development of good specifications which were rigorously adhered to during the procurement phase, and to base as much as possible on available commercial parts, adapted to the application. This placed a load on the CMS team to qualify parts and ensure that they met requirements, both before contracts were placed and during the procurement, but avoided the development of in-house parts in specialised technologies. Other projects approached this in different ways but the final cost of the CMS links seems to be equivalent to, or less than, that of alternatives.[41] Operational experience will be a future measure of the success of the choices of in-house or commercial routes.

CMS could have made different choices at several stages. The optical link was the dominant part of the readout system cost and, as it proved, could have been significantly reduced only by decreasing the number of links drastically. This would have required a different readout architecture with on-detector sparsification and sharing of links between a greater number of detector modules. Similarly, the early decision to use CMOS amplifier technology seems to have been vindicated in terms of performance, power and cost, but this could never have been proved to be the right choice during the development phase. Had commercial development of $0.25\,\mu$m CMOS technology not coincided with the LHC, the CMS tracker readout system might look quite different today.

6. Applications Elsewhere and Lessons Learned

One of the major benefits to the wider community from the tracker development has been the deployment of components in other CMS sub-systems, and other projects. The optical links have been used for 800 Mbps digital data transmission in the CMS ECAL and other projects, following the development of high speed serialisers in $0.25\,\mu$m CMOS. The family of ancillary chips, such as the PLL, CCU, monitoring circuits, and buffers, have been widely used elsewhere. The $0.25\,\mu$m process was rapidly taken up by other users following the demonstration of the analogue performance of the APV25 and the relative ease of design in the technology, and the APV25 itself has been successfully used by many other projects for readout of silicon and gas detectors. The expertise gained by the tracker team was put to good use in completely revising the ECAL readout at a very late stage, which was achieved very rapidly, partly by deploying tracker components. The experience gained in the use of FPGAs during the evolution of the system allowed the development of components such as the APVe.

Are there general lessons to be learned from this experience? It has certainly been beneficial to assemble a team who work closely with minimal competition and maximal co-operation; this takes time. Frequent meetings and close contacts are important for building trust and exchanging information efficiently; small working meetings are more effective than large public gatherings. During developments which are still maturing, self-confidence is needed even when things look uncertain.

While the LHC schedule has been longer than anticipated, the tracker electronics fortunately managed to stay in synchronisation and not slip behind it, and so early commissioning was possible. The discipline arising from knowing that the system was going to be inaccessible may have been an advantage. The tracker off-detector electronics was the first to be installed and operational in CMS, although it could not be fully commissioned until the tracker was in place and services connected.

The early estimation of cost was essential and there was considerable pressure from outside to design a system which could plausibly be constructed within a modest per channel budget, which was largely achieved. However, as with other projects, development costs are usually not easily accounted for but, for a large scale project, can be absorbed. In particle physics, it is rare for costs to evolve downwards but modern technologies occasionally offer such benefits, such as the switch to commercial $0.25\,\mu$m CMOS and the steady evolution of FPGAs. For a large system, there is benefit from the scale, but small errors in costs can become intolerable, while for small systems it is possible to overrun with tolerable impact. This account may have over-emphasised this aspect, but I believe that maintaining control of costs was a big factor in the success of the project.

On several occasions during the CMS tracker electronic development, it became clear that close attention to commercial trends was advisable and almost any progress could be undermined rapidly by shifts in market priorities. It is practically unavoidable that advanced technological developments are both costly and risky and it is difficult to ensure a safe fallback position in all circumstances. Probably the only answer to this is to maintain sufficient expertise in a few laboratories, such as CERN, and to ensure a high calibre of expertise on the project, while guarding against over-ambition. With hindsight, this may be the biggest achievement in building this system.

Acknowledgments

The CMS tracker is a vast and successful project, and all those who have contributed to it should be very proud of it. I am confident that the electronic system will perform well for many years. Many individuals made enormous efforts to bring the original vision to fruition and I am glad to be one of those who have seen it through its entire development up to LHC operation. This account has concentrated on the origins of the readout system, and rationalisation of the past should not be seen as the ultimate guide to the future, although there are certainly important lessons contained in it. It is impossible to acknowledge everyone individually but I would like especially to draw attention to major contributions from a few people who envisaged and laid foundations for the system: J. Coughlan, M. French, R. Halsall, A. Marchioro, M. Raymond, G. Stefanini and F. Vasey. I would also like to acknowledge with gratitude the constant, important support from the Tracker Project Managers: R. Castaldi, G. Rolandi and P. Sharp.

This article has not relied entirely on my memory but, even with the aid of many old documents, it is hard to confidently retrieve all past history. Mistakes or misapprehensions are mine.

References

1. RD20 Collaboration. RD20 status report (CERN/DRDC 92-28, 1992).
2. G. C. Messenger and M. S. Ash. *The Effects of Radiation in Electronic Systems* (Van-Nostrand-Reinhold, New York, 1986).
3. A. Holmes-Siedle and L. Adams. *Handbook of Radiation Effects* (Oxford University Press, 1993).
4. T. P. Ma and P. Dressendorfer. *Ionising Effects in MOS Devices and Circuits* (John Wiley, New York, 1989).
5. H. Spieler. *Semiconductor Detector Systems* (Oxford University Press, 2005).
6. H. Börner *et al.* Development of high resolution Si strip detectors for experiments at high luminosity at the LHC. RD 20 proposal (CERN-DRDC-91-10, 1991).
7. S. Gadomski, G. Hall, T. Høgh, P. Jalocha, E. Nygård and P. Weilhammer. *The deconvolution method of fast pulse shaping at hadron colliders, Nucl. Instrum. Methods A* **320** (1992) 217–227.
8. N. Bingefors, S. Bouvier, S. Gadomski, G. Hall, T. Høgh, P. Jalocha, H. vd. Lippe, J. Michel, E. Nygård, M. Raymond, A. Rudge, R. Sachdeva, P. Weilhammer and K. Yoshioka. A novel technique for fast pulse shaping using a slow amplifier at LHC, *Nucl. Instrum. Methods A* **326** (1993) 112–119.
9. J. Matheson *et al. Proc. Third Workshop on Electronics for LHC Experiments* (CERN/LHCC/97-60, 1997), pp. 168–172.
10. M. Raymond *et al. Proc. Third Workshop on Electronics for LHC Experiments* (CERN/LHCC/97-60, 1997), pp. 158–162.
11. E. Beuville *et al.* A mixed analog–digital radiation hard technology for high energy physics electronics: DMILL (Durci Mixte sur Isolant Logico-Linéaire) (CERN-DRDC-92-31, 1992).
12. CMS Tracker Technical Design Report (CERN-LHCC-98-006, 1998).
13. N. Saks *et al. IEEE Trans. Nucl. Sci.* **31** (1984) 1245.
14. N. Saks *et al. IEEE Trans. Nucl. Sci.* **33** (1986) 1185.
15. P. Winokur *et al. Proc. Third Workshop on Electronics for LHC Experiments* (CERN/LHCC/97-60, 1997), p. 48.

16. M. Shaneyfelt *et al. IEEE Trans. Nucl. Sci.* **41** (1994) 2536.

17. R. L. Pease. *IEEE Trans. Nucl. Sci.* **43** (1996) 442.

18. G. Anelli *et al. Proc. Third Workshop on Electronics for LHC Experiments* (CERN/LHCC/97-60, 1997), p. 139.

19. F. Faccio *et al. Proc. Fourth Workshop on Electronics for LHC Experiments* (CERN/LHCC/98-36, 1998), p. 105.

20. W. Snoeys *et al. Proc. Fourth Workshop on Electronics for LHC Experiments* (CERN/LHCC/98-36, 1998), p. 114.

21. W. Snoeys *et al. IEEE J. Solid-State Circuits* **35** (2000) 2018.

22. M. Raymond *et al. Proc. Sixth Workshop on Electronics for LHC Experiments* (CERN/LHCC/2000-041, 2000), p. 130.

23. G. Hall. *Nucl. Instrum. Methods A* **453** (2000) 353.

24. R. Bainbridge *et al.* Production testing and quality assurance of CMS silicon microstrip tracker readout chips, *Nucl. Instrum. Methods Phys. Res. A* **543** (2005) 619–644.

25. G. Hall. Recent progress in front end ASICs for high-energy physics, *Nucl. Instrum. Methods Phys. Res. A* **541** (2005) 248–258.

26. E. Noah *et al.* Single Event Upset studies on the CMS Tracker APV25 readout chip, *Nucl. Instrum. Methods Phys. Res. A* **492** (2002) 434–450.

27. J. Dowell *et al.* Optoelectronic analogue signal transfer for LHC detectors. RD 23 proposal (CERN-DRDC-91-41, 1991).

28. C. Da Via *et al.* Lightwave analogue links for LHC detector front ends, *Nucl. Instrum. Methods. A* **344** (1994) 199.

29. G. Hall, G. Stefanini and F. Vasey. Fibre optic link technology for the CMS Tracker CMS Note 1996/012 (1996).

30. F. Vasey, V. Arbet-Engels, G. Cervelli, K. Gill, R. Grabit, C. Mommaert and G. Stefanini. Development of rad-hard laser-based optical links for CMS front-ends, *Proc. Third Workshop on Electronics for LHC Experiments* (CERN/LHCC/97-60, 1997), pp. 270–275.

31. Publications of the CERN CMS Tracker Optical Link team can be found at http://cms-tk-opto.web.cern.ch/cms-tk-opto/tk/publications/default.htm

32. R. G. Hunsperger. *Integrated Optics: Theory and Technology* (Springer, 2002).

33. J. Troska, G. Cervelli, F. Faccio, K. Gill, R. Grabit, A. M. Sandvik, F. Vasey and A. Zanet. Optical readout and control systems for the CMS tracker, *IEEE Trans. Nucl. Sci.* **50** (2003) 1067–1072.

34. Addendum to the CMS Tracker TDR (CERN/LHCC 2000-016, 2000).

35. CMS Tracker Collaboration. Performance studies of the CMS Strip Tracker before installation. Accepted for publication by *JINST* (2009).

36. K. Gill, L. Mirabito and B. Trocmé. Synchronization of the CMS tracker, in *Proc. Ninth Workshop on Electronics for LHC Experiments* (CERN-LHCC-2003-055, 2003), pp. 289–293.

37. G. Iles *et al.* The APVE emulator to prevent front-end buffer overflows within the CMS silicon strip tracker, in *Proc. Eighth LECC Workshop* (CERN-LHCC-2002-34, 2002), pp. 396–401.

38. F. Drouhin *et al.* The CERN CMS tracker control system, *IEEE Nuclear Science Symposium Conference Record*, 2004, pp. 1196–1200.

39. J. Coughlan *et al.* The CMS tracker front-end driver, in *Proc. Ninth LECC Workshop* (CERN-LHCC-2003-055, 2003), pp. 255–260.

40. P. Alfke. Field programmable gate arrays in 2004, in *Proc. Tenth Workshop on Electronics for LHC and Future Experiments* (CERN-LHCC-2004-030), pp. 26–30.

41. K. K. Gan, F. Vasey and T. Weidberg. *Lessons Learned and to Be Learned from LHC.* https://edms.cern.ch/document/882775/3.8

Chapter 9

TileCal: THE HADRONIC SECTION OF THE CENTRAL ATLAS CALORIMETER

K. Anderson

Enrico Fermi Institute, University of Chicago
5640 S. Ellis, Chicago, IL 60637, USA
kelby@uchep.uchicago.edu

T. Del Prete

Dipartimento di Fiscia E. Fermi, Università degli Studi di Pisa
and Instituto Nazionale di Fisica Nucleare, Sezione di Pisa
Largo B. Pontecorva 3, I-56127 Pisa, Italy
delprete@pi.infn.it

E. Fullana

High Energy Physics Division, Argonne National Laboratory
Argonne, IL 60439, USA

J. Huston

Department of Physics and Astronomy
Michigan State University
East Lansing, MI 48824, USA
huston@pa.msu.edu

C. Roda

Dipartimento di Fiscia E. Fermi, Universita degli studi di Pisa
and lustitudo Nazionale di Fisica Nucleare, Segione di Pisa
Largo B. Pontecorva 3, I-56127 Pisa, Italy
roda@mail.cern.ch

R. Stanek

High Energy Physics Division, Argonne National Laboratory
Argonne, IL 60439, USA
bob@hep.anl.gov

This chapter describes the concepts behind the TileCal design, as well as the requirements and the constraints that have finally produced the ATLAS central hadronic calorimeter.

1. TileCal Design: Motivation and Requirements

The central hadronic calorimeter of ATLAS (the tile calorimeter, also known as TileCal) is a sampling calorimeter with steel as absorber and scintillating tiles as the active medium. The calorimeter has been described in detail previously in several papers; here, we describe primarily the manner in which the final TileCal design was determined.

In 1991, it was known that the calorimeter resolution in a sampling hadronic calorimeter is not critically dependent on the orientation of the active medium.[1] The understanding was that, at the end of a hadronic shower, low energy charged particles had velocities nearly isotropically distributed, as in a gas. Thus, a nonstandard orientation of the active medium could be exploited in order to optimize hermeticity, while maintaining a relatively good energy resolution. Scintillating tiles, for example, could be oriented inside a steel absorber in the r–ϕ plane (see Fig. 3), i.e. parallel to the direction of the incoming particles. The readout can then be performed via wavelength shifting (WLS) fibers coupled to the radial edges of the scintillator plates. Fibers could be routed in the radial direction, fitting in small slots in the absorber. This configuration is illustrated in Fig. 1. At a radius outside the absorber, the fibers could easily be grouped in order to obtain the needed calorimeter segmentation, and then coupled to photomultiplier tubes (PMTs).

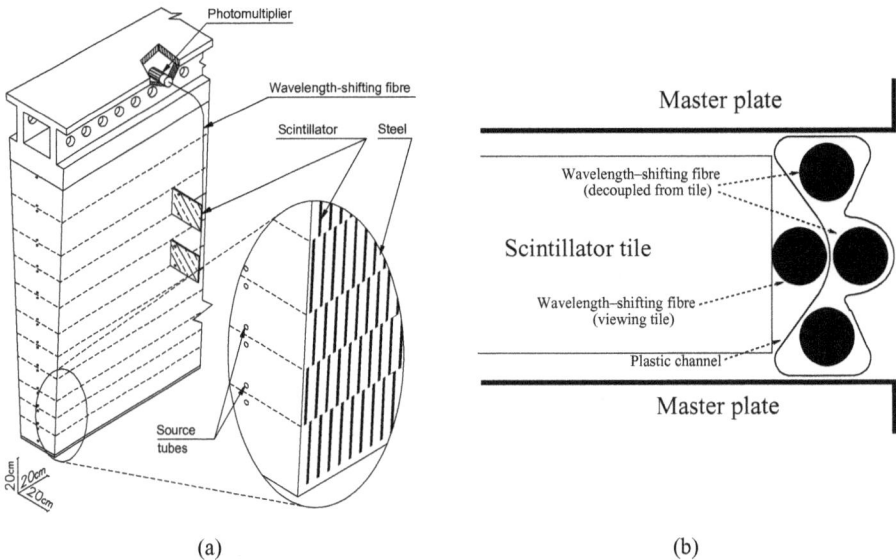

(a)   (b)

Fig. 1. (a) A schematic showing the sampling structure of the tile calorimeter and the integration of scintillator tiles and readout fibers with the absorber structure running radially from the interaction point. (b) A radial view schematic of the integration between the absorber structure, the scintillator tiles and the fibers.

At rapidities close to zero, the scintillating tiles are nearly parallel to the primary particles. Thus, the percentage of active material crossed by collimated showers depends strongly on the impact point. However, TileCal is placed downstream of an electromagnetic calorimeter; most of the particles from the interaction will have already started to shower, and thus any channelling effect will be limited. To prevent the particles from traversing too long a path inside the active material, the scintillator tiles can be made small and staggered with respect to each other.

This concept was validated and refined in detailed analyses of data acquired with prototype modules at test beams.[2,3]

Other conceptual designs had also been considered as possibilities for the central ATLAS hadron calorimeter. For example, a design using a lead–liquid-argon calorimeter (10 cm Pb plates and 4 mm Ar) was simulated in Monte Carlo. The resulting performance was found to be not superior to the TileCal design. Similar conclusions were reached for a lead-scintillating tile geometry.[1]

In addition, prototypes of an integrated electromagnetic–hadronic calorimeter using lead and scintillating fibers ("SPACAL") were built and tested.[4] This option was seriously considered for the complete calorimeter system but was finally rejected in favor of an electromagnetic section (now liquid argon) with a better performance and a less expensive hadronic section.

Ultimately, it was decided that the TileCal option for the hadron calorimeter, complemented by a lead/liquid-argon calorimeter (LAr) for the electromagnetic section, could not only achieve the physics goals but also be more modular and cost-effective.

The overall TileCal geometry was, in part, determined by the surrounding detectors: the liquid argon calorimeter inside and the muon spectrometer outside. The total thickness of the calorimeter system is a minimum of 10 interaction lengths (λ). Simulation studies indicate that a thickness of 9 λ provides sufficient containment of the hadronic cascades for precision measurements both of jet properties and of E^T_{miss}.

Another essential constraint was that the construction of TileCal was modular so that institutions around the globe could contribute with their resources, expertise and personnel to reduce the overall cost and the time needed to complete the work. TileCal elements were assembled in several laboratories in Russia, in the Czech Republic, Italy, Spain and the USA. The raw materials and finished elements were shipped back and forth across the globe over a four-year period. In spite of this dispersed effort, the strict tolerances required for the mechanical construction (about 100 μm) were maintained in all the modules and the tight schedule was fulfilled to within a few weeks. The mechanical assembly details are discussed in Subsec. 2.1.

In parallel with the design and construction of the mechanical components, R&D projects were carried out to optimize the optics. Since commercial tiles are both expensive and produced at a low rate, an alternative technique of scintillator production was developed based on injection molding.[5–7] This method proved both cost- and time-effective, but with the drawback of a lower light budget. The light budget was recovered by optimizing the light transmission to the PMTs: high quality optical fibers and efficient tile-fiber-PMT coupling.[8–10] Details of all the optics aspects are discussed in Subsec. 2.2.

It was demonstrated in the very first prototype modules that the TileCal design was capable of producing a light yield of more than 50 photoelectrons (PE) per GeV of energy deposited in the absorber.

2. TileCal Solutions

2.1. *Mechanics*

The tile calorimeter[11] for ATLAS[12, 13] is physically the largest component in this system, with the function of extending the depth of calorimetry to wholly contain hadronic showers from proton–proton collisions at the interaction point. The TileCal scintillators are located in pockets in the steel structure and read out using wavelength-shifting fibers coupled to PMTs located inside the outer support girders of the calorimeter structure.

In addition to its role as a detector for high energy particles, the tile calorimeter provides the direct support of the liquid argon calorimeter in the barrel region, and both of the liquid argon electromagnetic and hadronic calorimeters in the endcap region. Through these, it indirectly supports the inner tracking system and the beam pipe. The steel absorber, and in particular the support girders, provides the flux return for the solenoidal field from the central tracker solenoid. Finally, the end surfaces of the barrel calorimeter are used to mount services, power supplies and readout crates for the inner tracking systems and for the electromagnetic calorimeter section.

2.1.1. *The principles of the detector design*

Some of the considerations in the development of the design of TileCal can be summarized thus:

- The overall cost should be minimized;
- Good hermeticity;

- Good energy linearity and resolution;
- Modularity for parallel construction at several participating institutions;
- Implementation of the flux return for the central solenoid and shielding for the front-end electronics.

The tile calorimeter is constructed in three sections (one central barrel and two extended barrels), each comprising 64 individually constructed modules, which are stacked on one another to form each cylinder (Fig. 2). The central barrel covers the rapidity region $|\eta| < 1.7$, while the extended barrels have coverage of $1.4 < |\eta| < 3.2$. A barrel module weighs 20,020 kg and an extended barrel module weighs 9600 kg. The mechanical structure has a nominal outer diameter of 4230 mm and an inner diameter of 2288 mm. A barrel (extended barrel) module has a length of 5640 mm (2900 mm). The calorimeter itself is self-supporting and rests simply on support saddles placed in the lower regions of the structure. The instrumented region of the calorimeter has a radial length of 1.64 m, and contributes 7.4 λ for particles emitted at 90° to the interaction point.

2.1.2. *Module construction*

Each of the 256 modules is constructed in the following fashion. The absorber structure is formed using steel laminations, and has the unique feature that the steel plates run radially outward from the beam axis. This is the feature of the mechanical design which allows the fiducial acceptance of this calorimeter to be maximized. A schematic of the calorimeter structure is shown in Figs. 3 and 1. The structure is a glued and welded steel lamination

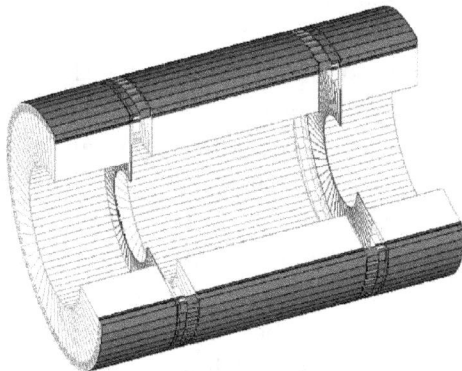

Fig. 2.   The tile calorimeter system in the ATLAS experiment at the Large Hadron Collider.

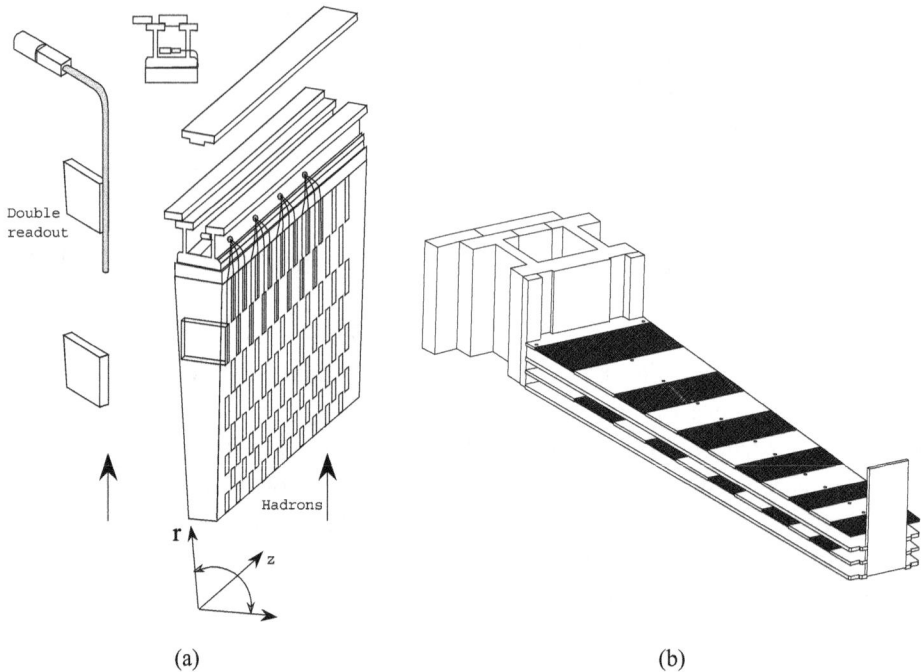

Fig. 3.　(a) The tile calorimeter module design with the placement of all the optics elements (tiles, fibers and phototubes). "Double readout" refers to the fact that both edges of the tiles are read out into separate phototubes. (b) This sketch shows the periodicity of the steel absorber as discussed below.

of full-length 5-mm-thick trapezoidal plates (master plates) spanning the full radial dimension of the module, and smaller 4-mm-thick trapezoidal plates (spacer plates) interspersed along their length to form pockets in which the scintillator tiles are inserted. The spacer plates are set back from the edge of the master plate outer envelope by approximately 1.5 mm to provide a slot in which the readout fibers are inserted. The scintillator tiles are inserted into the gaps between the spacer plates, resulting in an iron-to-scintillator ratio of 4.67:1 by volume. In addition, the plates have 22 precisely located holes through which the tubes for the Cs system (see Subsec. 2.4.2) calibration are inserted along the entire length of the structure.

In order to facilitate a relatively straightforward construction procedure for precision modules, the absorber structure for each of the modules is first fabricated into submodules, which are then stacked and welded along the length of a support girder to form each module. This is illustrated in Fig. 4(a).

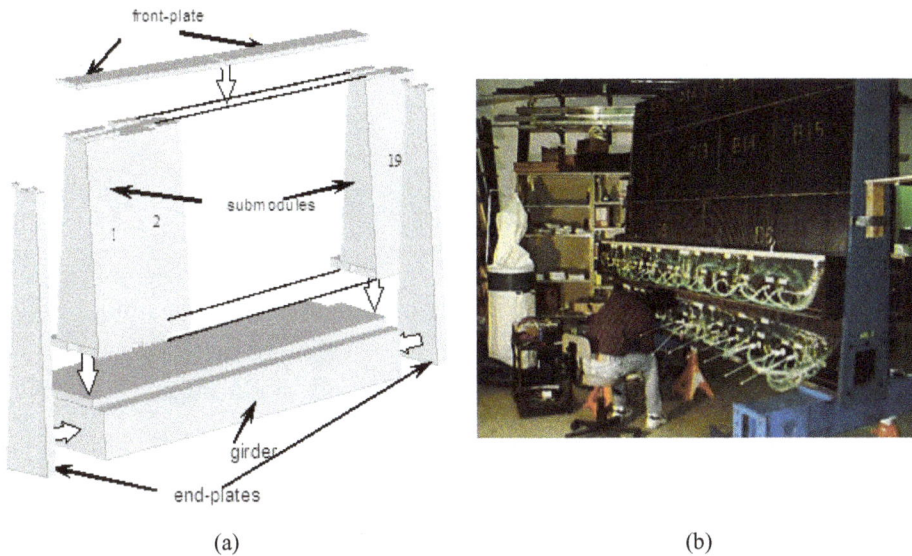

Fig. 4.   (a) A sketch of the assembly of submodules on the girder forming a module. (b) A partially completed extended barrel module, with details of the fiber routing clearly seen.

2.1.3. *Submodule construction and module production*

Eight standard submodules and two customized submodules are used to construct an extended barrel module, while 18 standard submodules and 1 customized submodule are used to construct a barrel module. Die stamping was used to produce the master and spacer plates, again to realize the precision needed for the module envelope and for cost-effective production.[14] A nearly finished extended barrel module is shown in Fig. 4(b).

The girder is the outer structural support of the module. In addition to providing the outer bearing surface, the support girder gives the largest magnetic flux return for the central solenoid and also serves as a convenient location for the phototubes and front-end readout electronics. Low voltage power supplies, cooling and readout cables enter at the end of each of the girders. These are all contained in extensions of the girder called fingers, which provide shielding from the return magnetic field as well as physical protection for cables and connectors.

In the region between the barrel and extended barrel calorimeters, the services to the inner detector systems and the LAr cryostat flange result in dead material. In order to help correct for the energy loss in this material, an intermediate tile calorimeter (ITC) and scintillators are installed in this region. This is illustrated in Fig. 5.

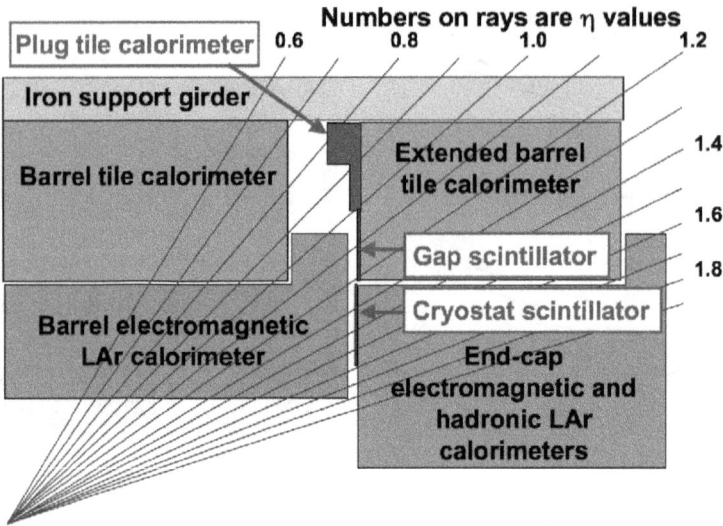

Fig. 5. A cut through the calorimeter system at the interface between barrel and extended barrel calorimeter.

Due to the wedge shape of the modules, and designing for a self-supporting structure, the calorimeters were constructed by stacking each module on top of the previous one, with appropriate shimming at the inner and the outer radius to ensure closure at the top with the last module. The installation of the ATLAS detector in the experimental cavern was started in 2004. The three TileCal barrels were first preassembled on the surface. This allowed a check that the strict geometrical constraints, a few millimeters over a few meters, were fulfilled, and that the design for the complicated integration of cable services could be successfully implemented. The experience gained in the preassembly phase permitted a solution to many installation problems before facing the assembly in the experimental cavern. TileCal was then partially dismounted to be lowered into the experimental cavern. The assembly in the cavern was completed in December 2005 and TileCal was the first ATLAS subdetector to provide cosmic muon data.

2.2. *Optics*

The driving concept behind the optics is to achieve and maintain a uniform, minimum light yield, such that the detector resolution is not compromised by lack of photostatistics or by nonuniformities. Key to the physics performance of TileCal is the optical budget; through the chain from scintillator to PMT, we require that a minimum ionizing particle (MIP) should produce at least

0.5 PE per tile at normal incidence. This response leads to 20 PE produced per GeV deposited in the calorimeter. The contribution from photostatistics is small relative to the intrinsic resolution endemic to hadron calorimeters. Based on past experiences with similar calorimeter systems, we expect an average 1–3% loss in light per year. A safety factor of 2 was added, leading to the requirement of at least 1 PE per MIP.

A nonuniformity of tile response can also degrade the physics performance of TileCal. Due to the large transverse size of hadron showers and the sharing of energy between the electromagnetic and hadronic sections of the ATLAS calorimetry, the nonuniformity requirements for TileCal are less stringent than for the LAr calorimeter. Monte Carlo simulations show that a nonuniformity of 10% RMS would result in an increase of the constant term of the calorimeter resolution of 1%. To realize this 10% goal, it is necessary to restrict the nonuniformity within a tile, with tile-to-tile fluctuations and fiber-to-fiber fluctuations each below 5%.

The mechanical structure and the scintillator orientation of TileCal necessitate the use of a very large number of scintillating tiles, of the order of 460,000, with 11 different radial sizes (59 tons of scintillator). Each of the scintillating tiles has a thickness of 3 mm. The standard technology for production of the highest quality plastic scintillator uses styrene polymerization between two high quality glass plates. This process is slow and relatively expensive, since large pieces of scintillator are cast and then cut into the shapes required and subsequently polished.

The light budget adopted for TileCal allows for the use of tiles produced by injection molding. Tiles produced by such a process typically have a lower light yield than the cast scintillator. The injection molding process uses commercially available optically transparent granulated polystyrene pellets mixed with a primary and a secondary scintillating dye (PTP and POPOP) adjusted to optimize the light yield and uniformity. As mentioned previously, there are 11 different radial sizes for the tile, but only 4 different tile widths. Thus, it was possible to produce the 11 different types of tiles using only 4 different molds with removable inserts, further reducing the cost and complexity. The mold surfaces were machined very smooth, so no additional polishing step for the tiles was required. Approximately 5% of the tile production was tested with a ^{90}Sr radioactive source, with the results used to characterize the light output of each batch of 20 tiles.

The surface of each tile is quite sensitive to scratching and crazing, so it was necessary to protect each tile. Welded Tyvek® sleeves were developed, into which each tile was inserted after fabrication. The reflectivity of Tyvek results in an increase in the tile light yield of approximately 20%. The light yield is larger near the edges of the tiles (since this is closer to the fibers), increasing the nonuniformity of the tile response to a level greater than that required for physics performance. Thus, masking for the Tyvek sleeves was introduced. The masks are printed on the Tyvek material prior to the production of the sleeves and reduce any nonuniformity of response across the tile to less than 5%.

The blue scintillation light from the tiles is collected by wavelength-shifting (WLS) fibers coupled to the ϕ edges of the tiles, as shown in Fig. 3. The mechanical and optical requirements are satisfied by the use of 1-mm-diameter, double-clad polystyrene optical fibers, produced by Kuraray [Y11(200)MS]. The double layer of cladding not only increases the light yield of the fibers but also adds to their durability. A fast (< 10 ns) WLS dye in the fiber shifts the scintillator light to wavelengths of ~ 490 nm, providing a long attenuation length for transmission through the fibers, as well as matching to the response of the photocathodes in the PMTs. In order to improve the light yield and uniformity, the ends of the WLS fibers are polished and aluminized.

The fibers sit inside grooves between the steel plates, as shown in Fig. 1(b). Each side of a TileCal module is read out by separate fibers leading to separate PMTs, thus resulting in a redundancy in the case of PMT/electronics failure for a particular channel. To hold the fibers in the groove and against the tile, a special reflective plastic profile was designed which could be easily pressed into the groove between the steel plates, as shown in Fig. 1(b). The fibers directly couple to the edges of the tiles which they are reading out, and are shielded inside the profile when traversing tiles they are not reading out. Only the portion of the fiber directly coupled to the tile needs to be wavelength-shifting and so consideration was given to splicing the WLS portion of the fiber to a clear fiber which would then be routed to the PMT. This would improve the light yield (clear fiber has a larger attenuation length), as well as reducing Cerenkov and scintillation light produced in the fibers themselves. This is particularly a problem for muons traversing the ϕ crack between modules, creating the potential for large nonuniformities. This problem is reduced by the addition of a small amount of UV absorber to the fibers, and it was decided not to pursue the option of splicing. Fibers and profiles arrived at the construction site as a unit premade by a robotic system for fiber insertion into the profiles.

There is a great deal of flexibility in the manner in which the WLS fibers could be routed to the PMTs. There were a number of considerations that led to the optimal solution:

- The nonuniformities within the cell should be minimized;
- The light output should be maximized;
- The fiber length should be minimized;
- The nonuniformities among cells should be small;
- The number of different fiber lengths should be kept as small as possible;
- Sharp bends (with the radius less than 5 cm) should be avoided in the fiber routing.

A picture of the fiber routing in progress for one of the extended barrel modules is shown in Fig. 4(b). Above the module fibers is a template used for guidance. Tile–fiber pairs with a response less than 75% of the average response for the tile row containing the cell being tested were repaired.

The average light output of a tile is inversely proportional to the area of the tile and linearly proportional to the tile–fiber coupling length. As the PMTs are located at the outer radius of each module, the tiles with the largest area have the shortest fiber path, leading to a better equalization of the light yields from the 11 layers of tiles. The fibers for each cell on each side are combined into a bundle and routed to the respective PMT positions. This position is as close to the geometric center of the cell as possible, in order to equalize the needed fiber lengths. Each bundle is glued and then the ends are polished after the glue hardens. A special cutting/polishing machine was developed that traversed the inside of the girder. The total number of fibers needed to read out the whole calorimeter is 640,000, amounting to a total length greater than 1000 km.

A pseudoprojective readout in η is achieved by the appropriate grouping of WLS fibers to the PMTs. The fiber grouping defines a three-dimensional structure in such a way as to form three radial sampling depths (1.5, 4.1 and 1.8 λ thick at $\eta = 0$) composed of cells having dimensions $\Delta\eta \times \Delta\phi = 0.1 \times 0.1$ (0.2 × 0.1 in the outer layer).

Due to space constraints, the TileCal ITC consists only of scintillators at high η values (1.0–1.6). The gap scintillators cover the η region from 1.0 to 1.2, while the cryostat, or crack, scintillators (located between the central and endcap electromagnetic calorimeter cryostats) cover the η region from 1.2 to 1.6. The energy deposited in the gap scintillators is used to correct the hadronic energy response, while the cryostat scintillators are crucial for correcting the electromagnetic energy response. As the gap and cryostat scintillators consist of single pieces of scintillator, the highest light yield scintillator from Bicron (BC408) was purchased for their fabrication.

The highest expected radiation dose for TileCal is approximately 40 Gy/year, at η of 1.2. Irradiation tests indicated that radiation damage for the first longitudinal segment in TileCal, for an integrated dose corresponding to approximately 10 years of running at the design LHC luminosity, would result in a light loss of less than 10%, and thus future replacement of tiles or fibers would not be necessary. The cryostat scintillators, given their location at more forward rapidities and near the electromagnetic shower maximum, will accumulate approximately 1 kGy/year, which will lead to significant light loss, and thus have been designed for easy replacement/installation.

2.3. *Readout*

The TileCal front-end electronics design allows calibration and monitoring of the readout system to better than 1%.[15] It measures cell energies from ~ 30 MeV to ~ 2 TeV for the $\sim 10,000$ channels of the front-end readout. The precision and noise levels of the electronics system do not significantly degrade the resolution of the calorimeter energy measurement, which is set by the fluctuations in the physics of the hadronic showers.

2.3.1. *Mechanics and overview*

The TileCal front-end electronics resides in the backbone support girders of the calorimeter wedges (see Fig. 6). This position has the advantage of low radiation levels, a low magnetic field, and places the electronics behind the active elements of the calorimeter where the hadronic showers occur. The electronics is located in an element (drawer) that slides into the girder. Limited space and accessibility make compactness and reliability important design criteria.

The front-end electronics must process the signals generated from the PMTs. The size of these signals ranges from a single MIP (e.g. a muon) traversing the detector to a 2 TeV shower. The latter corresponds to a signal of upto 800 pC from each of the two PMTs associated with a calorimeter cell. The requirement set by the resolution for a MIP and by the dynamic range demands a 16-bit range digitization. This requirement is met by passing the PMT pulse through a shaping stage to remove any shape fluctuations present in the raw PMT signal, amplifying the pulse in two gain scales and then digitizing the resulting signals in 40 MHz 10-bit flash ADCs. Calibration and trigger functions must also be carried out.

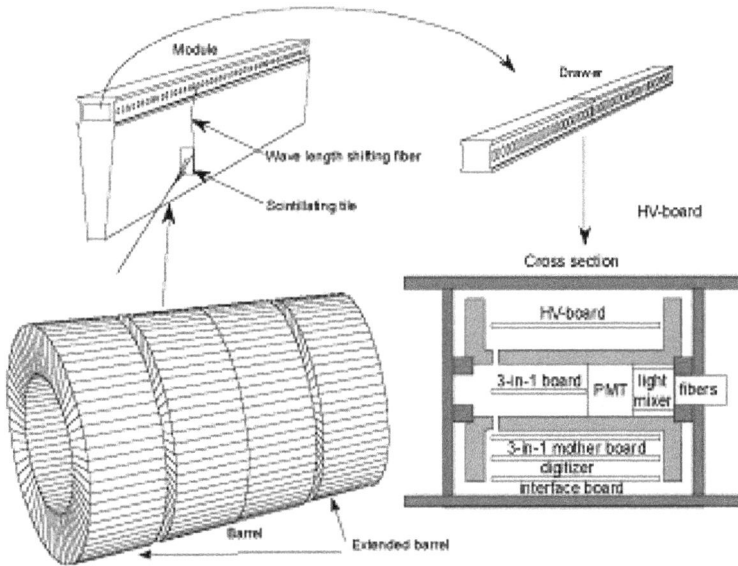

Fig. 6.   A schematic of TileCal, its wedge-shaped modules with removable drawers, along with a cross section of a drawer.

The front-end electronics consists of several subsystems:

(1) The 3-in-1 cards inside the PMT magnetic shielding can;
(2) The motherboard system, which controls up to 48 3-in-1 cards in the drawer and supports tower trigger sum cards and an integrator ADC card;
(3) Digitizer boards;
(4) The optical interface board.

2.3.2. *3-in-1 card*

Figure 7 is a functional diagram of the 3-in-1 card. Most analog functions of the front-end electronics are contained on a 7 cm by 4.7 cm printed circuit board, called the 3-in-1 card, located inside the steel shield of each PMT block.

The shaped PMT signals are produced in a passive LC shaping network. This network removes the pulse-to-pulse signal shape fluctuation in the raw PMT pulse, producing a standard signal shape for all channels of the Tile-Cal system. The extremely low noise of the passive LC shaping network is critical in maintaining the 16-bit dynamic range requirement. The shaped signal is passed to two operational amplifiers, which produces two signals for the high and the low gain range. The amplifiers have a gain ratio of 64 and a

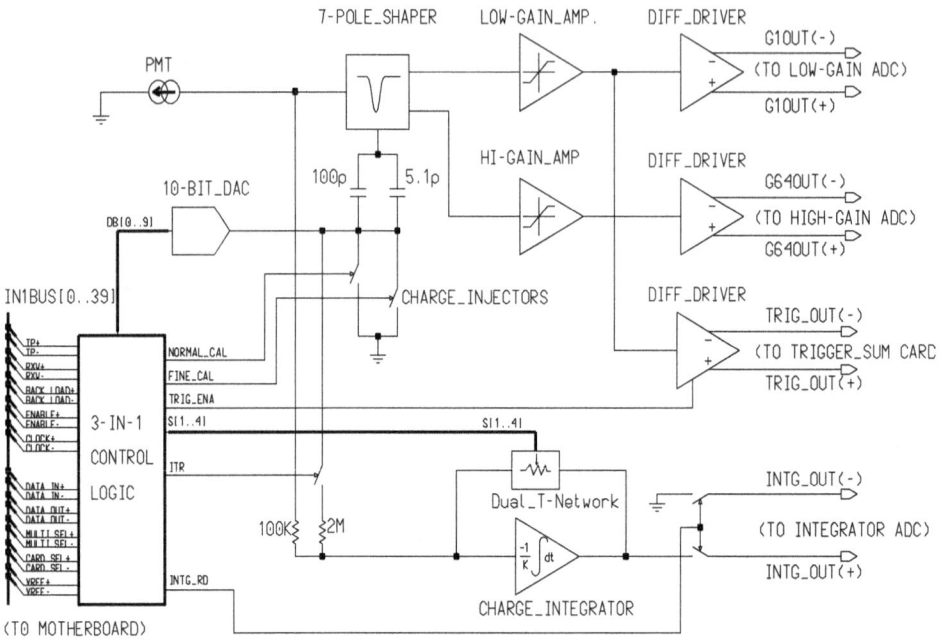

Fig. 7.   A block diagram of the TileCal 3-in-1 card.

clamping design. Clamping amplifiers provide fast recovery from saturation, which is particularly important for the high gain amplifier. The amplified signals are then passed to differential drivers, which send the signals to the digitizer boards.

In addition to the high and low gain outputs sent to the digitizer boards, a differential fast trigger signal, derived from the low gain output, is sent to the trigger sum boards mounted on the motherboards in order to produce trigger tower sums. The fine timing requirements for the analog sums are satisfied by using appropriate cable lengths. The outputs of the tower sum boards are sent to the level 1 trigger system.

A charge injection system is designed to calibrate the whole electronic channel. Two capacitors (5.1 pF and 100 pF) are connected through fast switches to the shaping network. When the switch is closed, a fast pulse is sent through the shaping network. The 100 pF capacitor can calibrate the system over the full 800 pC range, while the small 5.1 pF capacitor calibrates up to about 40 pC, giving a finer scale for calibrating the high gain channel.

The PMT signal is also sent to the slow integrator amplifier. This amplifier averages over a time period of 10 ms the DC level of the PMT signal. This signal is multiplexed onto a bus on the motherboard and sent to the integrator ADC card, which contains a microprocessor, ADC, and a CAN-bus interface. It digitizes the integrator amplifier signal level and transmits the data to the counting room via CANbus. Such signals are used in the radioactive source calibration of the PMT signals discussed in Subsec. 2.4.2.

The charge injection must be fast, because its phase relative to the digitization clock must be maintained to control the time of digitization relative to the peak amplitude.

2.3.3. *Motherboard control system*

The 3-in-1 cards hold control settings in a small EPLD. These control the various functions of the 3-in-1 cards: the DAC setting for charge injection and integrator calibration, switch settings for the integrator gain, selection of the charge injection capacitor, control of the connection to the charge integrator to the output analog bus, and the enable/disable of the trigger output. The trigger, timing and control (TTC) system is the primary means of controlling the drawers. It is used extensively in control of the LHC machine and experiments.[16] Commands are received via optical cable and decoded on the first motherboard. They are then sent on a serial differential bus to 3-in-1 cards. They can be sent to individual 3-in-1 cards or in parallel to any subset of cards. One exception is the charge injection command, which must be timed precisely with the digitization and bunch crossing clocks. This command is sent directly to the 3-in-1 cards. A CANbus control system is also connected to the drawer. Its primary function is to control the radioactive source calibration of the detector. It controls the cesium source position and reads out the integrators. It can also control all functions of the 3-in-1 cards except for charge injection, which is synchronized with the LHC clock. Unlike the TTC system, the CANbus system is bidirectional and also allows readback of 3-in-1 card configuration settings. This is very useful for monitoring the status of the readout system.

A more detailed description of the 3-in-1 card and motherboard system can be found in Ref. 17.

2.3.4. *Digitizers*

A functional diagram of the digitization system is shown in Fig. 8. The digitizer boards (eight per drawer) each receive the high and low gain differential signals from six 3-in-1 cards. These signals are digitized every 25 ns by 10-bit ADCs, using a TTC system clock that can be adjusted in units of 106 ps. In this way we can phase correctly the signal to time the ADC sampling time.

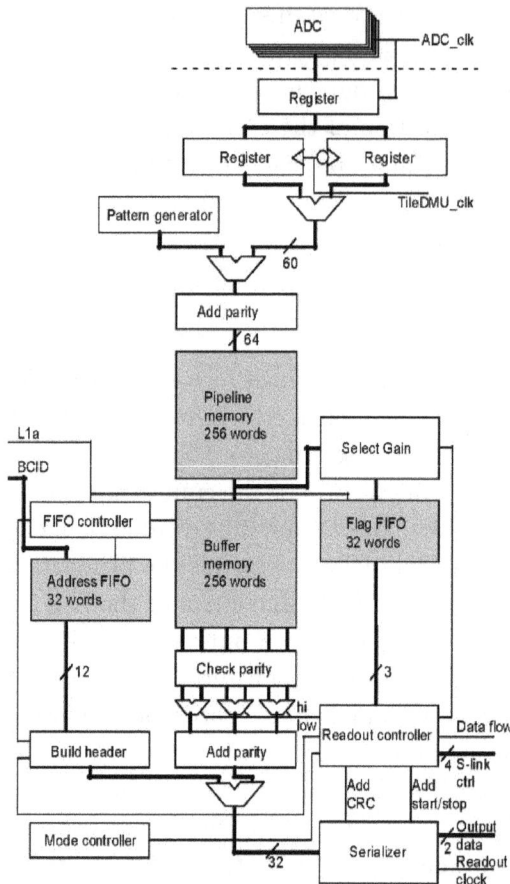

Fig. 8.   A simplified block diagram of the TileCal digitization system.

The TTC system clocks provide precise timing relative to the accelerator time structure. Each ADC has a programmable DC offset to prevent the ADC from receiving negative signals.

The digitized data words are processed by a so-called TileDMU custom ASIC chip, which temporarily stores the data in pipeline memories. The pipeline length is programmable up to 256 samples, giving a pipeline latency of up to $6.4\,\mu s$, which is considerably longer than the $2.5\,\mu s$ ATLAS requirement. The latency is needed to let the level 1 trigger system make a decision and to return a level 1 accept (L1A) trigger signal to the digitizer system.

At reception of an L1A signal, the TileDMU captures an event frame and sends it to an output storage memory. The position of the event frame along the pipeline and the number of data samples stored are programmable.

Typically, when running at an L1A trigger rate of 100 kHz seven samples are recorded. The captured event is all at either high or low gain, depending on the value of the maximum sample in the event frame. If the maximum sample is below a programmable maximum value in the high gain data, high gain samples are selected. Otherwise, low gain samples are sent to memory. Overlapping event frames are permitted and L1A trigger signals can be processed with a three-bunch-crossing (75 ns) separation.

With each event frame data block, a header word is stored in memory. This header word contains the bunch crossing identification number of the event, various diagnostic flags (e.g. parity words) and the gain scale flag bit.

The output memory can buffer 36 events at 7 samples. A control process monitors the output memory and sends events serially to the optical interface card along with control signals and the data clock.

A more detailed description of the digitization system can be found elsewhere.[18]

2.3.5. *Optical interface card*

The optical interface card receives the serial data streams from the 16 TileDMU chips in the drawer. It realigns the data to a common clock and packs it into 32 bit words and transmits data over optical links to the readout driver (ROD) crates in the counting room. Cyclic redundancy checks are performed on the data link between the digitizers and the interface card and between the interface cards and the RODs. The interface card memory holds 16 events at 7 samples.

2.4. *Monitoring system*

TileCal is complemented by four systems to monitor, test and calibrate the calorimeter readout at various stages. The use of four systems provides a reasonable redundancy to both control and monitor the performances and the overall calibration. Moreover, the calibration systems have been heavily used during the instrumentation and commissioning of TileCal, especially to check and eventually repair any defective channels.

2.4.1. *Charge injection system*

The linearity and stability of the front-end electronics, and the stability of the two gain circuits, are monitored by injecting a precise charge at the input stage of each electronic channel. With this method we can precisely calibrate the readout of each ADC channel from ADC counts to picocoulombs.

2.4.2. Radioactive ^{137}Cs system

The final calibration of the calorimeter at the electromagnetic scale can only be performed using electron beams of known energy. Given the nature of the detector, only the outermost cells can be directly calibrated with electrons. The calibration is transported to the inner cells with high precision using the Cs system. In this system the TileCal optics is monitored by exciting each scintillating tile with a moving radioactive source. Through this technique, we can monitor and equalize the scintillator light yield, the scintillator coupling to the WLS fibers and the integrity of these fibers to transport light to the PMT. TileCal is equipped with three calibrated ^{137}Cs sources — one for the barrel and one for each of the extended barrels. The sources travel in tubes passing through each scintillator tile, their motion controlled by a hydraulic fluid. The DC current from each PMT is measured by the integrator system described above. In this manner, an "x-ray picture" of the detector is obtained.[19, 20] Equalization is obtained by adjusting the voltage of each PMT. The resulting intermodule calibration is consistent within a few per mil. Figure 9 shows the working principle of the system and an example of the analysis of Cs data taken during the calorimeter commissioning. The accuracy in measuring the single tile response is about 2%.

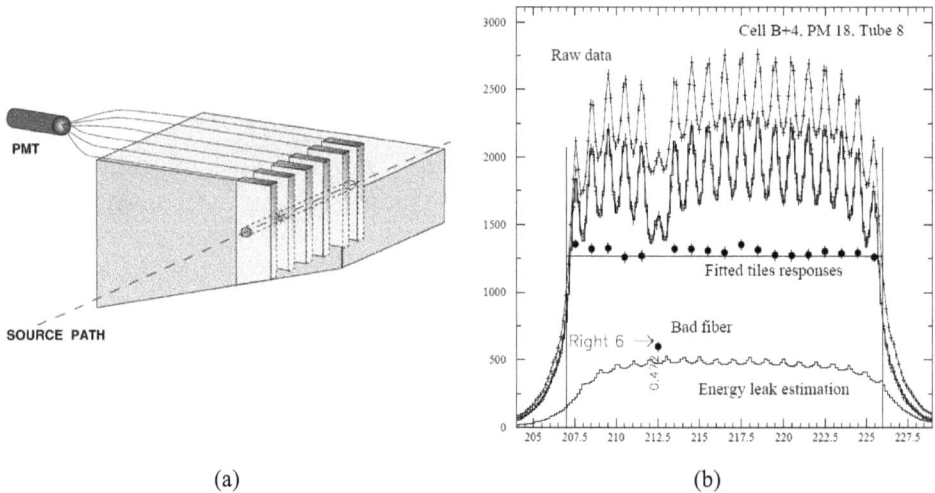

(a)   (b)

Fig. 9.   (a) The cesium source monitoring concept. (b) An example of calibration data analysis taken during TileCal commissioning. One tile-to-fiber coupling was found to be defective and was afterward fixed.

2.4.3. *Laser system*

The PMT performances and the fast readout chain are monitored with fast laser light pulses transported to the photocathode of each PMT by dedicated clear, plastic optical fibers. The shape of the laser signal is very similar to that of the scintillating light and its amplitude covers the whole dynamic range up to the saturation level of the PMTs. The PMT gain, stability and their linearity can be monitored at a level better than 1%.

2.4.4. *Minimum bias current system*

The slow integrator system will monitor, during LHC stable beam operation, the current in each PMT due to minimum bias events. The integration time is set to 10 ms with different amplifications to cope with the large dynamic range needed.

3. Performance of the Tile Calorimeter

The main requirements that drove the TileCal design were performance in energy measurement and practical considerations such as cost and the need for easy construction. While the latter two issues have been addressed in the previous sections, in this section the performance requirements are considered in more detail.

The key variables needed to evaluate the calorimeter performance are energy linearity and resolution. TileCal is required to have a maximum deviation from linearity of a few percent for energy deposits ranging from a few hundreds of MeV, corresponding to muons passing through the calorimeter, up to a few TeV. The linearity requirement has an impact on basically all jet measurements and it is particularly important for the sensitivity to new physics. As an example, a deviation of a few percent from linearity would mask a signal of compositeness at the scale of 20–30 TeV.[11]

The energy resolution is required to be of the order of $50\%\sqrt{(E(\text{GeV}))} \oplus 3\%$. This would give a mass resolution better than 10% for decays $W \rightarrow jj$. Clearly, the quoted global performances on the jet energy measurement are defined by the characteristics of both the electromagnetic and hadronic sections of the calorimeter.

TileCal is based on a semiprojective geometry where each tower is segmented in three longitudinal layers. While the transverse cell dimension is determined to be of the order of the hadronic shower lateral spread ($\lambda \simeq 23$ cm), the longitudinal cell dimension varies layer by layer. Various choices have been considered for the longitudinal dimension of the layers; the final design consists of two relatively thin layers interleaved by a thick central layer. The width of the first longitudinal layer is chosen to be sensitive to the energy losses in the passive material lying between the electromagnetic and the hadronic calorimeter sections. In fact the correlation between this energy deposit and the one in the last electromagnetic calorimeter layer is used to correct for the energy losses in the passive material. The width of the central layer is large enough to sample the bulk of the shower, while the last layer acts as a kind of tail catcher.

The length of each longitudinal layer is large enough to provide an integrated muon signal that could be well separated from noise, and used to form an alternative source for a muon trigger. This could be particularly useful for the low energy muons which do not reach the outer layers of the muon spectrometer.

At $\eta = 0$ the total length of the instrumented region is about 8.6 λ, including the 1.2 λ of the three electromagnetic sections and the 1.5, 4.1 and 1.8 λ of the three TileCal sections. The total material from the interaction point to the outside of the calorimeter is, however, 11 λ.

The longitudinal segmentation of the LAr and TileCal sections is very important for obtaining information on the shower shape and therefore on the shower composition. In fact, since both calorimeter sections are noncompensating, the shower shape information is used to apply software techniques to correct the calorimeter response. This procedure is essential in order to be able to meet both the linearity and resolution requirements for the energy measurement of hadronic showers.

3.1. *Measuring the performance with test beams*

The first TileCal R&D program dates back to 1991, and the final module construction started in 1998. The modular construction has allowed, during the construction years, the carrying out of various tests in beams on a total of about 12% of the calorimeter modules. The primary aim of these tests was to calibrate the energy scale of the calorimeter.[21] These measurements have also been very important for improving the GEANT4[22] description of hadronic showers in TileCal.

During these tests, the modules were placed on a moveable table, which allowed the beam to be directed to any point of the outer perimeter. The cell-by-cell response was then equalized using the cesium source, resulting in a uniformity of the order of a few per mil. During the equalization procedure, the PMT voltages are adjusted for cell response of about 1.1 pC/GeV for electrons. The value chosen for the signal equalization is determined by the need to measure, with the highest precision, both the low and the high end of the dynamic range.

The precise measurement of the electromagnetic calibration constant is then obtained using electron beams of momenta ranging between 20 and 180 GeV/c, directed at 20° on all the cells of the first calorimeter layer (A cells). The energy deposit is almost completely contained in a single cell. The cell-by-cell calibration uniformity, evaluated by the RMS of the calibration constants of all the calibrated cells, is $2.4 \pm 0.1\%$ [Fig. 10(a)]. This residual spread is due to local variation in individual tile and tile/fiber responses, as confirmed by Monte Carlo studies.[23]

The electron beam used to calibrate the outer cells of the modules does not give any information on the inner cells. The uniformity of the response of the inner cells is therefore verified with muon beams. Figure 10(b) shows the mean response to 180 GeV/c muons, normalized to the path length, as a function of the pseudorapidity. The calibration uniformity over the whole η range (averaged on different longitudinal layers) is less than 1%.

The overall performance of the calorimeter is measured with pions. During the test beam studies, a large data sample of pions with momenta ranging from 20 to 180 Gev/c has been acquired. Figure 11 shows the energy

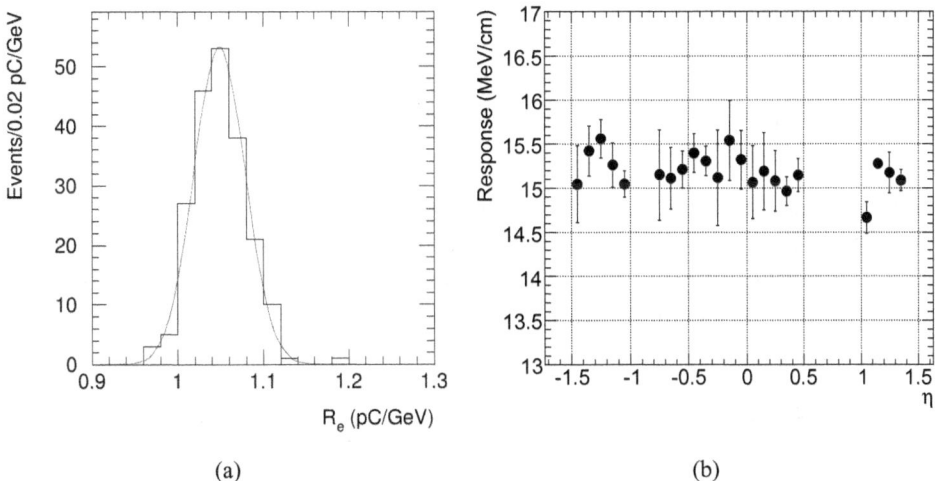

(a)   (b)

Fig. 10.   (a) The cell response of electrons entering the calorimeter modules exposed to the beam at an incidence angle of 20°, normalized to beam energy, with one entry for each A cell measured. The plot contains data at various energies ranging from 20 to 180 GeV. The mean value, 1.050 ± 0.003 pC/GeV, defines the TileCal electromagnetic scale factor. The RMS spread is $2.4\pm0.1\%$. (b) The total calorimeter response to 180 GeV muons, normalized to the path length, as a function of pseudorapidity.

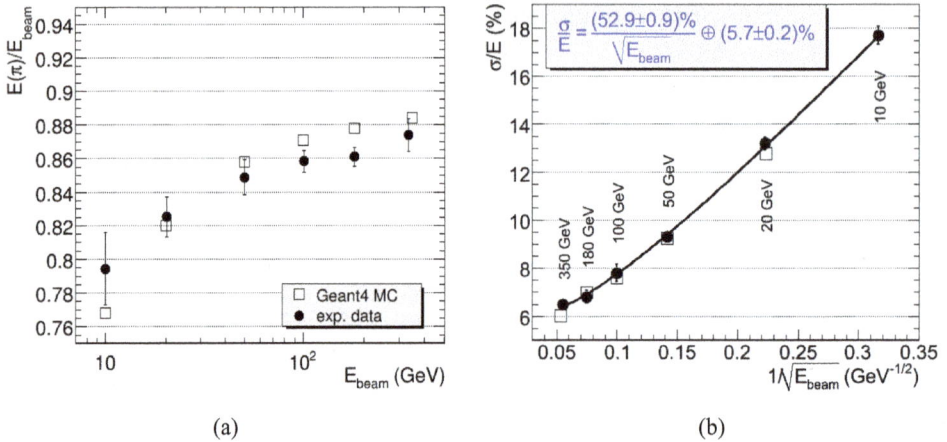

(a)

(b)

Fig. 11. Energy linearity (a) and resolution (b) at the electromagnetic scale, for pions impinging on the calorimeter at $\eta = 0.35$. Both plots show experimental data (filled circles) and Geant4-simulated data (squares).

linearity and resolution, at the electromagnetic scale, for pions impinging on the calorimeter at $\eta = 0.35$ for data (full circles) and Geant4[a] simulation (open squares). Since the plots are shown at the electromagnetic scale the noncompensation effects are clearly visible on the linearity plot. The Geant4 simulation describes both the linearity and the resolution within a few percent over the whole energy range.

3.2. *Experience with the installed detector: cosmic muons and single beam events*

The assembly in the cavern was completed in December 2005 and TileCal was the first ATLAS subdetector to provide cosmic muon data. The installation of all the other subdetectors was completed later and the whole ATLAS detector has been available for cosmic data taking for some time. An example of the events acquired with cosmic muons is given in the event display shown in Fig. 12.

Once the detector installation phase was completed, the work concentrated on improving the stability of the whole system and on developing a data quality framework that enabled one to quickly identify any problematic channels. Such a system should be automated as much as possible, in order to allow quick diagnostics on 10,000 channels. Data generated with random triggers, with the integrated TileCal control/calibration systems and with

---

[a]The Geant4 version is 4.8.3 and the hadronic list is the QGSP Bertini model.

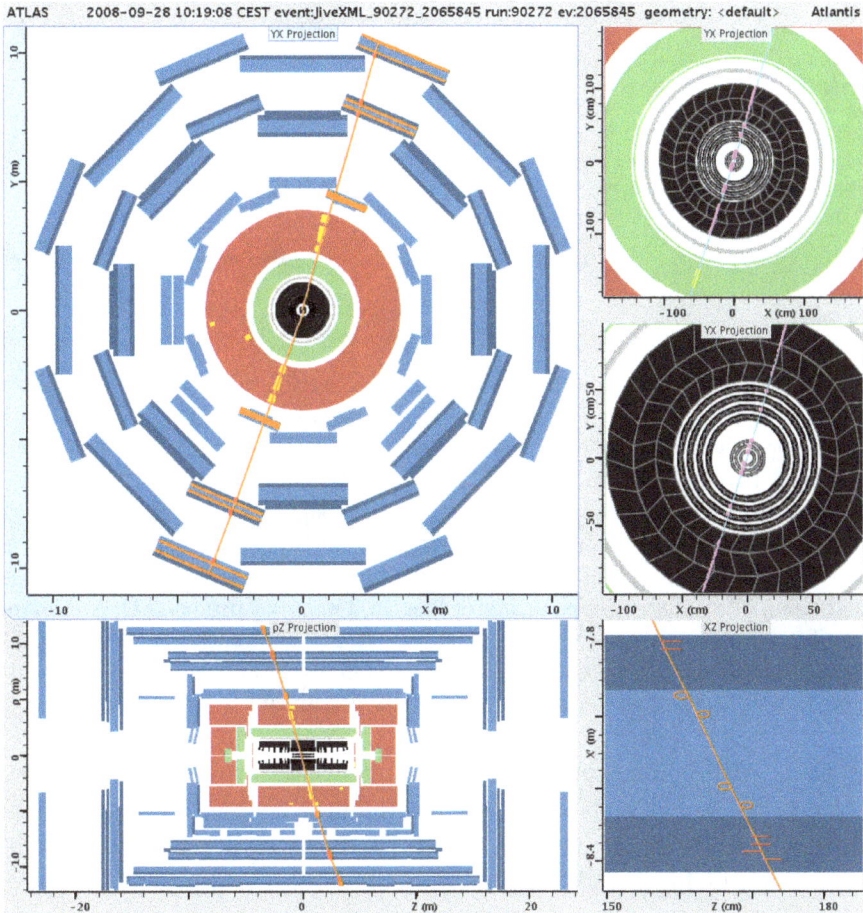

Fig. 12.   An event display of the signal produced by a cosmic muon crossing the central region of the ATLAS detector. Both the solenoid and toroid fields were on during this run.

cosmic muons, have been used both for diagnostic and for stability studies. As an example, the noise stability evaluated on randomly triggered events as a function of time for a period of three months is shown in Fig. 13(a). The electronic noise is evaluated by the RMS of the digital samples and is averaged over the whole calorimeter. The relative variation in time is within 1% (represented by the green dotted lines) and the average value is 1.44 ADC counts. The time dependence of the electronic noise for a single typical channel is also shown.

On September 10, 2008, the LHC began operation and single proton beams were successfully circulated in both directions. The interaction of

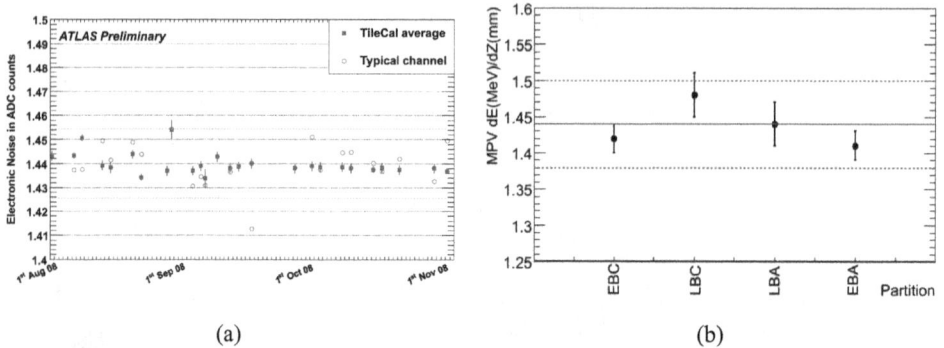

(a)

(b)

Fig. 13. (a) Time dependence of the electronic noise stability evaluated, on randomly triggered events, by the RMS of the digital samples averaged over the whole calorimeter. The green dotted lines represent the ±1% variation from the average values. (b) The most probable value (MPV) of dE/dx signals recorded by TileCal with horizontal muons from single beam data on September 10, 2008. The average response over all cells within a given partition to horizontal muons is shown for each partition. About 500 muons were selected by requiring a consistency to the expected signal along 12 m of the tile calorimeter length. This data provided the opportunity to verify the intercalibration of the TileCal cylinders, already calibrated with the cesium gamma sources, down to the 4% precision level. The red lines represent the average MPV of the four barrels and its 4% uncertainty.

this beam with collimators, lying 148 m upstream of the ATLAS detector, produced the so-called "splash" events. In these events, the many particles produced in the beam interaction with the collimator reached the ATLAS detector, producing signals in all the subdetectors. These events allowed us to carry out the first studies of TileCal performance with beam-generated particles.[24] The muons that traverse TileCal horizontally provide useful information on intercalibration of the three barrels. Figure 13(b) shows the most probable value of dE/dx signals recorded by TileCal when traversed by horizontal muons. Each point in the plot shows the average value calculated over all cells within a given barrel (the central barrel is divided into two parts). The barrel-to-barrel intercalibration is within 4%.

References

1. O. Gildemeister, F. Nessi-Tedaldi and M. Nessi. An economic concept for a barrel hadron calorimeter with iron scintillator sampling and WLS-fiber readout. Prepared for 2nd Int. Conf. Calorimetry in High-Energy Physics (Capri, Italy, 14–18 Oct 1991).
2. M. Bosman *et al.* Developments for a scintillator tile sampling hadron calorimeter with longitudinal tile configuration: R & D proposal. CERN-DRDC-93-3.

3. F. Ariztizabal *et al.* Construction and performance of an iron scintillator hadron calorimeter with longitudinal tile configuration, *Nucl. Instrum. Methods A* **349** (1994) 384–397. doi: 10.1016/0168-9002(94)91201-7.

4. P. Jenni, P. Sonderegger, H. P. Paar and R. Wigmans. The high resolution spaghetti hadron calorimeter: proposal. NIKHEF-H/87-7.

5. V. Semenov. *Proc. IX Conf. Scintillators* (Kharkov, 1986), p. 86.

6. M. Kadykov. Preprint JINR 13-90-10. (Dubna, 1990).

7. J. Abdallah *et al.* The production and qualification of scintillator tiles for the ATLAS hadronic calorimeter. Technical Report ATL-TILECAL-PUB-2007-010. ATL-COM-TILECAL-2007-026, CERN, Geneva (Dec. 2007). Note written by IHEP Group, Protvino.

8. M. David, A. Gomes, A. Maio, J. Pina and B. Tom. 15 years of experience with quality control of WLS fibers for the ATLAS tile calorimeter. Technical Report ATL-TILECAL-PUB-2008-003. ATL-COM-TILECAL-2007-022, CERN, Geneva (Dec. 2007).

9. M. David, A. Gomes and A. Maio. Radiation hardness of WLS fibers for the ATLAS tile calorimeter. Technical Report ATL-TILECAL-PUB-2008-002. ATL-COM-TILECAL-2007-021, CERN, Geneva (Dec. 2007).

10. F. Bosi, S. Burdin, V. Cavasinni, D. Costanzo, T. D. Prete, V. Flaminio, E. Mazzoni, C. Roda, G. Usai and A. Vasiljev. A device to characterize optical fibers, *Nucl. Instrum. Methods Phys. Res. Sec. A* **485**(3) (2002) 311–321. doi: 10.1016/S0168-9002(01)02067-8. http://www.sciencedirect.com/science/article/B6TJM-44CXYSS-9/2/9d17476b%1b7b60a95c21b163c2eee68a.

11. ATLAS Tile Calorimeter: Technical Design Report. CERN-LHCC-96-42.

12. ATLAS: Detector and Physics Performance Technical Design Report. Vol. 1. CERN-LHCC-99-14.

13. ATLAS: Detector and Physics Performance Technical Design Report. Vol. 2. CERN-LHCC-99-15.

14. B. A. Alikov *et al.* ATLAS barrel hadron calorimeter: general manufacturing concepts for 300,000 absorber plates mass production. JINR-E13-98-135.

15. K. Anderson *et al.* ATLAS tile calorimeter electronics. http://hep.uchicago.edu/atlas/tilecal.

16. M. Ashton *et al.* Timing, trigger and control systems for LHC detectors. CERN-LHCC-2000-002.

17. K. Anderson *et al.* Design of the front-end analog electronics for the ATLAS tile calorimeter, *Nucl. Instrum. Methods A* **551** (2005) 469–476. doi: 10.1016/j.nima.2005.06.048.

18. S. Berglund *et al.* The ATLAS tile calorimeter digitizer, *JINST* **3** (2008) P01004. doi: 10.1088/1748-0221/3/01/P01004.

19. E. Starchenko *et al.* Cesium monitoring system for ATLAS tile hadron calorimeter, *Nucl. Instrum. Methods A* **494** (2002) 381–384. doi: 10.1016/S0168-9002(02)01507-3.

20. N. Shalanda *et al.* Radioactive source control and electronics for the ATLAS tile calorimeter cesium calibration system, *Nucl. Instrum. Methods A* **508** (2003) 276–286. doi: 10.1016/S0168-9002(03)01700-5.

21. K. J. Anderson *et al.* Calibration of ATLAS tile calorimeter at the electromagnetic scale (2009). ATL-TILECAL-PUB-2009-002.

22. J. Allison *et al.* Geant4 developments and applications, *IEEE Trans. Nucl. Sci.* **53** (2006) 270. doi: 10.1109/TNS.2006.869826.

23. T. Davidek *et al.* Testbeam studies of production modules of the ATLAS tile calorimeter (2009). To be published in *Nucl. Instrum. Methods A*.

24. H. Okawa. Commissioning of the ATLAS tile calorimeter with cosmic ray and single beam data. ATL-TILECAL-PROC-2008-002.

Chapter 10

INNOVATIONS FOR THE CMS HCAL

J. Freeman

Fermilab, MS 205
Batavia, IL 60510-500, USA
freeman@fnal.ger

The CMS hadron calorimeter (HCAL) is a sampling calorimeter with brass absorber plates and sheets of a scintillator as the active medium.[1] The scintillator is divided into tiles that are optically summed into projective towers. The overall concept of CMS is a large high field magnetic volume instrumented by all-silicon tracking. A very high resolution crystal electromagnetic calorimeter follows the tracker. Finally, the HCAL is stationed at the outer radius of the magnetic volume, just inside the superconducting solenoidal magnet.

One of the tasks before us while designing the CMS HCAL was selection of the front-end signal preparation. The central part of the HCAL uses a photon detector and digitizing electronics. The choice/design of the photon detector and front-end electronics are very tightly tied together. The front-end electronics has to accommodate the sensitivity, capacitance, shaping, and other properties of the photon detector. In the development of the HCAL these two tasks of developing the front-end electronics and the photon detector were done in parallel.

The CMS HCAL consists of four regions. The barrel and endcap HCALs, HB and HE, use a scintillator as the active medium and are located in the central detector. The very forward calorimeter, HF, based on Cerenkov radiating quartz fiber, is located in the forward region outside of the magnetic field volume. The central HCAL sits inside the CMS solenoidal superconducting magnet. The final region — the outer calorimeter HO — sits outside the central magnet and, like the central calorimeters it has a scintillator as the active medium. Figure 1 shows the relative placement of the HB, HE, and HO.

The choice of placement of the HCAL inside the solenoid increases the bending path for the muon system. Additionally, placement of the HCAL

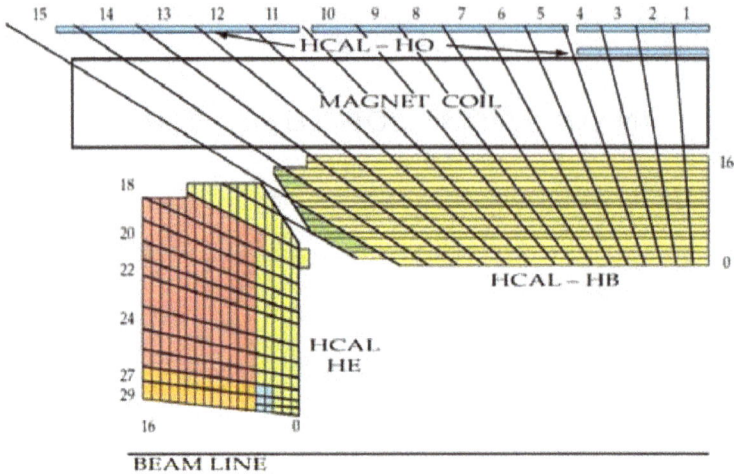

Fig. 1.   The relative placement of the central HCAL relative to the solenoid.

immediately adjacent to the electromagnetic calorimeter (as compared to being on the outside of the magnet) allowed for a very conservative robust magnet design. (The solenoid is about one hadron absorption length thick at 90°.)

The CMS central HCAL is a scintillator-based sampling calorimeter. It has thin layers of a scintillator interleaved between brass absorber plates. To maximize the absorber thickness in the small available space (about 1 m radially), the brass plates are relatively thick ($\sim$5.5 cm) and the scintillator is relatively thin (3.8 mm).

Figure 2 shows the layout of the optical design of the calorimeter. Light from the scintillator layers is carried to photodetectors at the back of the calorimeter.

The "art" of developing a detector concept for a new energy regime requires both physics judgment and guesswork as to what the interesting topics will be. Additionally, in the environment of more than one similar detector (as in the case of ATLAS and CMS) there is a wish for the detector to be complementary to the other and have unique abilities. We have seen this in the case of CMS, where electromagnetic calorimeter resolution and muon momentum resolution were emphasized. This concept directed us toward a very large high field magnet, and precise electromagnetic calorimetry. The effect of these choices on the HCAL was that it needed to be placed inside the large, thick solenoid. This caused the HCAL design to focus on high density (and a low active sampling fraction), to be nonmagnetic, and to operate in the 4 tesla field.

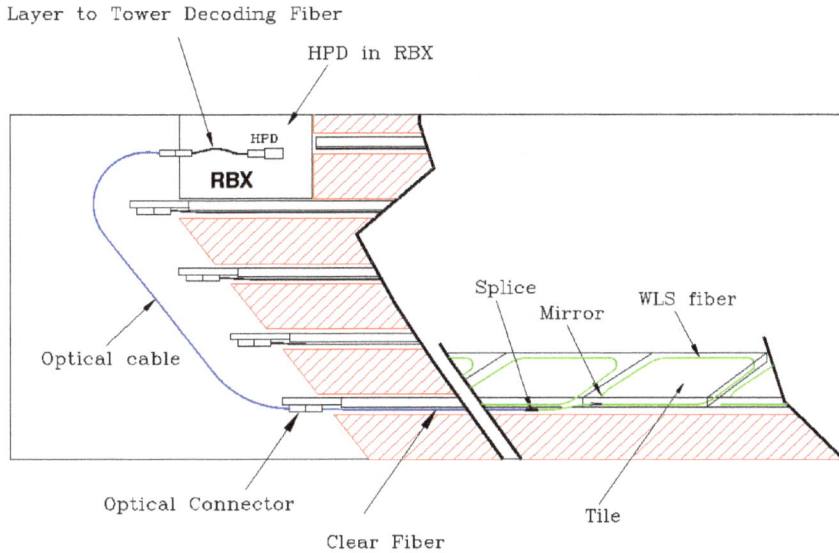

Fig. 2.   A schematic view of the CMS HCAL detector. Scintillating tiles are read out with wavelength shifting fibers, which are then coupled to clear fibers. The clear fibers carry the light to the outer radius of the HCAL where photodetectors and front-end electronics are located.

A unique feature of CMS is its moving-ring-based structure, allowing for very good access and maintenance of the detector elements. This design feature had the tradeoff for the HCAL in that the readout system (front-end electronics system) had to be placed inside the magnetic volume. Any alternative location for the photodetectors was so far away that there would have been prohibitively large light loss in the long clear optical fibers.

An investigation of possible thin active media that would work in a magnetic field and survive the radiation dose and particle fluxes of the LHC led us to the choice of layers of a scintillator. To read out the optical signal we searched for photodetectors that can operate in a 4 T field. Figure 1 shows the placement of the central barrel and endcap HCALs inside the solenoid magnet. It is interesting to note the relative thickness of the calorimeters compared to the magnet. Tower numbers are indicated. The photodetectors and front-end electronics were placed in small notches at the outer radii at eta tower 14 (HB) and tower 18 (HE).

In the mid-1990s, when we made our search for suitable photodetectors, there were a number of options: photomultipliers, photodiodes, APDs, microchannel plates, and various types of hybrid photodiodes (HPDs).

Our desired parameters for the device were:

(1) The device had to work in the 4 T magnetic field.
(2) Our anticipated light yield was 100–300 photons per GeV. We needed a device that had an acceptable quantum efficiency (at least 10%) in the 500 nm region, where our light signal would be.
(3) The front-end electronics noise was anticipated to be around 5000 electrons/25 ns sample. We did not want the electronics noise to profoundly degrade the calorimeter resolution, so this placed limits on the minimum acceptable amplification gain of the photodetector.
(4) The spread in variation of amplification of photoelectrons (excess noise factor) should be small enough that it did not seriously affect the calorimeter resolution.
(5) The radiation level was anticipated to be ~ 1 Krad and the device had to withstand this.
(6) The device had to have an operating lifetime of 10 years or more.
(7) The device had to be reasonably compact so as to fit in the region allowed for electronics. To fit our desired calorimeter segmentation we wanted the size of the photodetector per channel to be of order 20 cc or less.
(8) The cross-sectional area of the optical fibers that would constitute a calorimeter segment was a circle with a diameter of about 5 mm, so the photodetector had to accommodate this.
(9) A final requirement was that the photodetector should be sensitive to small DC light levels. One of the methods of calibrating the HCAL was to place a small radioactive source by the scintillators. The source generated a DC stream of photons in the scintillator. It was important for the photodetector to be able to measure this current. Accuracy of measuring the current corresponded to accuracy of calibration, and hence a-few-percent accuracy was required. The anticipated light level was about 3×10^7 photons per second at the photodetector.

Let us discuss a scenario with our parameters to understand their role in the choice of the photodetector. We consider a photomultiplier as the baseline. The ability of a photomultiplier to operate in a 4 T field is very dubious, to say the least. We will return to this point later. A photomultiplier can easily have a quantum efficiency of 10% in the green region, so parameter 2 is acceptable. This would then give us 10 photoelectrons (pe) per GeV of the signal. A typical photomultiplier gain can easily be 50 K, so a single pe signal would be 50 K electrons. The amplifier noise would then contribute at the level of 0.1 pe. With 10 pe/GeV, 0.1 pe is 10 MeV, which would be a completely acceptable noise level. Our design goal was less than 100 MeV.

The excess noise factor as a function of amplification M is defined as $\text{ENF}(M) = 1 + \sigma(M)^2/M^2$. A perfect amplifier with no noise would have $\sigma(M) = 0$, and the excess noise would be 1.0. If this were the case for a photomultiplier, then the single pe would appear as a delta function at 50 K electrons (gain $= 50$ K). This is in fact not the case for photomultipliers, which typically have an ENF of 1.3. This means that the single pe peak would have a sigma of about 30% or about 15 K electrons (at a central value of 50 K electrons). Since the pedestal sigma of our front-end electronics (centered at 0 electrons) is 5 K electrons wide, the single pe would be very well separated from the pedestal and would be clearly visible. The ability to cleanly see the single pe peak is useful for monitoring gain and performance, and hence would be a very desirable condition.

A typical small photomultiplier tube can have a size of 1 cm diameter and 10 cm length (including the base), for a volume of 8 cc, which fits our desired volume, and the desired active area for the fiber bundle. Most choices of phototube window will survive 1 Krad with no degradation.

So we conclude that a phototube would be a very satisfactory choice for our photodetector, with the proviso that it should work in a 4 T field. Unfortunately, phototubes lose their gain quickly in a magnetic field (due to inability to focus the electrons). Figure 3 shows the gain loss vs. magnetic

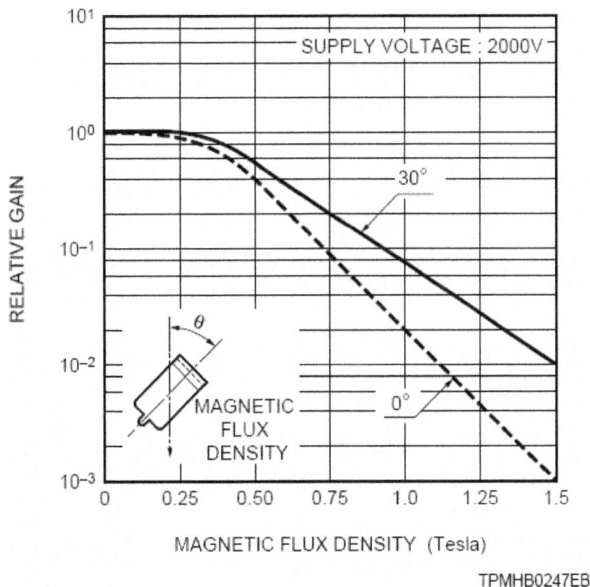

Fig. 3. Reduction in gain of a photomultiplier tube vs. magnetic field strength (Hamamatsu high magnetic field fine mesh phototube R6504).

field for a typical phototube. A survey of different structure phototubes yielded no candidate that could operate in fields much above 1 T.

Table 1 shows photodetectors available during the mid-1990s, with the exception of SIPMs, which are a recent development. The "Bias voltage" row indicates typical operating voltages. The Photon lifetime row displays the lifetime number of photons a typical device (under normal gain conditions) can survive. The "Expected electrons per GeV" row shows the number of electrons per GeV the HCAL would expect, folding in the device's gain and quantum efficiency.

Looking at Table 1, we can immediately eliminate photodiodes (PDs) from consideration. Although they have many desirable features, the unity gain implies that our electronics noise level of 5000 e^- would correspond to a 50 GeV signal.

Microchannel plates (MCPs) have a long history. Until recently they had a very limited charge lifetime. Additionally, until recently the large channel diameter limited their operational magnetic field limit to about 1–1.5 T. At the time of our decision (and even now) they were unsuitable due to the limited lifetime and magnetic field performance. (In the ensuing 10+ years since our evaluation of photodetectors, some devices, particularly MCPs, have improved their properties. Current MCPs can have a charge lifetime of up to 0.1 C and demonstrated operation in 2 T magnetic fields.)

As mentioned earlier, photomultipliers (PMTs) cannot operate in the 4 T magnetic field.

Silicon photomultipliers (SIPMs) are a comparatively recent development and have come into their own as realistic photodetectors only in the last few years. They did not exist in practical form in the mid-1990s. They are included in the table because they will have importance in future HCAL developments, as discussed below.

By elimination we were left with APDs and HPDs. Both of these were considered seriously. APDs have a marginal gain causing the lowest measurable signal to be of order 1 GeV. This would be an unacceptable feature if left alone. However, we speculated that perhaps some electronics developments could reduce the noise level, or perhaps we could operate at a higher gain. When operating at a gain of 50, APDs have a temperature dependence of gain of about 3% per °C. Therefore operating a detector with APDs would require a temperature stability of about 1°C to satisfy our desired energy resolution constant term.

We studied both of the APD and HPD options extensively in the laboratory. We also built prototype HCALs using these two photodetectors and

Table 1. A summary of key properties of photodetectors.

	Requirement	PD	APD	PMT	MCP	HPD	SiPM
Bias voltage (V_b)		>20 V	400 V	1–2 kV	1–5 kV	10 kV	50 V
Timing	1–3 ns	3 ns	3 ns	100 ps	50 ps	100 ps	30 ps
Sensitivity		100 pe	10 pe	1 pe	1 pe	1 pe	1 pe
Quantum efficiency	>10%	40%	80%	20%	20%	20%	30%
Expected e/GeV	>10^4	100	5000	10^6	10^5	2×10^4	3×10^5
Excess noise factor	<1.5	1	2.5–5	1.4	1.2	1.05	1.2
Dynamic range	10^4	>10^7	10^7	10^6	10^3	10^7	10^3–10^4
Gain (M)	>100	1	50–100	10^5–10^6	10^4–10^6	2×10^3	5×10^4–10^6
$\delta V_b / V_b$ for $\delta M/M = 1\%$	5×10^{-4}	—	5×10^{-4}	5×10^{-4}	5×10^{-4}	5×10^{-3}	10^{-3}
δT for $dM/M = 1\%$	1°C	10°C	0.3°C	3°C	3°C	3.5°C	0.3°C
Max magnetic field	>4T	>4T	>4T	~1T	~1T	>4T	>4T
Radiation tolerance (rad/n)	2 Krad 10^{11} n	1 Mrad 10^{14} n	1 Mrad 10^{13} n	10 Krad 10^{14} n	10 Krad 10^{14} n	10^{11} n	10^{13} n
Volume/channel	<20 cc	<1 cc	<1 cc	10 cc	1 cc	10 cc	<1 cc
Photon lifetime	5×10^{16}	No limit	No limit	10^{19}	>10^{11}	10^{17}	>10^{16}

response to muons

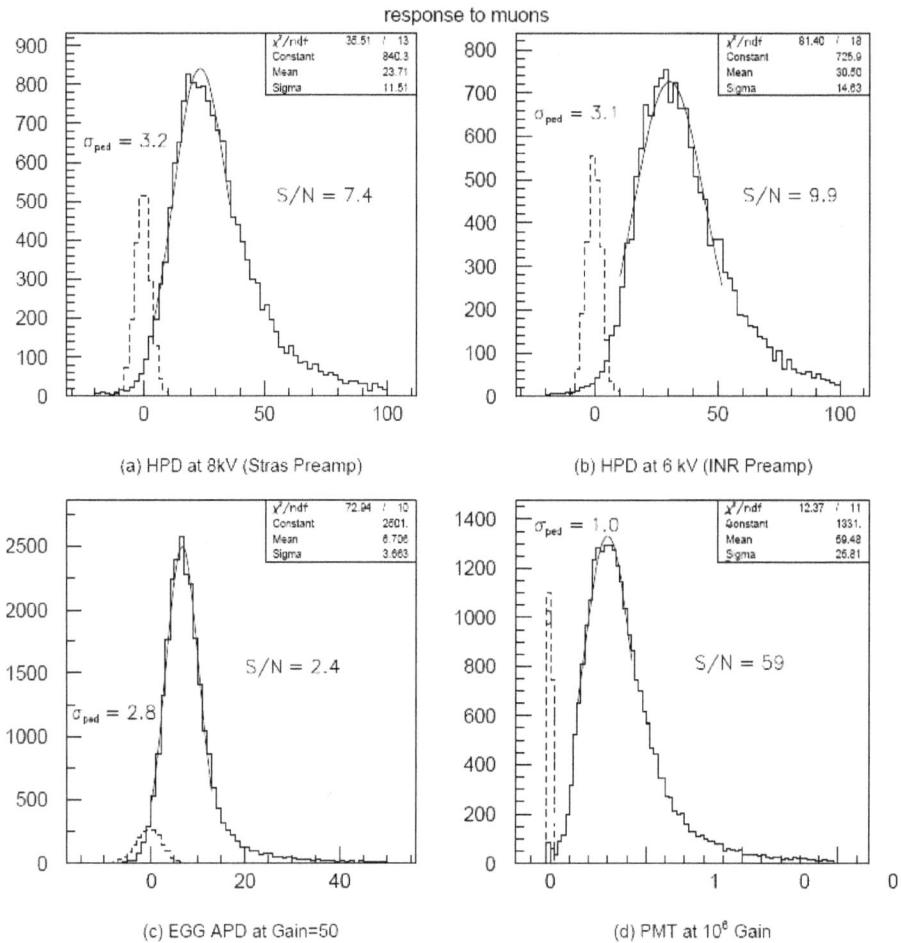

(a) HPD at 8kV (Stras Preamp)

(b) HPD at 6 kV (INR Preamp)

(c) EGG APD at Gain=50

(d) PMT at 10^6 Gain

Fig. 4.   Test beam comparison of APD vs. HPD vs. PMT for muon response.[2]

operated them in test beams. Figure 4 shows the test beam results for muons passing through the prototype calorimeter. The pedestal and the calorimeter muon response are shown for two types of HPDs (a, b), an APD (c), and a PMT (d). The signal-to-noise ratio S/N is shown for each case. The PMT has an outstanding value of S/N and if not for the magnetic field would have been a good choice. (In fact PMTs were chosen for the readout of the HF detector, which is not situated in a high magnetic field region.) Compared with the APD, the HPDs both show about 3× better S/N due to their higher gain and smaller excess noise factor. It is also worth noting that in this study an amplifier with an unrealistically low bandwidth was used for the APD. This caused the electronics noise to be about 1200 e^-, rather than our expected 5000. Use of a realistic amplifier (with a suitable bandwidth for the 25 ns bunch spacing of the LHC) would have resulted in substantially worse S/N for the APD. Finally, the APD required very good temperature stabilization. The combination of these issues led us to choose the HPD as the HCAL photodetector.

Figure 5 shows the cross section of the HPD. It is a vacuum device like the PMT. It has a transparent window followed by a photocathode that absorbs photons and emits photoelectrons. A high voltage ($\sim$10 kV) separates the photocathode from the $\sim$3 mm distant reverse-biased silicon diode. The kinetic energy of the accelerated photoelectrons ionizes the silicon and creates electron–hole pairs (3.6 eV/pair) which are the source of the gain of the device.

Having made the choice of HPD, we were still faced with some R&D to develop the device we wanted for the HCAL.[3] One key requirement was to have a multipixel device to reduce both the required photodetector volume

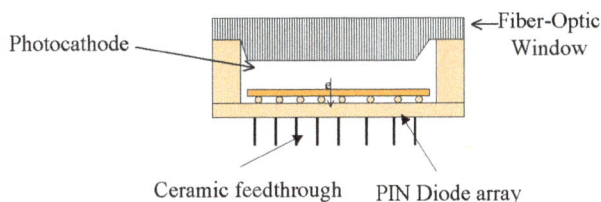

Fig. 5.   Cross section of the HPD.

Fig. 6.   An HPD cut in half.

and cost. The silicon diode was segmented into pixels, and the readout signals were carried through ceramic feedthroughs. The silicon diode was thinned from an initial 300 microns to 200 microns to speed the signal.

We observed the presence of capacitive crosstalk between neighboring pixels, which was traced to poor connection between the pixels and the diode bias supply.[4] We specified a layer of aluminization on the front of the diode and aluminization of the sides to carry the back-supplied bias voltage better to the individual pixels. This change eliminated the electrical crosstalk.

To prevent light from spreading into neighboring pixels when transiting the photocathode window, we specified that the window be made of fiber-optic glass. Photocathodes typical of phototubes and the HPD are semitransparent. Some of the light will transmit through the photocathode without interacting, and then hit the silicon diode. This light could reflect back to the photocathode and cause emission of additional photoelectrons. We found that this reflection caused crosstalk between neighboring pixels at the few-percent level. The vendor (DEP, Holland) was asked to develop an antireflective coating optimized for our light spectrum. A cross section diagram of the final diode is shown in Fig. 7.

Before starting mass production of the HPDs, we performed extensive tests. The tests included:

- Radiation damage studies, where we exposed them to 10^{11} neutrons/cm^2 and 15 Krad.

Diode Structure

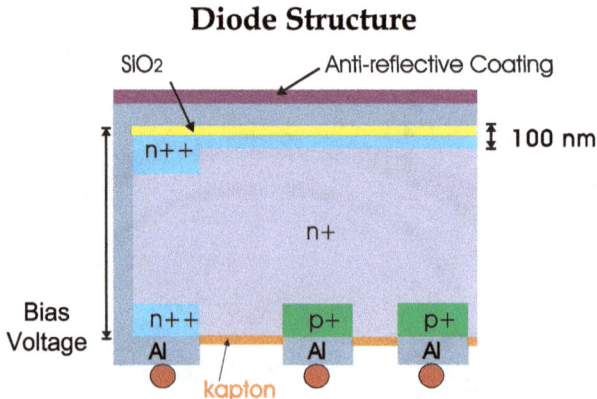

Fig. 7.   A diagram of the cross section of the HPD silicon diode. Aluminization is shown in gray. A SiO$_2$ layer prevents contamination of the silicon by the aluminum but allows high frequency coupling of the bias supply to the diode, eliminating the electrical crosstalk. A higher impedance path provides DC coupling of the diode to the bias supply.

- Magnetic field tests, where the devices were operated in 5 T fields. We also constructed a front-end electronics box, as would be placed on the calorimeter, and tested it in a 3.3 T magnetic field.
- Aging tests, where we measured any changes in performance for integrated charge of up to 10 C corresponding to our 10-year lifetime requirement.
- Test beam studies, to confirm the expected performance.

A total of about 600 HPDs were manufactured and passed our acceptance tests. The devices are working well in CMS central HCALs. As pointed out earlier, conventional photomultipliers were chosen for the HF.

Recently a new device, the SIPM, has become available.[5,6] It is an array of very small Geiger mode APDs in a single package. Typical array densities are 1000–10,000 per mm^2. These devices offer substantial advantages over the HPDs in that: they are much smaller, allowing finer calorimeter segmentation than was possible for the HPDs; they have substantially higher gain, reducing the noise floor of the calorimeter; and they operate at voltages of order 50 V, much smaller than the 10 KV of the HPDs. This last factor points toward less maintenance issues in the future. Important properties of the SIPM are included in Table 1. The CMS HCAL team is actively pursuing this possibility for future improvements.[7] To study possible upgrades we have installed 144 SIPMs in the HO calorimeter. As expected, their performance is very good. Figure 8 shows the energy distribution for cosmic rays in one HO tower read out with an SIPM. The energy deposition by the muon is cleanly separated from the pedestal.

Front-End Electronics

As we developed the choice of HPD for the HCAL photodetector, we also developed the requirements for the front-end ADC. Some of the requirements were driven by the environment which the electronics would be in, some by the physics goals of our detector, and some by constraints or properties of the HPD.

The electronics was to be placed close to the HPD on the back of the calorimeter. This choice of placement implied that the electronics needed reliability, low power, and radiation tolerance.[8]

The signals from the scintillator have a rise time of about 10 ns and a fall time of about 30 ns. Because the LHC will have high luminosities and large physics occupancy, we desired that the electronics would react to the signal as quickly as possible. The LHC operates at 40 MHz (25 ns buckets) and we wanted the signal to be in as few buckets as possible.

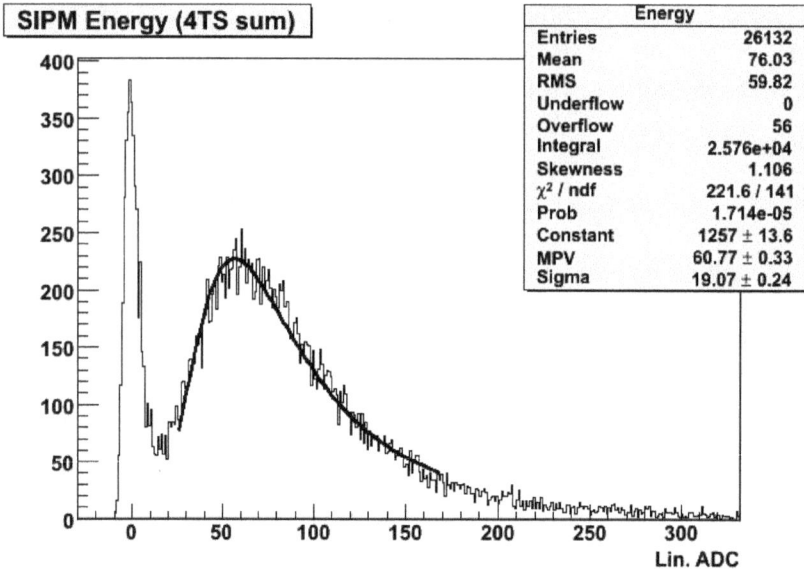

Fig. 8. Cosmic ray energy pulse height distribution for a CMS HO tower read out with an SIPM.

We designed and built a flash ADC (FADC) exactly tailored for the HCAL. We chose to make a totally custom FADC, because of the unique requirements we had in the detector. The FADC we developed is the QIE8 (charge Integrator and Encoder), the eighth generation of a successful FADC that has been developed and used at Fermilab.[9]

For small energy depositions into the calorimeter, the signal shape is not very regular. For this reason we felt that integrating the charge of the signal, rather than measuring a voltage, would give us the best performance. Because the signal for low energies was small, we wanted to have as quiet an amplifier as possible, with a target of a few thousand electrons. With a signal of about 10–20 photoelectrons per GeV and the HPD gain of 2000 (corresponding to 3–6 fC/GeV), we desired a sensitivity of about 1 fC (6000 electrons) and a noise level (pedestal width) of the same order. With the 1 fC minimum, we looked at physics simulations to find the maximum realistic energy deposit per channel. We found that for the 7-on-7 TeV LHC collider, a maximum energy per channel of 3 TeV was adequate. This corresponded to about 10,000 fC. Thus we specified the maximum scale of 10,000 fC and a dynamic range of 10^4.

A conventional linear ADC would require 13+ bits to supply the required dynamic range. Taking into account the typical differential nonlinearities of an ADC, we would need 14 bits of resolution. Because the LHC operates at 40 MHZ, the ADC would also have to sample at 40 MHz. The power required for this size of the flash ADC would be large. Another important consideration in our concept was to minimize the number of bits of information sent off detector to the counting house. (The available volume for the readout fibers was very limited.)

A standard trick to reduce the required size of the flash ADC, and at the same time reduce the amount of data sent off detector, is to put a multirange amplifier in front of the FADC, with each range having a different amplification factor. The signal from each range is sent to circuitry to detect the proper range. The signal from the lowest nonsaturating range is then sent on to the FADC for digitizing. The output data are then the result of the FADC, plus the selected range. Logically one can think of this compound number as the mantissa and exponent of a floating point number. By appropriate choice of the number of ranges, their amplification factors, and the number of bits of resolution of the FADC, the dynamic range can be satisfied and have a large reduction in the number of bits sent off detector.

Figure 9 shows an example of a typical four-range amplifier/FADC. Each range starts at zero and has a full scale of A^n, with n the range, 0, 1, 2, or 3. A is the gain of the amplifier, and for this example the gain for each range is A times the range of the preceding one. The input charge is on the horizontal scale and the output mantissa is on the vertical scale. The dotted lines indicate unused codes. For example, an input charge of $1.5\,AQ_0$ would generate an output code in range 2. The redundant possible code from range 3 would not be used, because it would have coarser resolution. The result of an algorithm such as this is that there would be many unused codes in order to cover the required dynamic range. This means that more bits are needed to get the required resolution and more data have to be sent off detector.

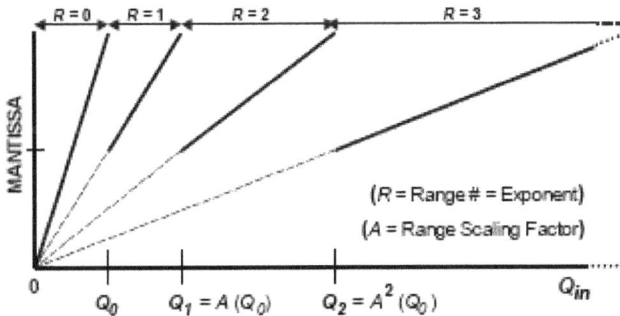

Fig. 9.   An example of a four-traditional-range FADC.[9]

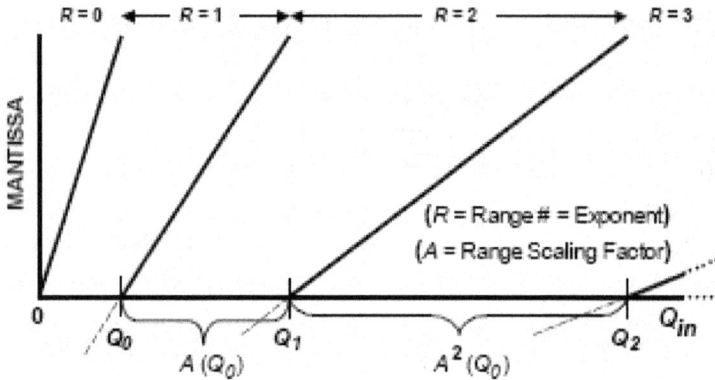

Fig. 10. A four-range FADC where each range has an offset of the maximum scale of the range before it.[9]

In Fig. 10 we see an improved design. Here each range does not start from 0 but rather from an offset that is (a little less than) the full scale of the range before it. In this design there are no wasted codes and one gets the highest resolution possible for the number of bits allocated to the FADC. (We note that to avoid having values of charge that generate no code we require a small overlap between the ranges and hence generate a few unused codes.) This is the design we chose for the QIE.

The energy resolution of the HCAL has been measured to be about $85\%/\mathrm{sqrt}(E) + 5\%$ for isolated pions. This energy resolution is shown as the smooth curve in Fig. 11, as a function of the energy. A required feature of our FADC was that it not should contribute significantly to the overall energy resolution of the calorimeter. The effect of the finite bin width must then lie significantly below the smooth curve in Fig. 11.

An FADC works by having a ladder of resistors and a voltage comparator at each point along the ladder, comparing the input voltage and the voltage of the point on the ladder. An encoder then senses which is the (for example) highest voltage comparator still less than the input signal. A conventional (linear) FADC ladder has the same value of the resistor all along the ladder. Therefore it has the same voltage drop between comparator ladder points and thus the same bin width for all bins. However, nothing requires that the same resistance value be used everywhere on the ladder. We tailored our ladder (and hence the bin values of the FADC) very carefully to reduce to a minimum the number of bits of resolution needed and at the same time keep the bin width contribution to a resolution well below the native energy resolution of the calorimeter.

Bins: 15*1 + 7*2 + 4*3 + 3*4 + 3*5 (total of 68 units = 510 mV, 1 unit = 0.3 GeV)
Ranges: *1, *5, *25, *125; Pedestal is in bin "3".
Calibration uses additional subset of comparators *3.

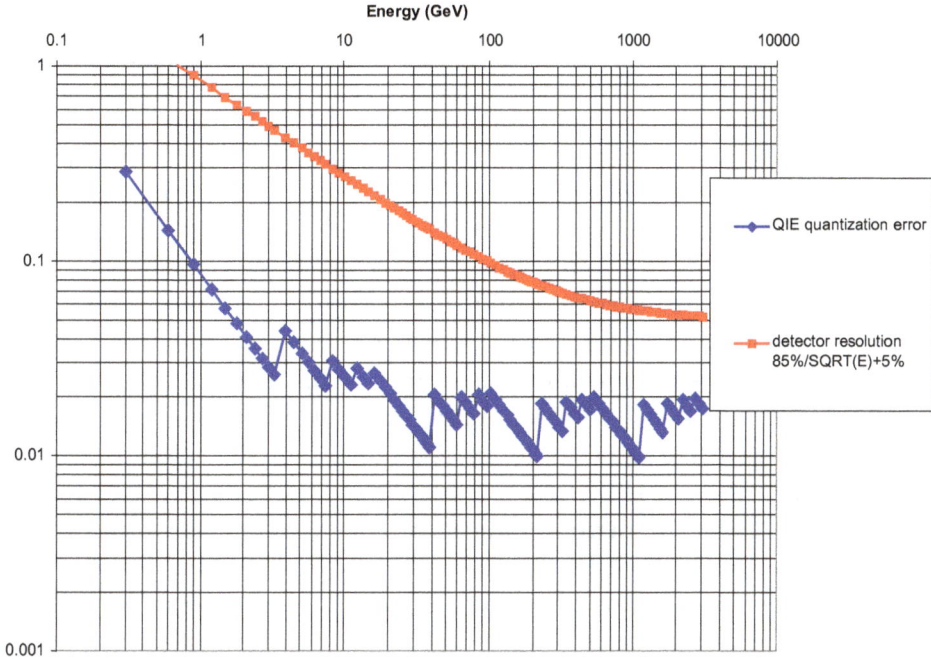

Fig. 11.   The HCAL resolution as a function of energy, and the FADC resolution as a function of energy. The choice of the FADC ladder is shown in the figure heading.[1]

Our development led to an FADC with four ranges (*1, *5, *25, and *125) and five bits of (a very nonlinear) ADC. Figure 11 displays the details of the choices. The figure heading shows the selected bins and ranges. We see that the dynamic range requirement of 10^4 is satisfied.

We also satisfied our requirement that the FADC bin width not contribute to the energy resolution. The jagged curve in Fig. 11 shows the bin width of our choice of the FADC ladder and the gain of the four ranges. Looking carefully, one can see that the bin width pattern repeats itself four times. The 5% calorimeter resolution constant term means that at very high energies the energy resolution will asymptotically approach 5%. We see that at very high energies the FADC bin width contribution is about 2%. Added in quadrature with the constant term the binning causes about a 6% worsening of the ultimate calorimeter energy resolution.

To satisfy our dynamic range requirement the design needed 5 bits of the ADC value and 2 bits of the range value, for a total of bits 7. This is about half the number of bits we would have expected from a single range linear FADC. The number of comparators doubles for each bit of the FADC resolution. In the QIE8 design there are 31 comparators; in the 14-bit FADC there would be 16,000.

A final aspect of the design was drawn from our desire to measure radioactive source currents as a means of calibration. As mentioned earlier, the presence of the radioactive source caused the scintillator to generate about 3×10^7 photons per second at the HPD. After the HDP quantum efficiency and gain, this corresponded to about 1 nA or about 150 electrons per 25 ns bucket. We wanted to measure this level accurately because it was a primary means of monitoring the gain of the calorimeter. A 5% error in measuring the radioactive source current would then correspond to a 5% error in the energy scale of the detector. To reach this accuracy we designed the QIE to have a high sensitivity calibration mode. In this mode of operation, the resistors in the ladder were three times higher in value than the nominal ones and all were the same value. In this mode the QIE acted as a linear FADC with 0.3 fC per bin. A control bit allowed us to switch from regular to calibration mode. To perform radioactive source calibrations, we read out the FADC repeatedly and formed histograms with 65,000 entries. We then calculated the mean of the histogram. Figure 12 shows the mean value of the histograms as a function of time. In the middle of the measurement the source was removed. Each dot corresponds to the mean value of the 65,000-entry histogram. Using this technique (of massive oversampling) we were able to measure shifts of the mean by as small as a few 1/1000s of a bin. This gave us the accuracy required for the radioactive source calibration.

The QIE is a four-stage pipelined device that runs at 40 MHz. The four range (approximately 100 MHz bandwidth) preamplifier is constantly at work, amplifying the input signal. At pipeline stage 1, integrators for the four ranges integrate the charge in a 25 ns gate; at stage 2, the lowest range that is not full-scale is selected, and the voltage presented to the FADC; at stage 3 the FADC digitizes the voltage; at the final stage, stage 4, the capacitor is reset; and then back to stage 1. In the QIE there are four independent capacitors, each at a different stage of the four-stage cycle. Thus every 25 ns a new 5-bit + 2-bit number is generated.

The penalty for the novel arrangement of the QIE amplifier gains and offsets is an abundance of calibration constants. Each of the four capacitors in the "round robin" can (and does) have slightly different constants. There are four ranges with slope and offset for a total of 32 calibration constants. Additionally, there can be variation in the ladder that can be measured.

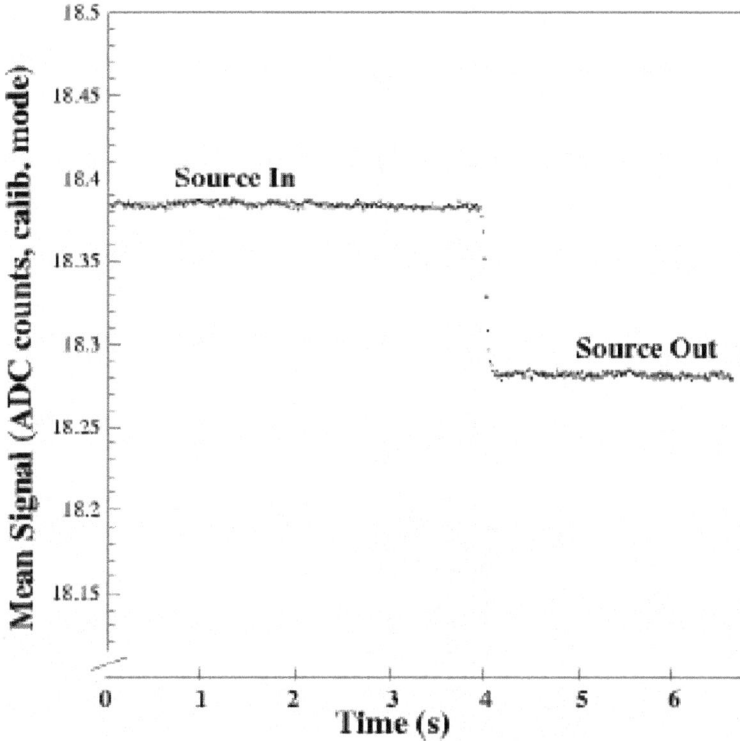

Fig. 12. The mean ADC value in calibration mode for the source in position by the scintillator, and out. The source caused about a 0.1 bin shift, corresponding to about 200 e$^-$/25 ns.[9]

About 18,000 QIE chips were calibrated and tested in an automated robotic testing station. The robot sorted the chips based on test results. After the good chips were mounted on electronics boards, they were given a final, detailed test. An example of a test result, the slope for the set of chips for one of the ranges, is shown in Fig. 13.

Summary

The CMS HCAL photodetector and front-end electronics were developed simultaneously. The high magnetic field where the HCAL is located caused a severe limitation on the choice of photodetector. In fact no suitable photodetector existed and we were forced to develop our own, a pixilated proximity-focused HPD. The low gain of the HPD (1000–2000) and the limited light yield of the sampling scintillator calorimeter (10–20 pe/GeV) necessitated the development of a very sensitive and low noise front-end readout. We

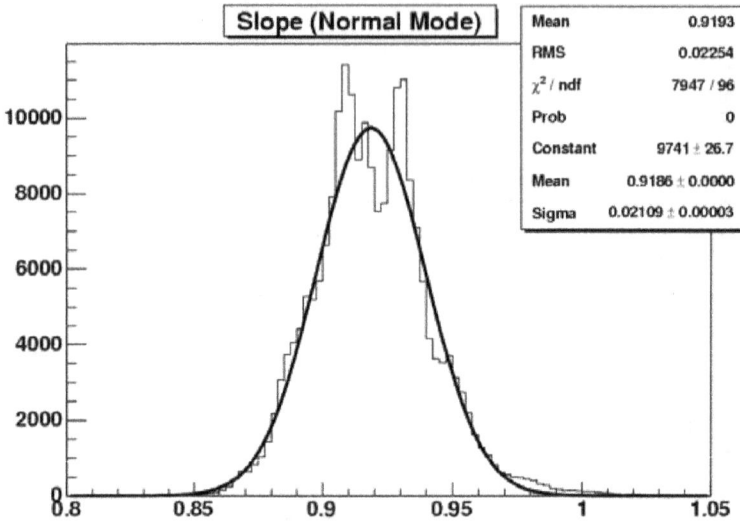

Fig. 13.   Slope measurement for a set of QIE chips for one of the four ranges.[10]

developed a novel range-switching nonlinear FADC, the QIE8, for this purpose. It has a sensitivity of 1 fC and a noise level of about 0.7 fC. The required 10,000–1 dynamic range was achieved using a very nonlinear response function of the 5-bit FADC and automatic switching between four ranges of preamplifier gain. The readout of the calorimeter channel requires only 7 bits of data rather than 14 bits for a completely linear single range readout. The data compression reduces front-end power requirements and minimizes the cable volume leaving the detector.

References

1. CMS Collaboration. The CMS detector at the CERN LHC, *JINST* **3 S08004v** (2008) 122.
2. P. B. Cushman. Test beam performance of HPDs and APDs for the CMS HCAL, *Nucl. Instrum. Methods A* **387** (1997) 107.
3. P. B. Cushman and A. H. Heering. Problems and solutions in high-rate multichannel hybrid photodiode design: the CMS experience, *IEEE Trans. Nucl. Sci.* **49** (2002) 963–970.
4. P. Cushman, A. H. Heering, N. Pearson, J. Elias, J. Freeman, D. Green and A. Ronzhin. Crosstalk properties of the CMS HCAL hybrid photodiode, *Nucl. Instrum. Methods A* **504** (2003) 62–69.
5. Z. Sadygov *et al.* Three advanced designs of micro-pixel avalanche photodiodes, *Nucl. Instrum. Methods A* **567** (2005) 70–76.

6. B. Dolgoshein *et al.* Large area silicon photomultipliers: performance and applications, *Nucl. Instrum. Methods A* **567** (2005) 70–76.

7. A. Heering *et al.* Performance of silicon photomultipliers with the CMS HCAL front-end electronic, *Nucl. Instrum. Methods A* **576** (2007) 341–349.

8. T. M. Shaw *et al.* Front end readout electronics for the CMS Hadron Calorimeter, in *IEEE 2002 NSS* (Norfolk, Virginia, USA; 10–16 Nov 2002).

9. T. Zimmerman and J. R. Hoff. The design of a charge integrating, modified floating point ADC chip. *IEEE J. Solid State Circuits* **39** (2004) 895–905.

10. E. Hazen *et al.* Radioactive source calibration technique for the CMS hadron calorimeter. 2003. 17pp. Published in *Nucl. Instrum. Methods A* **511**(2003) 311–327.

11. J. Damgov *et al.* HB Performance and Validation of Calibrations using CRAFT, and LHC Beam data of September 2008, CMS Note 2009/999, May 24, 2009.

Chapter 11

ATLAS SUPERCONDUCTING TOROIDS — THE LARGEST EVER BUILT

Herman H. J. ten Kate

Physics Department, CERN
CH-1211 Genève 23, Switzerland
Herman.tenKate@cern.ch

1. Introduction

Three huge superconducting toroids and a relatively small solenoid together give shape to the ATLAS superconducting magnet system exhibiting record dimensions of 22 m in diameter and 25 m in length. The four magnets provide the magnetic field for particle bending in, respectively, the muon detectors and inner detectors in the ATLAS experiment at the Large Hadron Collider.[1] The magnet technology used for this largest toroidal magnet system in the world is based on winding large size aluminum-stabilized NbTi superconducting cables into impregnated coils housed in aluminum alloy coil casings which are conduction-cooled by forced flow liquid helium flowing in aluminum cooling tubes attached to the various coil casings guaranteeing a 4.6 K operating temperature.[2]

Figure 1 shows the ATLAS experiment. Four types of detectors as well as the four superconducting magnets can be recognized. The central solenoid provides an axial magnetic field of 2 T for the inner trackers. The barrel toroid and the two end-cap toroids makeup the toroidal magnetic field of about 1 T for, respectively, the layered muon detectors in the radial direction and the detector chambers installed on big flat wheels in the forward and backward directions.

The magnetic field is generated in the usable detector area of some 16,000 m³, and the peak field in the toroid coil windings is 4.1 T. The magnet system stored energy is 1.6 GJ or, more for the reader's imagination, equivalent to the kinetic energy of a French high speed TGV train of 385 tons on its way from Geneva to Paris speeding at 330 km/h and the brakes do not work when it is arriving in "Gare de Lyon"! More selected key parameters of the ATLAS magnet system are listed in Table 1.

Some 17 years ago, in 1992, the first predesign sketches were published, and in 1997 the technical design report on the magnet system was approved.

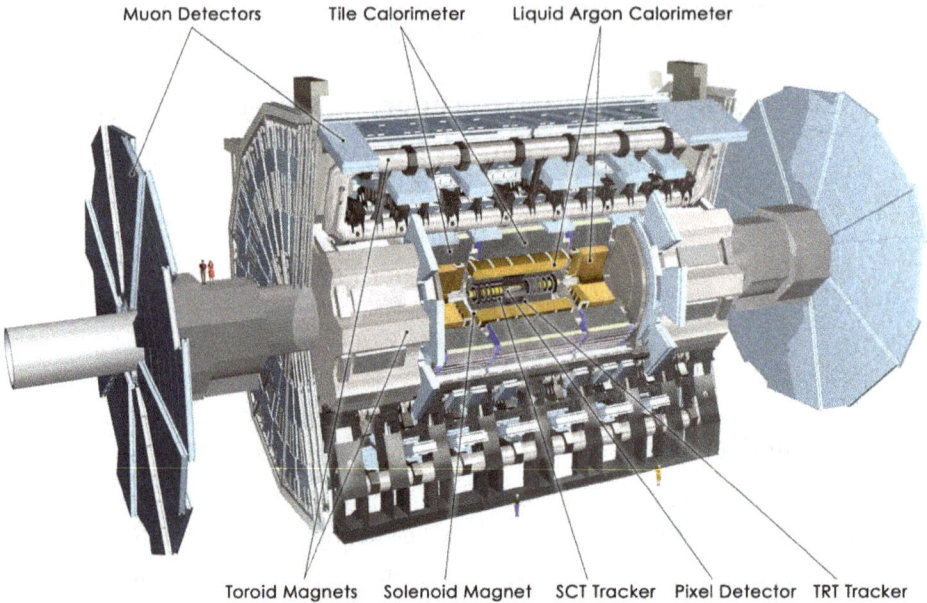

Muon Detectors   Tile Calorimeter   Liquid Argon Calorimeter

Toroid Magnets   Solenoid Magnet   SCT Tracker   Pixel Detector   TRT Tracker

Fig. 1. Schematic of the ATLAS detector showing the barrel toroid and the two end-cap toroids on either side of the central solenoid.

The first construction contracts were awarded in the winter of the same year, and in the years thereafter bare coils, coil casings and vacuum vessels arrived at CERN. In 2001 cold mass and cryostat integration on the CERN site started.

The solenoid was ready for installation in 2004, followed by the barrel toroid in 2005–2006 and the end-cap toroids in 2007. In summer 2008 the system was fully operational, giving reason to party over the completion of this very challenging and largest toroid-based superconducting magnet system successfully constructed so far. In September of the same year, the first bunch of protons passed the detector and gave exciting bent particle trajectories in the inner and muon detectors.

2. Why a Toroid?

Superconductors need to be used when a high magnetic field is required in a large volume for which normal conducting and water-cooled copper-based conductors are not feasible or economical. In modern particle detector magnets, typically a magnetic field of a few teslas is required but in a volume of many thousand cubic meters. Essentially, detector magnets can be solenoid or toroid-based. The majority of detector magnets are using

Table 1. Main specifications of the toroids and solenoid in the ATLAS magnet system.

Property	Unit	Barrel toroid	End-cap toroids	Central solenoid
Size: Inner diameter	m	9.4	1.65	2.46
Outer diameter	m	20.1	10.7	2.63
Axial length	m	25.3	5	5.8
Number of coils	—	8	2×8	1
Mass: Conductor	Tons	118	2×20.5	3.8
Cold mass	Tons	370	2×140	5.4
Total assembly	Tons	830	2×239	5.7
Coils: Turns/coil	—	120	116	1154
Nominal current	kA	20.5	20.5	7.7
Magnet stored energy	GJ	1.08	2×0.25	0.04
Peak field	T	3.9	4.1	2.6
Conductor: Overall size	mm^2	57×12	41×12	30×4.25
Ratio Al:Cu:NbTi	—	28:1.3:1	19:1.3:1	15.6:0.9:1
Number of strands	—	38	40	12
Strand diameter	mm	1.3	1.3	1.22
Critical current @ 5 T, 4.2 K	kA	58	60	20.4
RRR Al	—	> 800	> 800	> 400
I/Ic margin @ 4.5 K	%	30	30	20
Temperature margin	K	1.9	1.9	2.7
Number of units × length	# × m	32×1730	32×800	4×2290
Total length	km	56	2×13	10
Heat load: at 4.5 K	W	990	330	130
at 60–80 K	kW	7.4	1.7	0.50
Liquid He mass flow	g/s	410	280	7

solenoids enclosed by an iron yoke to return the flux and provide a structure for the muon detectors, as this obviously is a simple and straightforward concept. An example is the CMS experiment,[3] the second general purpose detector at the LHC.

The ATLAS muon physics community, however, opted for the best possible muon bending in the forward directions in the shower of after proton–proton collision products and consequently asked for the construction of a toroidal magnet system that can fulfill this demand. The main reason for a toroid for muon detection is to have the best possible measuring acceptance at high rapidity (being the important variable in proton–proton collisions).

Solenoids do not allow for a high resolution measurement of muons and other charged particles going forward, since the muons traverse only a small part of the magnetic field in a region where the magnetic field is not uniform, and not in the transverse direction to the muon trajectories. Another important advantage of the toroid is that it allows one to measure momenta, which for forward-going particles is crucial, since they carry high momenta but low transverse momenta. The ATLAS inner detector community agreed on an axial-magnetic-field-producing solenoid within the first 1.2 m radius around the collision point, and so the hybrid magnet system of the ATLAS experiment was born.

Obviously, the magnetic field is nearly perfectly contained in the torus and consequently no iron return yoke is necessary. In practice a small stray field is present, whose value depends on the uniformity of the toroid coil winding and thus the number of separate coils in the toroid. No iron also means that the magnetic field is very predictable and easy to calculate with high precision. On the other hand, the magnetic field efficiency of a toroid is much worse.

The magnetic field inside the torus decreases in the radial direction with $1/r$, and is therefore very nonuniform and relatively low compared to the peak magnetic field in the windings that determine the peak forces in the structures. For generating the same magnetic field, an ideal toroid needs about 2.5 times the current of a solenoid and about 2.5^2 more forces and complexity to worry about. The quest for optimal muon bending and resolution and thus toroidal magnets in detectors has a price. On the other hand, for magnet scientists and engineers the complex hybrid system of a solenoid and three toroids is a dreamed of project.

The important advantages of the toroidal geometry were already well known to the physics community preparing the LHC experiments in the early 1990s. Two toroid-based experiments were proposed — called EAGLE and ASCOT[4] — and their primitive layouts are sketched in Fig. 2. EAGLE was based on a central solenoid surrounded by a warm iron toroid, while ASCOT featured a central solenoid surrounded by a 12-coil barrel toroid and 2 iron-cored end-cap toroids. In 1992 the two communities agreed to merge and gave birth to the principle layout of ATLAS, a central solenoid with a relatively thin cylindrical iron return yoke and three air core toroids.

The overall size of ATLAS is unprecedented and required scaling-up of magnet technology accordingly. Few-meter coils for toroids were constructed, such as for the Tore Supra fusion experiment,[5] the Large Coil Task fusion coils construction experiment[6] and for the CLAS/CEBAF detector.[7]

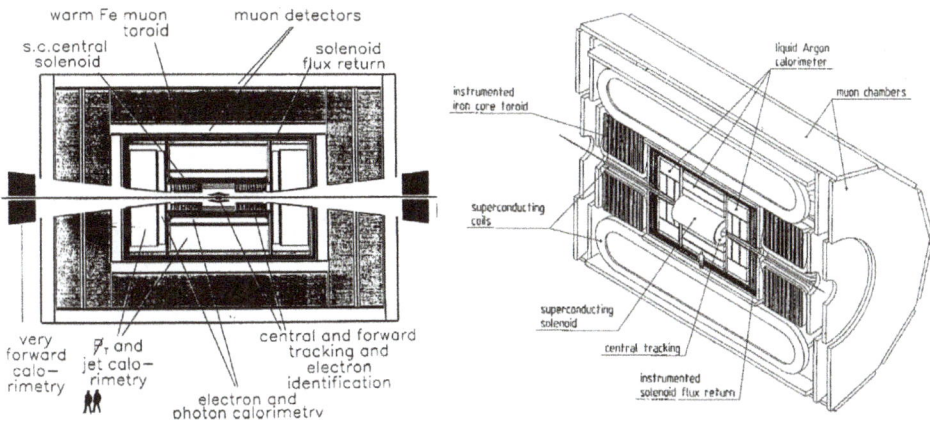

Fig. 2. EAGLE (left) and ASCOT (right) in 1992 merging into ATLAS (Fig. 1).

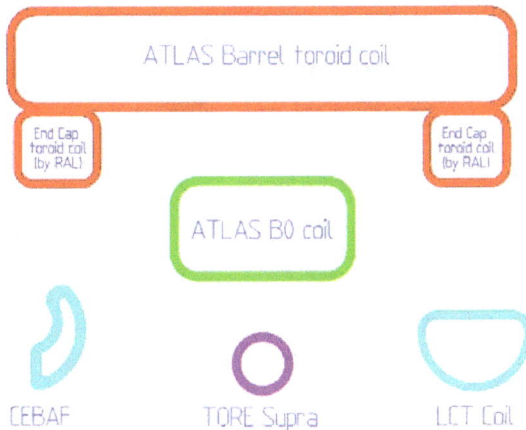

Fig. 3. Dimensions of large superconducting coils illustrating the scaling-up from few-meter coils (bottom) to the 25 m barrel toroid coils (top). The 9 m long ATLAS B0 model coil was constructed to bridge the gap and learn on the job.

Figure 3 shows these coils in perspective of the 25 m ATLAS toroid coils; obviously, a big step forward had to be taken. To accommodate this, thereby reducing the technical risks, the ATLAS magnet laboratories decided to construct an intermediate but still 9-m-long and full cross-sectional B0 model coil — a very sensible move to test the technology and exercise the production steps with the companies involved.

When taking away the various detectors and support structures, we see the bare windings of the ATLAS magnet system in Fig. 4.

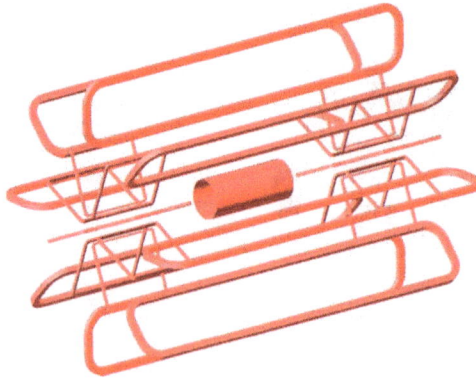

Fig. 4. Layout of the bare windings of the ATLAS magnet system. One can distinguish the solenoid in the center enclosed by the eight barrel toroid coils and the two sets of eight end-cap toroid coils at either end of the solenoid generating bending muon power in low angle forward and backward directions.

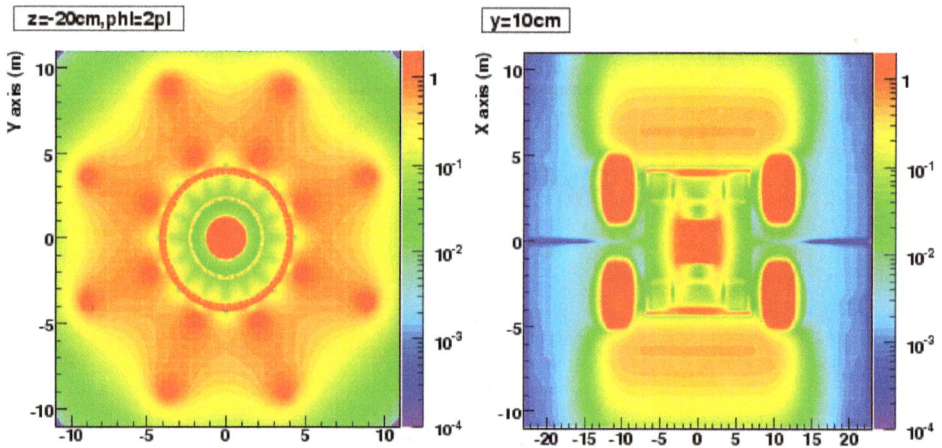

Fig. 5. The magnetic field configuration generated by the single central solenoid (generating 2 T in the center and a saturated cylindrical return yoke), the eight barrel toroid coils (0.5–0.8 T within a torus outside the return yoke), and the powerful end-cap toroids (1–1.5 T on either side of the solenoid and providing bending power for a low angle in the forward/backward direction).

The current in the windings of the solenoid and toroids together generate at nominal the spatial distribution of the magnetic field, as shown in Fig. 5.

3. Superconductors for Detector Magnets

A primary concern in designing a superconducting magnet is to define and engineer, given the cooling mode, a fully functional conductor with sufficient superconducting material, sufficient stabilizer and sufficient parallel path for quench protection. For detector magnets there is a long tradition of using conduction-cooled windings by which the superconductor is kept at 4.6 K by circulating liquid He in cooling tubes attached to the coil casing or support cylinder. This implies a thermal path that conducts the heat from the NbTi through several layers of the conductor and ground insulation as well as coil casing structures to the helium in the cooling pipes. An example of such a cold mass is given in Fig. 6, showing the cross-section of a barrel toroid coil. Details of this construction will be explained later.

The evolution of the detector type of conductors is nicely shown in Fig. 7. With increasing size, magnetic field and stored energy the conductor has to grow in size accordingly. For the ATLAS magnets the conductors used are shown from left to right for the central solenoid CS, end-cap toroids and barrel toroid coils. Their main parameters are included in Table 1 as well. The concept is a NbTi/Cu-strand-based Rutherford cable, which is coextruded (or plated) with pure Al. Figure 8 shows the conductor used in the barrel toroid coils.[8] The section of Al is present to take over the superconductor transport current in the case of a quench for a few minutes, which allows one to detect a quench reliably and to take measures to release the stored energy safely. The amount of Al is determined by the amount of time needed. It is crucial that during the coextrusion or plating process an intermetallic bond

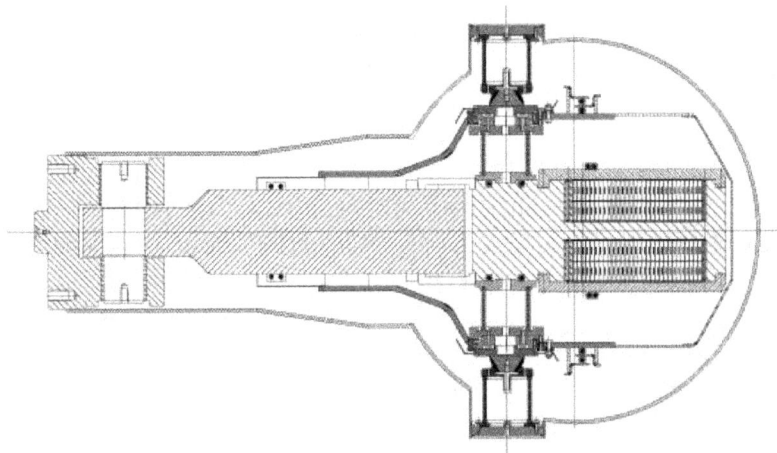

Fig. 6. Cross-section of the barrel toroid coil in the position of a tie rod. From inside out we see the two double pancake windings wound from rectangular super-conductor, Al alloy coil casing, the cooling tubes attached to the top and bottom surfaces, thermal shield, two sets of cold mass lateral supports and a tie rod, all assembled within the 1.1-m-diameter vacuum vessel.

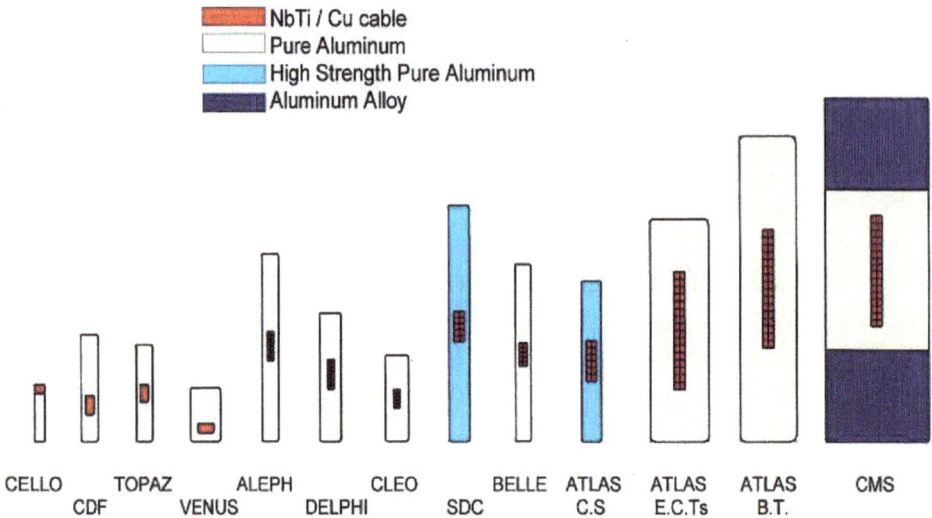

Fig. 7.   Cross-sections of various Al-stabilized conductors used in conduction cooled detector magnets. Evolution from small, older magnets to large modern LHC detector magnets for the ATLAS central solenoid, the end-cap and barrel toroids and the CMS solenoid.

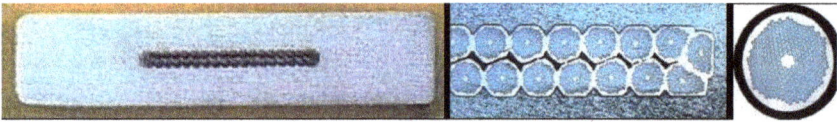

Fig. 8.   Cross-sections of the conductor used in the barrel toroid. On the left is the full conductor sized $12 \times 57\,\text{mm}^2$, in the middle a detail of the cable and the bonding between Al and copper, and on the right the NbTi/Cu strand of 1.3 mm diameter.

is formed between Al and Cu in order to ensure unhindered heat and current flow. To get this bonding right along the entire production of about 100 km of ATLAS conductor in three versions was the main challenge in conductor manufacturing.[9]

For the toroids a pure Al stabilizer (with RRR values of 1500–3000) was acceptable given the stress in the coil windings. An important challenge for the solenoid was to make the radiation length in the radial direction (the amount of high density materials) as small as possible by using a relatively thin support cylinder in combination with reinforced pure Al in the conductor. By microalloying pure Al with Ni (or Zn) the yield strength could be raised from the 30 to some 120 MPa while keeping the RRR value at a reasonable level of 400.[9] This special development for ATLAS is a true innovation and allows maximizing the conductor strength and thus a minimum thickness of the obligatory support cylinder for the windings. It is foreseen that in the near future we will follow this strategy and further improve the yield strength of highly conductive Al to the 200–400 MPa level. For the CMS coil an alternative conductor-reinforcing strategy was followed by using Al alloy side bars that are eb-welded to the Al extruded conductor; see the rightmost conductor in Fig. 7.

4. Barrel Toroid Construction Challenges

The barrel toroid is an assembly of eight $25\,m \times 5\,m$ size racetrack coils that are equally spaced and mechanically supported by two sets of eight rings of beam elements bolted to connection flanges on the coils, as can be recognized in Fig. 9. The beauty of symmetry is remarkable in perspective of the person standing between the two bottom coils in front. The central solenoid bore surrounded by calorimeters is visible at the back and waiting to be moved to its final position in the center.

Very challenging in the coil production is the coil winding and vacuum impregnation of the 16 double pancakes of size $25 \times 5 \times 0.12\,m^3$, the gluing of the sets of double pancakes under prestress using the bladder technology, as well as the high quality welding and vacuum tightness of the all Al cooling circuits.

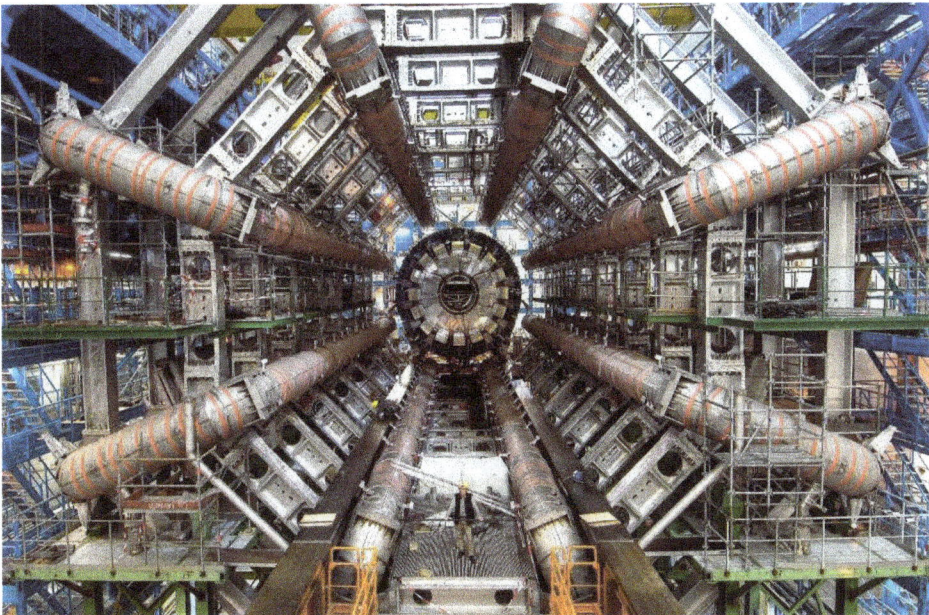

Fig. 9.   Picture of the naked barrel toroid taken at the end of the assembly in the underground cavern. The depth of the toroid is 25 m and the outer diameter 22 m.

Fig. 10. Production stages of the barrel toroid coils. From top to bottom, left to right: Finishing of a vacuum-impregnated double pancake; assembly by welding of the Al-5083 coil casing; insertion gluing under prestress of the double pancakes in the coil casing; installation of segmented thermal shield panels around cold mass; cold mass insertion in the bottom part of the vacuum vessel; finished coil being prepared for acceptance testing.

The various coil components were built in industry and then transported to CERN, where cold mass assembly,[10] cryostat integration[11] and single coil tests[12] took place. A few pictures of the production and integration steps are shown in Fig. 10.

The 8 coils consist of an Al-5083 coil casing enclosing 2 double pancake and vacuum-impregnated windings, each showing 60 turns in 2 layers of the aluminum-stabilized Rutherford type of NbTi cable of 40 strands, as shown in Fig. 8.

The layout of the cold mass and its positioning in the vacuum vessel are shown in Fig. 6. The cold mass and so the superconductor is conduction-cooled by aluminum alloy cooling tubes glued onto the cold mass. The cold mass is surrounded by an aluminum alloy thermal shield, blankets of 30 layers of superinsulation, and positioned in a 10-mm-thick 304 L stainless steel vacuum vessel.

The net Lorentz force of about 1200 tons acting on the single cold mass is transferred by 8 titanium tie rods to the warm structure and reacted by compression in the 8 warm structure inner rings. The lateral movement of the cold mass inside the vacuum vessel is blocked by 2 sets of 16 so-called cryostops on either side of the cold mass. These stops take on average a load of about 3 tons in compression and 1 ton in shear. They also include a sliding interface that allows the shrinkage of the cold mass with respect to the cryostat. The axial position of the cold mass in its vacuum vessel is frozen by a fixed point at the central rib supporting 10 tons of axial force between cold mass and vacuum vessel.[13]

The double pancakes are glued in slots in the coil casing under prestress generated by a pressurized bladder system, to reduce the shear stress in the bond layers when cold and charged, thereby minimizing the risk of epoxy cracking and delamination.[10]

After completion of the cryostat integration, the eight coils underwent, before the decision was made to install the coil in the cavern, on-surface tests including a full cooling cycle to 4.6 K and powering up to nominal current with some margin.[12] The forces on the cold mass suspension system were arranged by putting the tie rod side of the coil across 25 m in contact with an iron mirror. This gave a realistic test and the tied rods could be loaded up to some 180 tons like in the real toroid once fully assembled. All eight coils passed the on-surface test.

The next challenge was the assembly of the toroid in the underground cavern. Obviously, after installation the shape of the toroid should be nearly perfect, meaning that the bore is cylindrical. Since the toroid will sag due its weight, the deflections had to be calculated as precisely as possible. On the vertical axis a deflection of 27 mm was estimated based on a full 3D finite element calculation. To be on the safe side a margin of 3 mm was anticipated and the toroid was assembled with +30 mm ovality in the vertical direction. The assembly method employed is to put the 8 coils in their calculated position using individual supports for all coils, to measure in this stress-free configuration the gaps between the coils at all the 64 connection flanges, to fill the gaps with struts and shims and to bolt all together with sufficient

Fig. 11.   Assembly of the barrel toroid in the ATLAS cavern: arrival of a BT coil in the cavern while hanging on the crane hook (upper left); manipulation of a 100-ton coil with mm precision during the insertion in the toroid structure (upper right); assembly status after six coils are installed (bottom). Note the various supporting platforms and pillars for keeping the coils in their calculated positions during the assembly.

prestress to avoid opening of gaps. The assembly method is illustrated in Fig. 11.

The struts are typically 5–6 m long and the largest one has a connection flange section of $500 \times 800\,\mathrm{mm}^2$. The entire warm structure is made of Al alloy and produced by stamping, forging and extrusion. The warm structure design and production features are explained in Ref. 14.

Once the toroidal structure is closed, the supports holding the coils in their calculated position can be removed and the structure will deflect and seek its natural shape under its weight. Surprisingly, the sag measured is indeed 26–27 mm, which means that given the complexity of the structure and uncertainties in materials properties and production tolerances, the theoretical analysis was very precise and well within the expected ±10% range. Note that the additional deflection under the Lorentz forces when the toroid is charged is in the 1–2 mm range. The installation procedure is explained in more detail in Ref. 15.

Finally, the electrical, vacuum and cryogenics connections between the coils have to be installed. Bus bars and cooling lines linking the eight coils are routed through a ring of eight transfer line segments and exit at the top of the toroid, where the current lead cryostat is positioned.

The first cool-down and test of the barrel toroid took place in 2006. Given the imposed maximum thermal gradient within the cold masses, it took 5 weeks to bring the 360-ton cold mass to 4.6 K. A test sequence is to ramp the magnet with a ramp rate of 3 A/s to the 20.5 kA nominal current, followed by a plateau to prove stability and then either a normal slow dump in the dump resistors or provoking a fast dump by the magnet safety system that fires 4 heaters in all 8 coils by which the energy is dissipated in the cold mass within 2 min and the cold mass temperature rises to 60 K, a very safe value. After recooling to 4.6 K, which requires 50 h, the cycle is repeated to prove full reversibility.

Due to the many quench protection heaters, two in every double pancake, the entire coil is in the normal state within 2 s, which causes a uniform coil resistance and thus uniform absorption of the magnet-stored energy, and the toroid internal voltage is kept at about the 70 V level.

The internal coil suspension system — for each coil 8 titanium tie rods, 32 cold stops and a fixed point — behaved as expected, with no sign of yielding or hysteresis.

During the various tests no training or any other spontaneous quenches were observed in this new record size toroid — a convincing validation of the mechanical and thermal design as well as the production techniques used.

5. End-Cap Toroids

The two end-cap toroids that makeup the magnetic field in the forward direction are much smaller as they have to fit in the bore of the barrel toroid, and are therefore constructed as single cold masses in a vacuum vessel.[16] Figure 12 shows the two end-cap toroids during their integration.

Fig. 12.   The two end-cap toroids during their cold mass integration and cryostating. The overall dimension is about 11 m in diameter and 5 m in length.

On the left we see the completed cold mass, and on the right the second cold mass already in the vacuum vessel before closure of the end-plate.

The cold mass is assembled from eight coil modules and eight keystone boxes, which support and separate the coils.

The conductor has a slightly smaller Al-stabilizing section, since the stored energy is about one-quarter of the barrel toroid, but the superconducting cable is the same. A coil module comprises two double pancake coils, like in the barrel toroid coils, and they are sandwiched between a set of Al alloy coil casing plates, which also incorporates the cooling pipes. The coil modules are vacuum-impregnated; however, contrary to the barrel toroid, no bladders are used to prestress the coils and the design relies in full on the resin-bonding properties.

Cold mass modules, thermal shield sections, superinsulation, coil suspension parts and vacuum vessels were produced in industry and the integration of cold masses and cryostats was performed at CERN.[17] Apart from the coil suspension tie rods made of stainless steel, all other metal structures in cold mass, shields and vessels are of Al alloy. Though much smaller than the barrel toroid, still the end-cap toroids are huge and unique toroids.

In 2007, after their on-surface test, the two toroids of 240 tons each were transported, inserted in the ATLAS cavern by a 280-ton crane and moved into their operating position; see Fig. 13.

Fig. 13. End-cap toroids on the move: leaving the integration facility (upper left); 240 tons on the trailer to ATLAS point 1 (right), and insertion in the shaft (lower left).

6. Play of Forces between Barrel and End-Cap Toroids

A remarkable feature of ATLAS is the central rail system, which allows opening the experiment for maintenance and repairs by axially moving out the end-cap calorimeter, the small muon wheel and the end-cap toroids respectively, while the solenoid with the inner detectors inside its bore and the barrel calorimeter stay in their central position; see Fig. 14. The consequence of this design is that three toroids are not solidly connected. After every opening the end-cap toroids must be repositioned and the force transfer points well aligned and shimmed, by which axial symmetry and tangential repositioning is important to avoid unbalanced forces, undesired displacements affecting the magnetic field distribution and overloading of the coil suspension tie rods and brackets.

The second aspect is that the magnetic field should be continuous locally in the forward region where the end-cap toroids and barrel toroid overlap. This overlap is achieved by inserting the end-cap toroid coils between the barrel toroid coils, practically by offsetting in the tangential direction the

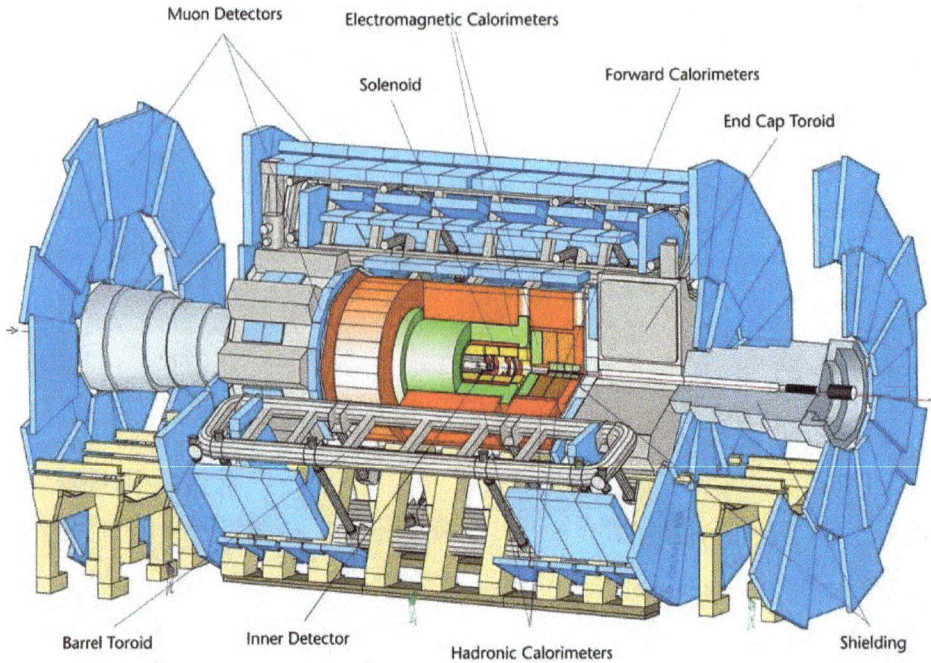

Fig. 14.   Schematic layout of the ATLAS detector, nicely showing the central rail system which allows the axial opening of the detector by moving in and out the end-cap calorimeters, small muon wheel and end-cap toroids using air-pad supports.

eight-coil toroids by 22.5°. This can be recognized in Figs. 1, 4 and 14, and the castellated shape of the vacuum vessel in Fig. 13. The implication of the magnetic field overlap is that the forces working between the three toroids are huge.

The Lorentz forces acting on the various cold masses and how they are reacted are schematically shown in Fig. 15. When charging the toroids, the end-cap toroids are sucked into the barrel toroids with an average force of 240 tons, provided that symmetry is kept. The path of the force transfer is as follows. The end-cap toroid cold mass is single and practically infinitely stiff. The accumulated force of 240 tons is transferred by eight sets of two axial tie rods of stainless steel to the thick vacuum vessel outer flange (see Fig. 13, right). At these eight positions, sets of brackets have meeting parts welded on the end-cap toroid vessel and at the corresponding points on the barrel toroid vessel.

The axial force transfer brackets guide the Lorentz force acting on the end-cap toroid cold mass into the barrel toroid vacuum vessel. And this happens from both sides of the barrel toroid, which means that the barrel

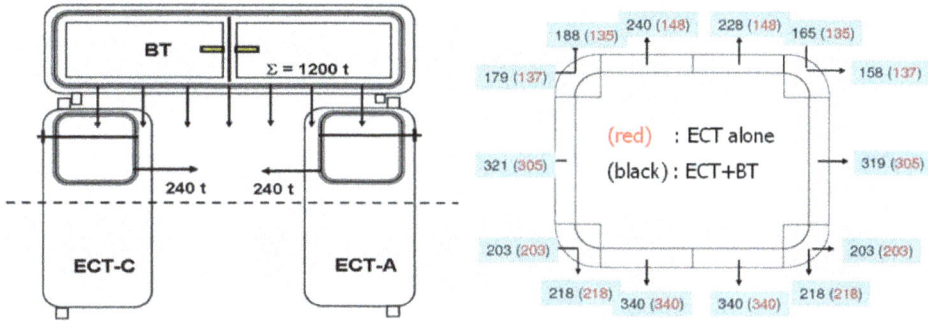

Fig. 15. Forces working between the three toroids and how they are transferred (left); forces per meter on the end-cap toroid coil (right) in stand-alone (red) and combined operation (black).

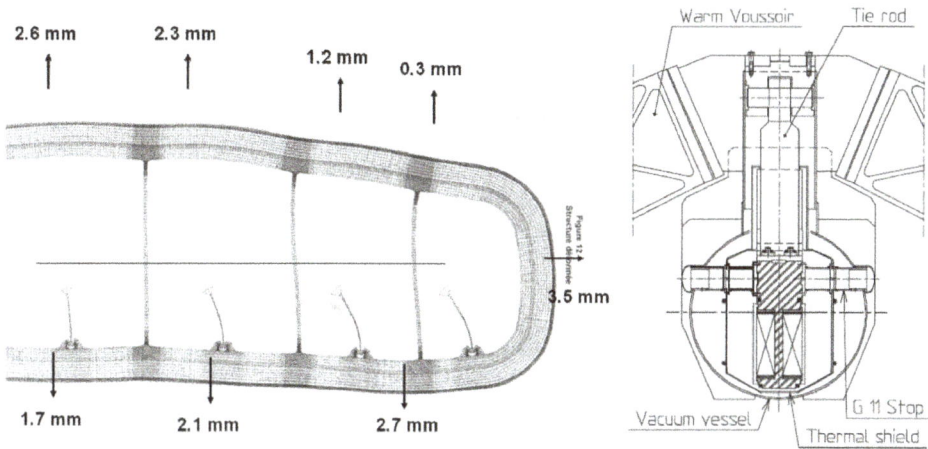

Fig. 16. Deflection of the barrel toroid cold mass and the bending-down effect due to the presence of the end-cap toroid magnetic field (left); barrel toroid coil cross-section showing the tie rods and how they are working to transfer the force into the warm support structure (right).

toroid structure has to sustain axial compression of 2×240 tons. When the alignment is correct the force transfer is uniform and every barrel toroid vessel is subjected to 2×30 tons compressive force.

The effect of the presence of the barrel toroid field in the end-cap toroid windings is large as well; see Fig. 15 (right). While the local Lorentz force per meter remains about the same on the inner leg, on the outer leg the forces and stress increase by 60%. The Lorentz forces acting locally on the barrel toroid cold mass change as well, which is shown in Fig. 16.

The deflections in the cold mass are in the 1–4 mm range. The combined load of the two toroids clearly provokes an extra bending-down effect in the cold mass. The load in the rightmost tie rod goes up by 200% to 188 tons, equivalent to 340 MPa Von Mises stress in the Ti tie rods. The additional stress induced locally in the windings due to the combined magnetic field is about 60%.

The 240-ton axial force working on the end-cap toroid introduces a displacement of the end-cap toroid of about 4 mm with respect to the barrel toroid and is composed of deflection of the brackets and spring effect of shims and rubbers at the interfaces. After many cycles the remnant deflection is less than 0.2 mm. The repeatability of the toroid positions after a few openings and closings was better than 1 mm — a good result given the weight and size of the objects to be moved.

7. Central Solenoid in the Barrel Toroid Bore

The solenoid pictured in Fig. 17, provides a 2 T axial magnetic field and is designed for minimum material in the radial direction to minimize particle scattering.[18] It is a single layer coil wound inside a thin Al alloy cylinder using an Al-stabilized NbTi conductor; see Table 1. The coil package is only 45 mm thick and thus extremely lightweight. The 5.6-ton light 39 MJ featuring magnet has an energy-to-mass ratio of 7 kJ/kg, which was made possible by enhancing the yield stress of the Al stabilizer instead of increasing the support cylinder thickness, as discussed in Sec. 2. The coil was entirely manufactured in industry[19] and integrated in the LiArgon calorimeter cryostat

Fig. 17. The 2-tesla central solenoid of ATLAS, featuring an extremely lightweight design to minimize particle scattering in front of the calorimeters. The winding structure against a thin outer support cylinder is shown in the inset.

at CERN. To house the solenoid in the calorimeter cryostat is a novel feature as well, and is done for the same minimum-material reason. Another novel feature is the use of triangle coil supports, fixed at one end, sliding at the other coil end, and that maintains the position of the coil with respect to the cryostat. Cooling is provided by a serpentine cooling circuit welded to the outer support cylinder. The coil requires a one-week cool-down, and the ultimate test in the ATLAS cavern up to a nominal field was successfully completed in summer 2006.[20]

8. Magnet Operation Necessities

Apart from coil construction, essential operational services required to run the magnets are a pumping system for creating and maintaining isolation vacuum in the cryostats, liquid helium coolant circulation in the cold mass cooling pipes, an electrical circuit that brings current to the coils and, last but not least, a magnet safety and control system.[21]

The isolation vacuum in the ten toroid coils is maintained by a total of 20 nonmagnetic diffusion pumps with $3000\,\mathrm{m^3/h}$ capacity, 2 per coil for redundancy reasons. The pumping capacity installed is actually about five times of what is needed when cryostats are perfectly sealed. The overcapacity is nonetheless installed to cope with eventual leaks and therefore to maximize the detector on time.

The ten toroid coils and the solenoid are kept at about 4.6 K by forced helium flow in the Al alloy cooling tubes that are connected to the cold masses. The cryogenic system of ATLAS is illustrated in Fig. 18.

The concept is to split helium production from helium circulation in the coils and to put a buffer in between. In the case of power cuts and other interruptions of services, the buffer guarantees a safe rundown of the magnets during a 3 h period. This is very beneficial in avoiding forced fast dumps and consequently 4–5-day interruption of detector operation, the period needed to recool the magnets after a provoked quench.

Helium is supplied by a 6 kW main refrigerator, while cooling down from ambient and thermal shield cooling at around 50 K is provided by a 20 kW shield refrigerator. The compressor station and gas buffers are located on-surface and the two refrigerators underground in the adjacent service cavern. The main refrigerator supplies helium to an 11 kL storage dewar and produces the 12 g/s coolant flow for the four pairs of current leads as well. Liquid helium circulation in the cold masses is provided by a 1200 g/s cold helium pump that takes the helium from the buffer in the case of a slow dump. Again for redundancy reasons, a second helium circulation pump is on standby. He

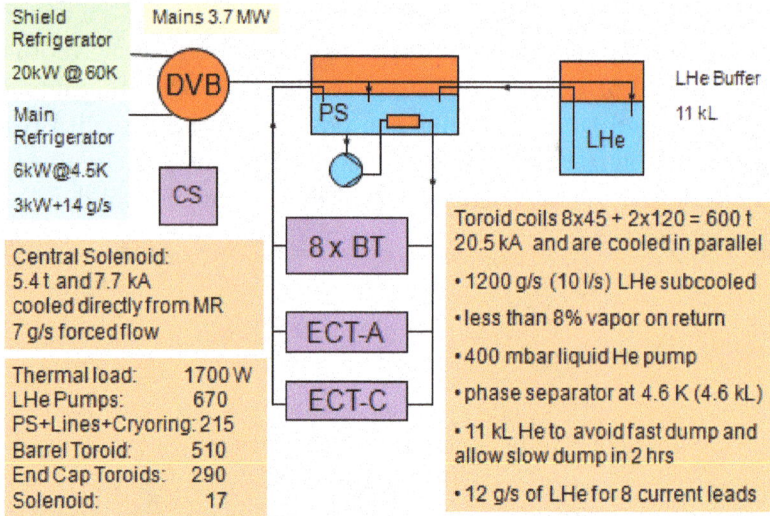

Fig. 18.   Schematic layout of the ATLAS magnet cryogenic system.

flow in the solenoid and ten toroid cold masses is controlled individually to cope with variations in flow resistance and to guarantee helium quality in all coils. The proximity cryogenic systems, like the storage dewar, distribution valve box and pump cryostats and transfer lines to the current lead cryostats, are located in the vicinity of the magnets in the ATLAS experimental cavern, as shown in Fig. 19.

Electrically, the three toroids are connected in series to a 20.5 kA–16 V power converter, see Fig. 20. The solenoid is a similar circuit laid out for 7.7 kA. The power converter, switches and diode/resistor units are located in an adjacent service cavern, and 200 m of aluminum bus bars make connections. A ramp-up of the toroids takes 3 h and de-energizing through a slow dump takes 2 h 40 min. Across the barrel toroid and the set of end-cap toroids, diode/resistor units guarantee safe de-energizing or a slow dump of the coils and limit the voltage to 40 V maximum across the current leads in the case of a fast dump provoked by the detection of a quench.

The magnet control and magnet safety system provides the process controls for the vacuum, cryogenics, electrical systems and the quench protection system triggering a fast dump. Quench detection is by bridge connections across the entire barrel toroid, across the end-cap toroids and across the solenoid. There is a sixfold redundancy in the toroid quench detection, grouped in two physically separated units. Quench protection is arranged by firing heaters in all coils so that a uniform distribution of the cold mass

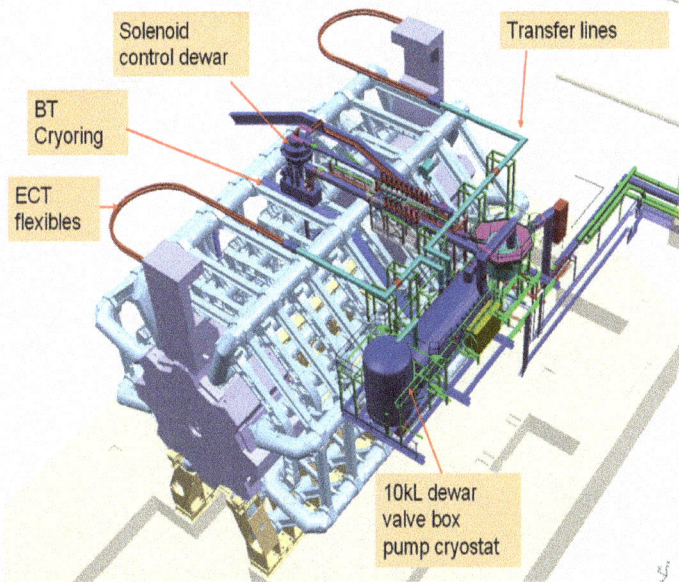

Fig. 19. The magnet's proximity cryogenic system located next to the top of the experiment. Transfer lines bring helium to the current lead cryostats and control dewars of the toroids and solenoid, respectively. Since the end-cap toroids are retractable, the service lines run through flexible chains.

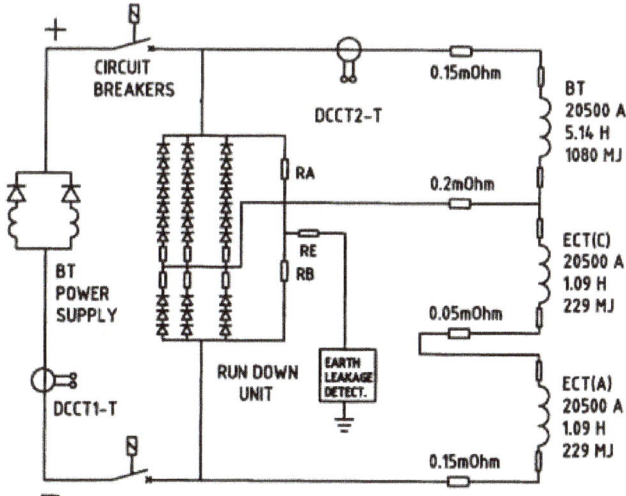

Fig. 20. The electrical circuit of the three toroids connected in series.

heating is achieved. Given the normal zone propagation of 10–15 m/s, a toroid coil is switched to the normal state within 1–2 s. The quench protection heater circuits and power supplies, cabling and heaters show a twofold redundancy. During a quench, the cold mass temperature rises to about 60 K in the toroids and about 100 K in the solenoid, which are very safe values.

9. First Operational Experience

In 2006–2008 the three toroids and the solenoid were commissioned in the cavern, first in stand-alone mode and later in various combinations, after which the experiment was prepared for the first full-dress run. After closing the experiment and all magnets in position, as a last step the LHC beam pipe was put in position (see Fig. 21) to allow the first bunches passing ATLAS.

The three toroids together were ramped up to nominal current for the first time in August 2008 and the historical registration in Fig. 21 was made. The ultimate proof that such huge toroids can be built was delivered.

In September 2008 the first bunches were circulated in the LHC and the magnet system was on and proved to have stable operation during the one-month data-taking period. The presence of a magnetic field in the toroids and the solenoid is nicely shown in Fig. 22, where bent tracks of beam halo events in the muon detectors and bent trajectories of cosmic showers in the inner detector are made visible.

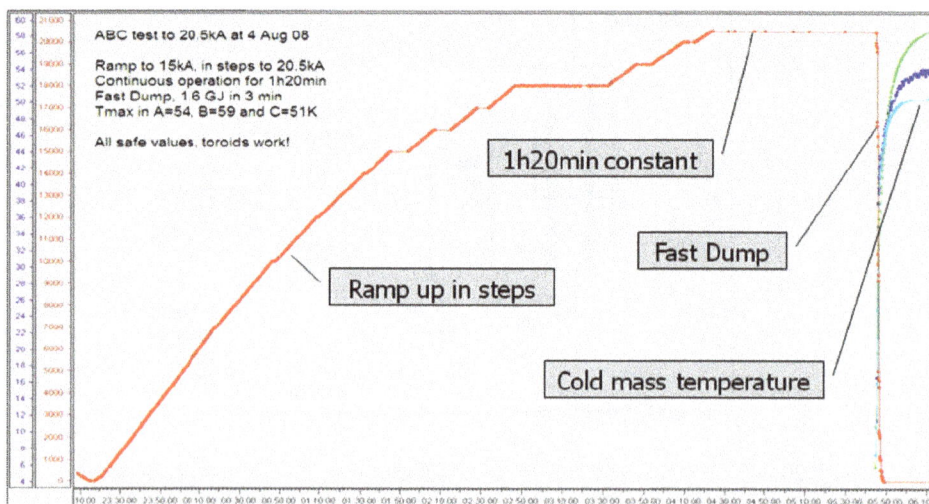

Fig. 21. The first ramp-up to 20,500 A nominal current in the three toroids in August 2008, followed by a provoked fast dump.

Fig. 22. Closing the beam pipe to prepare for running with the beam. On the left is the front plane of the barrel toroid and end-cap toroid on one side of the experiment, in the center are the beam pipe and radiation shielding segments, and on the right is the first large muon wheel.

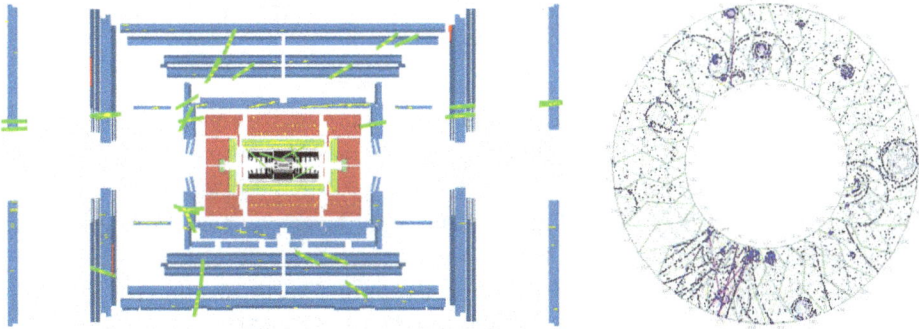

Fig. 23. The first operation of the full detector with the toroids and solenoid on: beam halo bent trajectories in the muon detectors (left) and bent trajectories of cosmic showers in the TRT inner detector.

10. Conclusion

The ATLAS barrel and end-cap toroids and central solenoid have been installed in the cavern, fully tested, and behave as expected. After five years of engineering design and ten years of construction and installation, the ultimate system test in the cavern and the proven stable operation in 2008 have demonstrated that superconducting toroids of this size $(10{,}000\,\mathrm{m}^3)$, complexity, peak magnetic field $(4\,\mathrm{T})$ and stored energy $(1.6\,\mathrm{GJ})$ can be built and operated safely, according to expectations.

Acknowledgments

Thanks are due to the engineers and technicians from the participating magnet laboratories who dedicated their energy and enthusiasm to this project for more than 15 years, CEA–Saclay and INFN–LASA for the barrel toroid, RAL and NIKHEF for the end-cap toroids, KEK for the central solenoid, and JINR–Dubna for their integration and installation works, as well colleagues from CERN service groups for realizing the magnet vacuum, cryogenic, electrical, safety and control systems. This project is funded by the ATLAS Collaboration, presently comprising about 2500 scientific authors from about 169 institutes in 37 countries worldwide, and is supported in part by CERN.

References

1. http://atlas.ch/index.html
2. H. H. J. ten Kate. The ATLAS superconducting magnet system: status of construction & installation, *IEEE Trans. Appl. Supercond.* **16** (2006) 499.
3. http://cms.web.cern.ch/cms/index.html
4. http://en.wikipedia.org/wiki/ATLAS_experiment
5. B. Turck and A. Torossian. Operating experience of Tore Supra superconducting magnets, in *15th IEEE/NPSS Symp. Fusion Engineering 1993*, **1** (1993) 393–398.
6. L. Dresner *et al.* Results of the international large coil task: a milestone for superconducting magnets in fusion power, *Cryogenics* **29** (1989) 875–882.
7. B. A. Mecking *et al.* The CEBAF large acceptance spectrometer (CLAS), *Nucl. Instrum. Methods Phys. Res. Sec. A* **503** (2003) 513–553.
8. G. Volpini, G. Baccaglioni, G. Cartegni, D. Pedrini, E. Baynham, C. Berriaud, B. Blau, I. L. Horvath and H. H. J. ten Kate. Production review of 85 km of Al-stabilized NbTi Cable for the ATLAS Toroids, *IEEE Trans. Appl. Supercond.* **16** (2006).
9. A. Yamamoto *et al.* Design and development of the ATLAS central solenoid, *IEEE Trans. Appl. Supercond.* **9** (1999) 852.
10. J.-M. Rey, M. Arnaud, C. Berriaud, S. Cazaux, M. Humeau, R. Leboeuf, C. Mayri, P. Vedrine, A. Dudarev and H. H. J. ten Kate. Cold mass integration

of the ATLAS barrel toroid magnets at CERN, *IEEE Trans. Appl. Supercond.* **16** (2006).

11. P. Védrine, F. Alessandria, M. Arnaud, C. Berriaud, R. Berthier, A. Dudarev, A. Leone, B. Levesy, C. Mayri, Y. Pabot, J.-M. Rey, L. Z. Sun, H. H. J. ten Kate, G. Volpini and Y. Zaitsev. Manufacturing and integration progress of the ATLAS barrel toroid magnet at CERN, *IEEE Trans. Appl. Supercond.* **14**(2) (2004) 491–494.

12. A. Dudarev, J. J. Rabbers, S. Junker, R. Pengo, H. H. J. ten Kate, C. Berriaud, M. Arnaud, P. Vedrine, F. Broggi and G. Volpini. On-surface test of the ATLAS barrel toroid coils: overview, *IEEE Trans. Appl. Supercond.* **16** (2006).

13. C. Mayri, P. Vedrine, C. Berriaud, M. Reytier, Y. Pabot, S. Cazaux, A. Foussat, A. Dudarev, Y. Zaitsev and H. H. J. ten Kate. Suspension system of the barrel toroid cold mass, *IEEE Trans. Appl. Supercond.* **16** (2006).

14. Z. Sun, B. Levesy, M. Massinger, C. Mayri, Y. Pabot, P. Vedrine, I. Zaitsev, A. Dudarev and H. H. J. ten Kate. ATLAS barrel toroid warm structure design and manufacturing, *IEEE Trans. Appl. Supercond.* **16** (2006).

15. A. Foussat, M. Raymond, H. H. J. ten Kate, B. Levesy, C. Mayri, P. Vedrine, Z. Sun and Y. Pabot. Assembly concept and technology of the ATLAS barrel toroid, *IEEE Trans. Appl. Supercond.* **16** (2006).

16. D. E. Baynham *et al.* Engineering design optimization of the super-conducting end-cap toroid magnets for the ATLAS experiment at LHC, *IEEE Trans. Appl. Supercond.* **9**(2) (1999).

17. D. E. Baynham, F. S. Carr, E. Holtom, J. Buskop, A. Dudarev, P. Benoit, R. Ruber, R. Pengo, G. Vandoni and H. H. J. ten Kate. ATLAS end-cap toroid integration and test, in *Proc. Applied Superconductivity Conference 2006.*

18. A. Yamamoto *et al.* Progress in the ATLAS central solenoid, *IEEE Trans. Appl. Supercond.* **10** (2000) 353.

19. S. Mizurnaki *et al.* Fabrication and mechanical performance of the ATLAS central solenoid, *IEEE Tr. Appl. Supercond.* **12** (2002) 415.

20. R. Ruber *et al.* ATLAS superconducting central solenoid on-surface test, *IEEE Trans. Appl. Supercond.* **15** (2005) 1283.

21. P. Miele, F. Cataneo, N. Delruelle, C. Geich-Gimbel, F. Haug, G. Olesen, R. Pengo, E. Sbrissa, H. Tyrvainen and H. H. J. ten Kate. ATLAS magnet common cryogenic, vacuum, electrical and control systems, *IEEE Trans. Appl. Supercond.* **14** (2004) 504–508.

Chapter 12

CONSTRUCTING A 4-TESLA LARGE THIN SOLENOID AT THE LIMIT OF WHAT CAN BE SAFELY OPERATED

A. Hervé*

Institute for Particle Physics, ETH Zürich
Schafmattstrasse 20, 8093 Zürich
Switzerland
alain.herve@cern.ch

The 4-tesla, 6 m free bore CMS solenoid[a] has been successfully tested, operated and mapped at CERN during the autumn of 2006 in a surface hall and fully recommissioned in the underground experimental area in the autumn of 2008. The conceptual design started in 1990, the R&D studies in 1993, and the construction was approved in 1997. At the time the main parameters of this project were considered beyond what was thought possible as, in particular, the total stored magnetic energy reaches 2.6 GJ for a specific magnetic energy density exceeding 11 kJ/kg of cold mass. During this period, the international design and construction team[b] had to make several important technical choices, particularly mechanical ones, to maximize the chances of reaching the nominal induction of 4 T. These design choices are explained and critically reviewed in the light of what is presently known to determine if better solutions would be possible today for constructing a new large high-field thin solenoid for a future detector magnet.

---

*Previously at CERN, CH-1211 Genève 23, Switzerland.

[a]A full bibliography for the CMS magnet in general and the coil in particular, together with the list of all collaborators, can be found at http://edms.cern.ch/document/CMS-SG-MG-0002.

[b]The author would like to thank the members of the CMS Magnet Collaboration: CEA Saclay (F. Kircher *et al.*), ETH Zürich (I. "Steve" Horvath *et al.*), Fermilab (R. Smith *et al.*), INFN Genova (P. Fabbricatore *et al.*), ITEP Moscow (V. Kaftanov *et al.*), University of Wisconsin (R. Loveless *et al.*) and CERN (D. Campi *et al.*) for their dedicated work, as well as the CMS collaboration for their constant support. A special mention goes to the Technical and Accelerator Sectors of CERN for their invaluable technical support.

1. The Role of the Coil in the Experiment Layout

The most important aspect of the overall detector design is the configuration and parameters of the magnetic field for the measurement of muon momenta. The requirement for good momentum resolution, without making stringent demands on the spatial resolution of muon chambers, leads naturally to the choice of a high solenoidal magnetic field. A long superconducting solenoid (12.5 m) has been chosen, with a free inner diameter of 6 m and a uniform magnetic induction of 4 T. The muon spectrometer then consists of a single magnet allowing for a simpler architecture of the detector. The inner coil radius is large enough to accommodate the inner tracker and the full calorimetry. The magnetic flux is returned via a 1.5-m thick nearly saturated iron yoke instrumented with four stations of muon chambers, see Fig. 1. The CMS magnet[1] is thus the backbone of the CMS experiment as all subdetectors are supported from it, see Fig. 2. It must be noted that the thickness of the CMS yoke is minimal as a nonnegligible stray field is acceptable; this

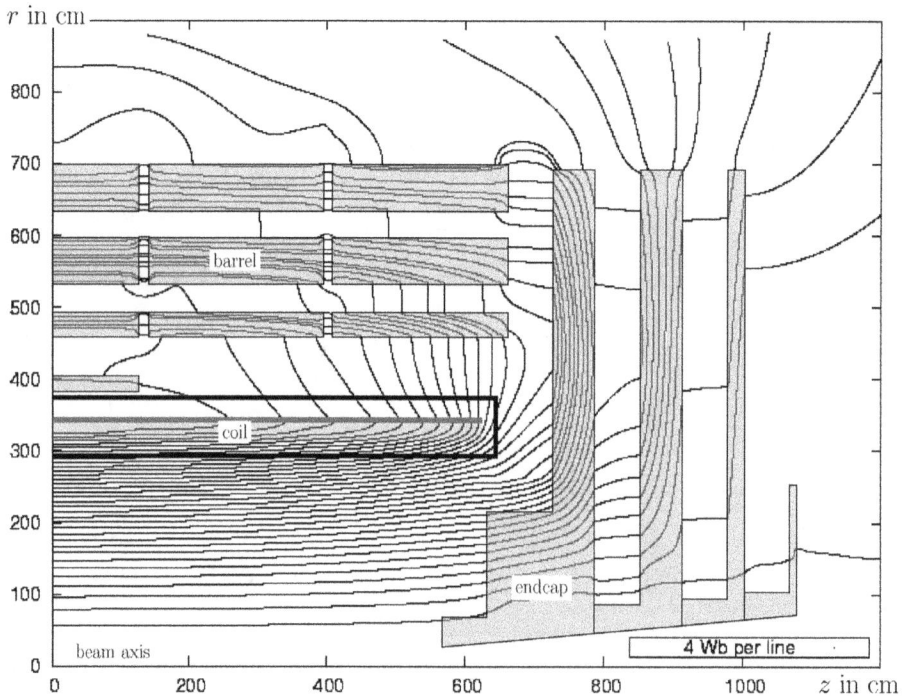

Fig. 1. Field lines of the CMS magnet. The coil is shown inside the cryostat, together with the iron masses of the return yoke leaving large pockets in the barrel section and large gaps in the endcap section to insert the four stations of muon stations.

Fig. 2. View of the coil before the installation of any detector; one can see the pockets in the barrel return yoke to receive the muon stations, and the two side rails imbedded in the inner vacuum vessel to support the hadronic calorimeter barrel.

may be different for future projects which plan to use a push–pull scenario accommodating two nearby experiments.

The magnet has been assembled,[2] tested[3,4] and mapped[5] in a surface hall during the autumn of 2006, and it has been lowered in the underground area by heavy lifting means, see Fig. 3. This bold choice has decoupled the work on the magnet assembly and test from the construction of the experimental area. The magnet has been fully recommissioned underground during the autumn of 2008.

2. Coil Description

The coil is an indirectly cooled, aluminum-stabilized, four-layer superconducting thin solenoid built in five modules. Its main parameters are given in Table 1.

For physics reasons, the radial extent of the coil had to be kept small, and thus the CMS coil is in effect a "thin coil." The thickness of the coil is

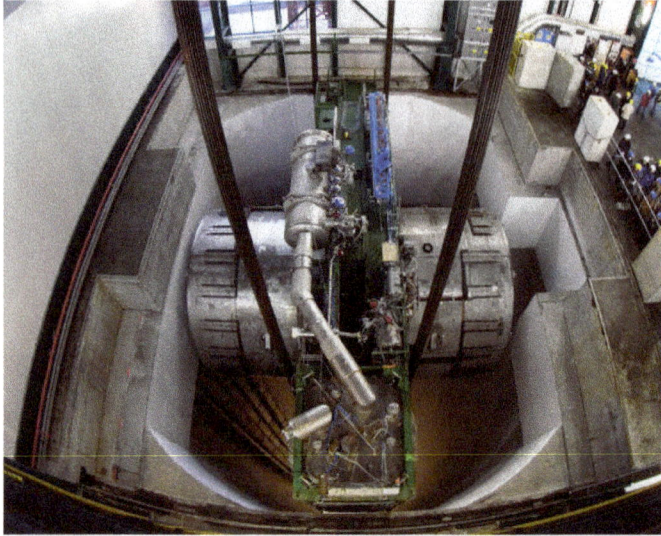

Fig. 3.   View of the coil supported from the central section of the barrel yoke, a 2000-ton load, being lowered in the experimental area by heavy lifting means through the 15 m × 21 m shaft opening. It took 10 h to reach the destination 100 m below.

Table 1.   Main parameters of the CMS coil.

Parameter	Value
Magnetic length	12.5 m
Free bore diameter	6.0 m
Central magnetic induction	4.0 T
Maximum induction on conductor	4.6 T
Nominal current	19,140 A
Average inductance	14.2 H
Stored magnetic energy	2.6 GJ
Stored energy per unit of cold mass	11.6 kJ/kg
Operating temperature	4.5 K
Temperature stability margin	1.8 K

3.9 X_0 (radiation length) and the specific energy ratio is 11.6 kJ/kg of cold mass. As seen in Fig. 12 of Ch. 1, this parameter compares favorably with other thin coils.

The coil has been wound using the inner winding method, and the external mandrels are used as quench-back cylinders providing intrinsic protection. The coil is indirectly cooled by saturated helium at 4.5 K circulating in the thermosiphon mode through a network of pipes welded to the external mandrels.

The first large superconducting detector magnets appeared at the beginning of the 1970s at CERN and DESY; the Cello and TPC magnets were the first thin solenoids ever built, culminating with the construction of the Topaz, Aleph and Delphi magnets at the end of the 1980s (see Ref. 6 for a full review of the detector magnets). Thus, very naturally, at the inception of the CMS project, in 1990, the Aleph solenoid[7] was taken as a demonstrator, and several of its good points were directly adopted:

(i) Use of NbTi with pure aluminum stabilization, and aluminum alloy construction to prevent problems with differential thermal contraction.
(ii) Targeting 1.8 K for the stability margin.
(iii) Use of an external mandrel to support an indirect cooling loop, and inner winding of the coil insulated with fiberglass and epoxy resin, potted using vacuum impregnation.
(iv) Take advantage of the external mandrel to benefit a passive protection scheme based on the quench-back effect.
(v) Keep the mean temperature of the coil below 130 K if all the magnetic energy is taken by the enthalpy of the cold mass, in case of malfunctioning of the active elements of the protection system, and 80 K in normal cases.

3. Challenges of Large High-Field Epoxy-Potted Superconducting Coils

3.1. *Quench protection method*

The protection of the coil in the case of a quench is considered the most critical issue because of the large stored magnetic energy. The coil protection system is based on the so-called quench-back effect and discharging the coil on outer resistors.

The winding cylinder configuration is basically a current transformer in which the winding acts as the primary and the cylinder as the secondary. During normal coil operation or slow discharge, the induced current in the coil cylinder is negligible due to the very low rate of current variation. When a fast current decrease occurs the induced loss is sufficient to heat the mandrel and drive the coil everywhere to the normal state within a few tens of seconds.

3.2. *Thermosiphon cooling method*

The thermosiphon cooling method[8] has been chosen because it operates continuously without the need for active pumps. In the case of a quench it spontaneously switches off, and gives rise to a limited pressure rise because of the small quantity of liquid helium involved in the cooling circuit.

3.3. *Enthalpy stabilization*

A superconducting coil is stable if the local thermal capacity is sufficiently large to prevent excessive temperature rise. In contrast to total cryostability, enthalpy stabilization only allows limited thermal disturbances. This is acceptable for detector magnets, as these perturbations can only be mechanical, epoxy cracking or conductor slippage.

   The choice of a low current density ($\sim$30% of today's state-of-the-art critical current density of $3000\,\mathrm{A/mm^2}$ at 4.5 K under 4.6 T) results in a relatively high critical temperature of 6.3 K, giving a temperature safety margin of 1.8 K for a working temperature of 4.5 K at nominal current. Because of the presence of the low resistivity stabilizer and the heat diffusion within the winding, even short local transient temperature excursions into the resistive domain are supported.[9, 10]

3.4. *Mechanical stresses and strains*

The Lorentz force $\mathbf{f} = \mathbf{I} \wedge \mathbf{B}$ creates magnetic forces that increase quadratically with the field because B is proportional to I. As a first approximation the solenoid can be considered as a thin tube submitted to an internal pressure $B^2/2\mu_0$. For large coils up to 1.5 T, this was not a real concern, however, as B goes over 2.5 to 3 T, the electrical conductor alone is not sufficient to safely contain this pressure. Structural material (aluminum alloy) has to be added either as a thick mandrel or more efficiently on the conductor itself to create a compound conductor.

   In addition to the internal pressure due to the interaction of the current with the axial component of $\mathbf{B}$, one must not neglect the large compressive forces inside the coil due to the interaction of the current with the radial component of $\mathbf{B}$. This effect increases with B as the magnetic permeability of the end cap iron is decreasing rapidly with B; more flux tends to cross directly the coil without going to the end cap (see Fig. 1). This axial stress has to be added quadratically to the hoop stress (alternatively, the equivalent von Mises stress has to be considered) to determine the actual stress of the material and compare it to the allowable stress compatible with construction codes. In addition, everywhere inside the coil, the local strain on the stabilizer must be kept under control and, also for this, the use of a compound conductor has been considered beneficial.

3.5. *Glass-epoxy insulation*

The elegance and efficiency in the design and construction of this type of coils is coming from the use of glass-fiber potted with epoxy to ensure electrical insulation from turn to turn, layer to layer and toward the mandrel. This electrical insulation is also used as "glue" between all these elements to make the cold mass behave as a mechanically solid structural element. In addition this glue is used as "thermal conducting compound" to ensure good thermal conduction between turns and between layers toward the mandrel, which is the only element cooled by circulating helium (see Fig. 4).

Needless to say, the good quality and stability with time of the glass-fiber-epoxy compound is fundamental to the good behavior and functioning of this

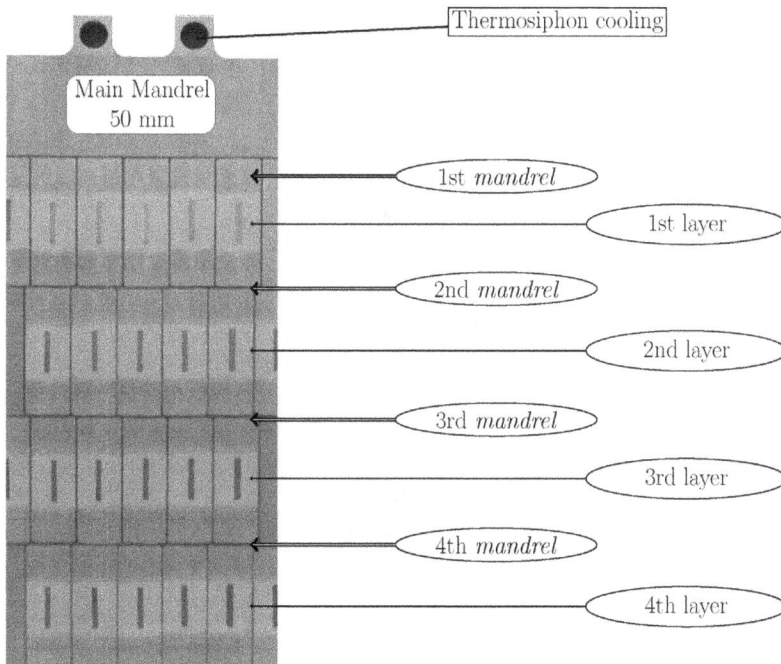

Fig. 4. View through a cut or the CMS coil. It must be noted how the glued combined reinforcement sections provide in effect individual *mandrels* for each layer of current carrying elements where the magnetic forces are created. The external mandrel is only 50 mm thick, and cooling is insured by conduction from the 4.5 K helium circulating in the pipes at the outside of the mandrel.

type of coil, and the designer is constantly looking for solutions that decrease the shear stress on glass-fiber-epoxy boundaries. Not only any cracking of the insulation would lower the mechanical and thermal behavior of the coil but, in addition, it must be realized that the sudden cracking of a small portion of insulation when ramping the field is sufficient to generate enough heat to quench the coil. When the central field increases, the general strain increases and the glass-fiber-epoxy compound is submitted to higher and higher dangerous stresses.

This is worrying, as the integrity of the glass-fiber-epoxy compound depends on its good sticking to the surfaces of the conductor and mandrel, and this sticking depends on the good cleaning and degreasing, sandblasting, geometrical quality of the previous layer and a perfect vacuum impregnation process — all operations that are difficult to control for a large coil. The design specification of CMS mentioned that no part of the insulation should be in tension (only compression stress is allowed) and the shear stress should not exceed 30 MPa. In fact this is possible using the compound conductor and orienting the design to keep the shear stress under control everywhere, in particular near module-to-module interfaces.

4. Departing Parameters

Although the general Aleph design has been adopted, numerous improvements were needed to match the challenge of increasing the central induction to 4 T, namely:

(i) Even when one chooses a much higher current (20 kA), a four-layer construction in five modules is needed. This imposes new constraints on the quality of the winding process to keep the geometrical quality, layer after layer.

(ii) The equivalent magnetic pressure of 64 bars acting on the coil necessitates the use of high-strength aluminum alloys for structural elements.

(iii) The founding assumption of the CMS design, introduced by J. C. Lottin,[11] was to position this high-strength aluminum alloy in direct contact with the conductor to create, in effect, a compound-reinforced conductor.

(iv) The circumferential hoop-strain has to reach 0.15% to stay within a reasonable thickness for the cold mass (310 mm), so that the radial extent of the coil system including the vacuum vessel is limited to 850 mm.

(v) In addition to the hoop-stress, a very large axial magnetic force inducing axial stresses has to be transmitted from module to module.

(vi) The stored magnetic energy being 2.6 GJ, the energy density per unit of cold mass reaches 11.6 kJ/kg, and this placed the coil away in the parameter space compared to previous superconducting solenoids (see Fig. 12 of Ch. 1), and, clearly, innovative solutions were needed to deal with this challenging positioning.

5. The Compound Conductor

The classical aluminum-stabilized conductor is a soft component, not able to sustain high working stresses. In the previous solenoids the magnetic stresses were low enough to be transferred to the aluminum alloy external cylinder. For CMS, this force containment method would lead to a thick external cylinder and to excessive shear stresses, especially at the winding–mandrel interface. By associating directly with the conductor most of the required structural material, the winding becomes a self-supporting structure (see Fig. 4).

6. Design Challenges at the Inception of the Project

The main design challenges have been identified, very early on, as the following:

(i) How to obtain a stabilized conductor having the necessary section to provide the 1.8 K stability margin? One must consider that the section of the stabilized superconducting element (the insert in Fig. 5) is an order of magnitude larger than what had been produced up to 1990.

(ii) How to get a compound conductor having the necessary mechanical strength? Using a compound-reinforced conductor has always been considered, by the design team, the only way to allow controlling strains, and keep low shear stress in the insulation. The fact that there is only one element to wind (with an excellent geometry due to the final in-line milling) eases the challenges of producing a high-quality multilayer winding. Also, if well chosen, the reinforcing alloy sections even harden during the curing process; meaning that the ultimate elastic limit is obtained only after the winding process is completed.

(iii) How to wind a very stiff conductor in four layers? The inertia of the compound conductor is very large compared to what had been done so far.

(iv) How to build precise mandrels? Because of the high stress, it is not possible to thermally stress-relieve the mandrels so as to stabilize them; nevertheless, good cylindricity is required to obtain a precise coil.

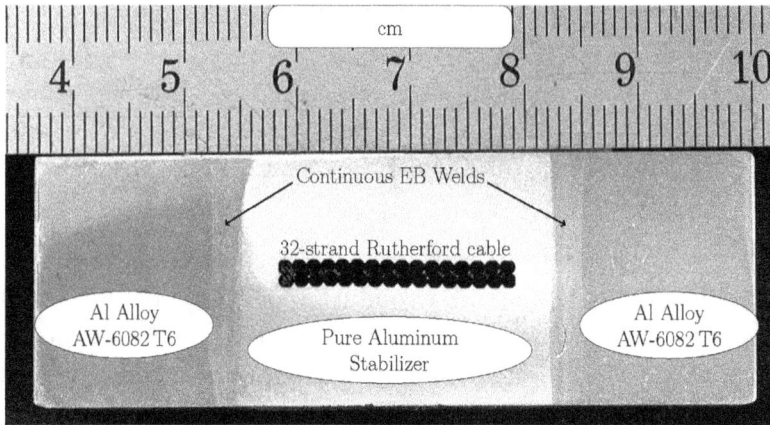

Fig. 5. Cross section of the CMS reinforced compound conductor. The inner part, the so-called insert, is a conventional conductor, Rutherford cable inside a pure aluminum matrix, to which is added by the continuous electron beam process the two aluminum alloy sections.

(v) How to limit shear stresses inside the coil? The use of a reinforced conductor solves this problem inside the modules; however, in the region between two modules the large compression force must be transmitted without creating undue shear stresses, particularly inside the insulation layer.

(vi) How to mount such a heavy coil (220 tons) with a horizontal axis inside the vacuum vessel?

7. How the Design Challenges Have Been Met

7.1. *Superconducting insert*

Building on the pioneering work done for the first thin solenoids, the super-conducting element of the conductor — the insert (the inner part of Fig. 5), a NbTi Rutherford cable stabilized with pure aluminum in a continuous extrusion with the needed section — was industrialized in the early 1990s,[12] and the process was subsequently used not only for CMS but also for the conductor of the Atlas toroids.

7.2. *Compound conductor*

Identification of the best geometry for the compound-reinforced conductor occupied most of the early effort and finally, as is often the case, the simplest geometry proved to be the best one (see Fig. 5).

How to attach the two reinforcement sections to the insert was the next challenge. Ambitious solutions involving parallel coextrusion were tried and abandoned. Continuous electron beam (EB) welding[13, 14] was adopted because it satisfied fully the requirements, there was expertise in the team, and a competent medium size firm interested in the manufacturing challenge was found nearby.

One difficulty has been to obtain the reinforcing sections in 2.6 km lengths with good mechanical uniformity, although they were not produced by continuous extrusion.

Several high-strength aluminum alloys in various tempers have been considered for the reinforcement. The typical alloy EN AW-6082 has been selected for its high toughness, good extrudability and weldability. The underaged state (T51) allows winding to proceed with a not-too-stiff conductor, nevertheless resulting after the curing cycle to a T6 temper with further overall increase in the tensile yield strength, reaching at 4.2 K an $R_{p0.2}$ of 400 MPa and an R_m of 670 MPa (see the right of Fig. 6), largely sufficient for the application.[15, 16]

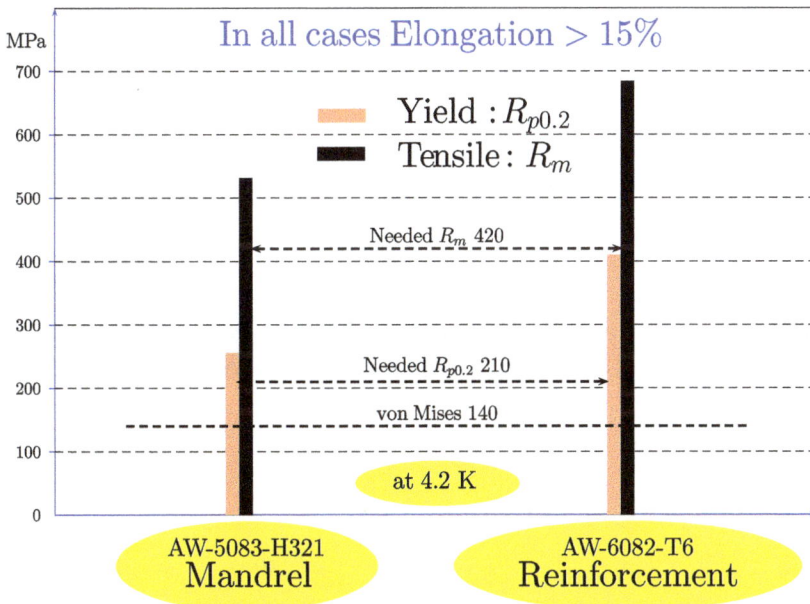

Fig. 6. Yield and tensile strength at 4.2 K of AW-5083-H321, used for mandrels, and AW-6082-T6, used for the reinforcement of the conductor. It must be noted that the 5083 in the fully annealed O state, or in plates thicker than 100 mm, does not satisfy the requirements in weldments.

7.3. *Selection of material for the mandrels*

To get aluminum plates with the necessary mechanical properties has been difficult. In fact, it seems impossible to build mandrels by welding together plates, with the required mechanical properties and thicker than 100 mm. This proved to be an *a posteriori* justification for the use of the compound-reinforced conductor. An extensive and comparative characterization of the low-temperature properties of different aluminum alloys and their welds resulted in the selection of the general purpose EN AW-5083-H321 as the base material for the welded fabrication of the shells. The strain-hardened and stabilized temper H321 guarantees an $R_{p0.2}$ of 254 MPa and an R_m of 510 MPa measured on plates at 4.2 K (see the left of Fig. 6). The MIG weldability of the alloy has been proven and specified. Moreover, the $R_{p0.2}$ of the joint at 4.2 K, produced with an EN AW-5556 filler, has been measured to be between 210 MPa and 233 MPa, which is satisfactory for the application.[17]

Large plates 75 mm thick in EN AW-5083-H321 (to be reduced in thickness to 50 mm by machining after welding) are industrially available and certified according to ASTM B209M. Nevertheless this standard covers, for plates in H321 temper, a thickness of up to only 80 mm guaranteeing a minimum room temperature (RT) $R_{p0.2}$ of 200 MPa. Since the strength decreases with increasing thickness, the properties associated with this temper would no longer be certified or adequate for the flanges or the shoulders thicker than 80 mm.

As an alternative to a welded construction, the ten end-flanges are obtained from seamless rings in the same alloy.

The selection of the ring rolling technique for thick components resulted in fully satisfactory properties,[18] with improved levels of homogeneity and isotropy of tensile properties compared to thick plates (minimum $R_{p0.2}$ measured at 4.2 K in the axial direction over 270 MPa), and exceptional ductility in the longitudinal direction, even increasing with decreasing T (from 18% at RT to 28% at 4.2 K).

7.4. *Manufacture of mandrels*

Almost all large solenoids constructed to date have used mandrels made up of bent plates in EN AW-5083-O (fully annealed). Several intermediate stress-relieving thermal heat treatments were applied during manufacture to relax stresses and fulfill geometrical tolerances. This is not possible anymore if one wants to maintain the mechanical properties of EN AW-5083-H321 up to the end. The firm in charge of the coil winding has employed a technique of

1-Machining inner surface   2-Machining outer surface

n times

1

2

3-Winding operation   4-Completed winding
ready for impregnation

3

4

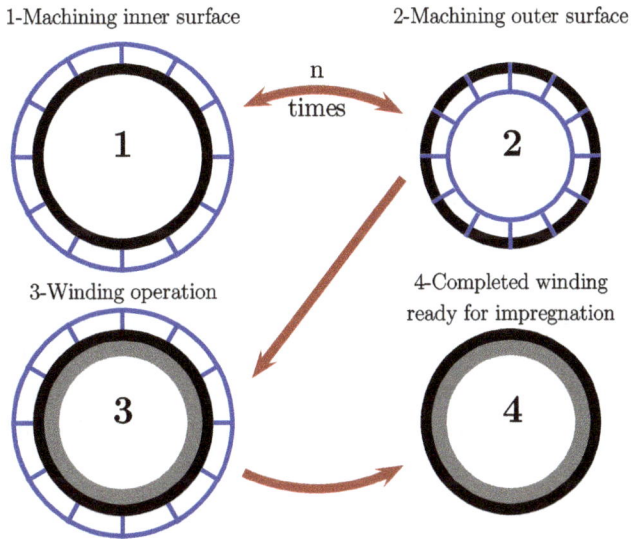

Fig. 7. Various phases of manufacture of mandrels and winding to obtain a cylindrical coil module without heat-treating the mandrel during or after machining.

construction in which referencing tooling is used, either inside or outside the mandrels, to allow machining respectively outside or inside. Then, keeping the outer tooling, winding can proceed. Once the winding is completed, its stiffness is sufficient to maintain the cylindricity of the ensemble within the 10 mm range, and the outer referencing tooling can be removed (see Fig. 7). A construction foreseeing seamless flanges welded to the shells helps avoid stress-relieving, compared to a fully welded construction, since the seamless rings represent also a good dimensional reference for each module.

7.5. *Winding a stiff conductor*

The inner winding of such a stiff conductor was a worry at first, but, in the end, after a dedicated preindustrialization activity, it proved easy to execute using an ad hoc winding machine.[19]

7.6. *Limiting the shear stress inside the coil*

Controlling the strain and keeping the shear stress low everywhere inside the cold mass has been a continuous driving goal of the design.[20]

Regions for potential high shear stress lie between modules because the full compression force must be transmitted from module to module in a very discontinuous region. To alleviate this difficulty a 220-ton coil in one unit was even considered for some time, but this design was abandoned as being too ambitious. In particular, it was not possible to obtain Rutherford cables longer than 3 km without junctions.

A novel design has been suggested by Ansaldo-Superconductori. It uses a set of wedges to terminate each layer so that the end of each module is already fairly flat. The subsequent gluing, after impregnation, of 5-mm thick G11 plates and the final machining of these plates, using a vertical lathe, allowed improvement of the flatness, thus permitting a perfect coupling from module to module.[21]

In fact, the use of a dedicated workshop to construct the mandrels near the winding area has proven to be a tremendous asset. Not only did it save numerous difficult transports of large pieces, but it also allowed the machining of the top flange region of each 50-ton coil module after impregnation.

7.7. *Coil assembly*

Coupling of the modules with their axes horizontal was abandoned because the shape of each module was difficult to control during this operation. Finally, it was decided to stack the modules with their axes vertical to ensure an easy coupling between modules, thus allowing assembly of the full coil in the vertical position without any deformation (see Fig. 8, left). Then, taking

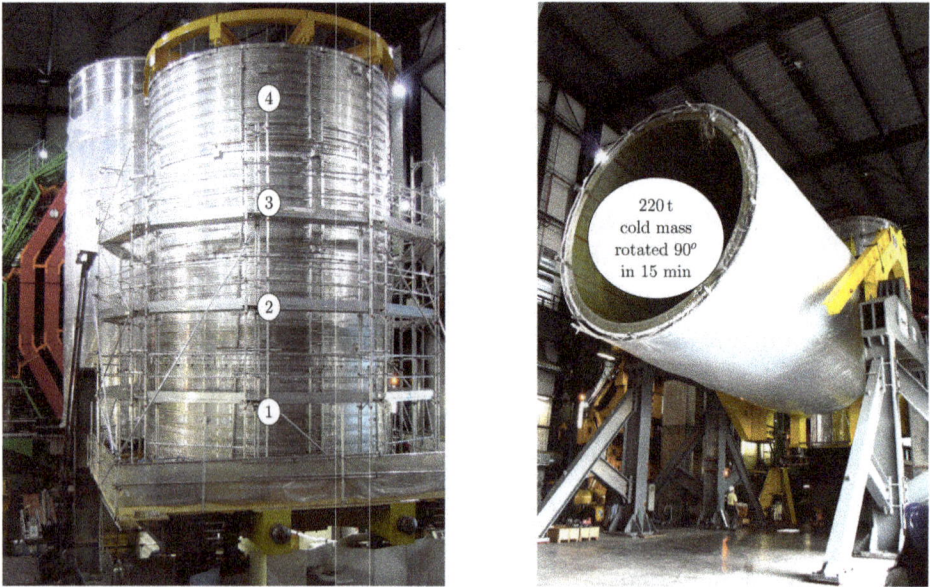

Fig. 8.   Left: View of the cold mass with four out of five modules stacked. Right: View of the cold mass with the horizontal axis ready to be inserted in the cryostat.

advantage of the very large stiffness of the coil, the complete coil was rotated by 90°, and cantilevered from one end (see Fig. 8, right), to allow insertion into the outer vacuum vessel. The same large tooling was used to insert the inner vacuum vessel into the coil.[2]

7.8. *Quality control*

Having to reject any element far down in the manufacturing process has so dramatic an implication for such large projects that it must be avoided at all costs.

Structured quality control, traceability and supervision have been important and continuous actions, in particular for all conductor elements, the finished conductor, the construction of the mandrels, the insulation, the winding and the impregnation.

7.9. *Project organization*

Constructing such a magnet is beyond the capabilities of any high-energy physics laboratory. The solution retained by CMS has been the creation of the *ad hoc* Magnet Collaboration (see footnote b), allowing regrouping international laboratories, having complementary expertise in the necessary fields, and leading working groups for each main task of the project. The management structure has been adapted to reflect it, and all technical and procurement decisions have been proposed by the Project Manager and endorsed by the Magnet Technical Board (MTB), on which all participating laboratories were represented. This has allowed a very integrated project in the framework of a truly international collaboration.

8. Critical Review of Retained Solutions and Their Relevance to Designing a New 4 or 5 T Coil

After the successful operation of the CMS coil, it is interesting to review the main technical choices made by the CMS Magnet Collaboration to maximize the chances of reaching 4 T. As several similar, even larger and more ambitious, coils are being actively proposed,[27–30] it is important to examine if better solutions would be possible today for constructing a new large high-field thin solenoid for a future detector magnet, with central induction in the range of 3.5–5 T; indeed, applying the same general design, possibly with some variations, seems feasible.

8.1. *Critical current and stability considerations*

The electrical performances of NbTi[22] are such that there is no problem in getting the needed current at 4.5 K under the maximum field which may reach 5.8 T for a 5 T coil. However, the temperature safety margin must be sufficient.

For a coil having a central field B_0 and a maximum field on the conductor $B_{\max}$ for the nominal current I_0 at the operating temperature T_{op}, it is customary to consider that one continues to increase the current above I_0 (assuming also that the central field and the maximum field on the conductor increase linearly with the current) until the current I reaches the critical current $I_c = kI_0$ at T_{op} and under the then maximum field on the conductor $kB_{\max}$. This is called the load line of the coil and it is usual to design the superconducting cable and choose the amount of superconducting wires so that $I_0 \sim 2/3 I_c$ (see Fig. 9), i.e. choose $k \sim 3/2$. The nominal current I_0 is then also roughly $1/3$ of the critical current at T_{op} and under $B_{\max}$.

This large safety factor on current is a simple way of specifying the temperature stability margin ΔT_s, which is not so easy to specify. The operation point of the coil is at a fixed operating temperature T_{op} and at a fixed applied maximum field on the conductor $B_{\max}$. The perturbations are local perturbations that generate energy locally and one can easily imagine that the coil remains stable if the local temperature $T_{\max}$ does not exceed the critical

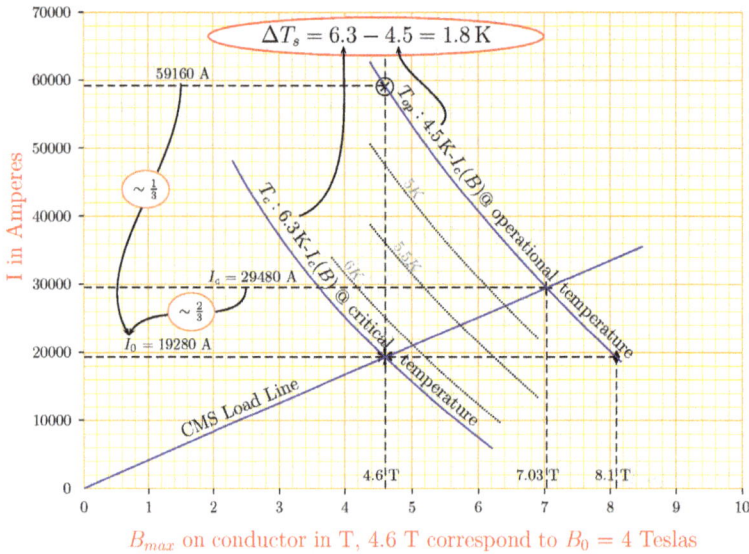

Fig. 9. CMS load line, and computation of the temperature safety margin ΔT_s.

temperature of the conductor carrying a current I_0 under the magnetic field B_{max}. The value $\Delta T_s = T_{max} - T_{op}$ is called the temperature safety margin. It was $2\,K$ for Aleph at $1.5\,T$ and it is $1.8\,K$ for CMS at $4\,T$. When designing for a higher central field (and a higher maximum field on the conductor), one has to be prepared to reduce the temperature safety margin and increase the amount of superconducting wires respecting the present limit of 40 strands for the cabling machine. One possibility of recovering some margin is to operate the coil at $4.2\,K$, accepting having the gas return circuit at a subatmospheric pressure.

If one excludes accidental heating due to a bad internal joint, the remaining perturbations that may generate heat locally are basically a slippage of the conductor or crack development in the fiber-glass-epoxy insulation layer; hence the importance of ensuring by design a monolithic cold mass preventing local movements, in which shear stresses are under control so that the cracking of the insulation layer is excluded. Involved studies have been carried out to identify possible local accidents, compute what energy can be released locally and the corresponding local increase in temperature.[9, 10] However, this remains fairly theoretical, as the possible perturbations are mainly due to manufacturing defects that are difficult to quantify for such large objects; but experience tends to show that if we exclude coils that exhibit an important manufacturing or impregnation defect, the coil systems look more stable than expected. Clearly, other defects may appear like sections of superconducting wire below specification, or bad bonding of the Rutherford cable, and thus the real temperature stability margin is one of the most difficult parameters to understand, specify and keep under control.

One can see from Fig. 10 that the actual CMS Rutherford cable would still exhibit a temperature safety margin of $\Delta T_s = 1\,K$ if used for a $5\,T$ solenoid at $4.5\,K$ ($+$). However, to keep the same temperature safety margin of $1.8\,K$, at least 2.6 times the amount of superconducting wires would have to be used ($\oplus$), and at least 1.7 times the amount for $\Delta T_s = 1.5\,K$ ($\otimes$).

In fact, during the testing of the CMS coil, the central field being constant at $4\,T$, the temperature of the coil has been allowed to drift by $0.8\,K$, up to $5.3\,K$, without inducing any quench,[4] showing that the coil has a sufficient stability margin, but it does not demonstrate that this margin can be safely reduced in new projects.

The conclusion is that no clear design recommendation exists for the temperature safety margin, and the present understanding between experts is that it must be as high as possible and certainly greater than $1\,K$, possibly $1.5\,K$.

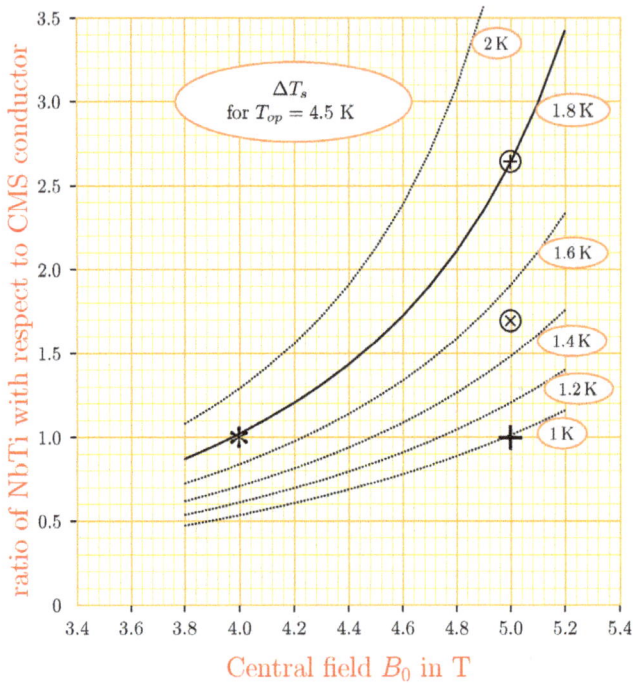

Fig. 10.   Ratio of the superconductor to be used with respect to the CMS conductor to reach a given central field B_0 keeping a given temperature safety margin ΔT_s.

The stability with respect to sudden localized disturbances depends also on the amount, geometry and electrical resistivity of the stabilizer under field and under stress. However, for large conductors the rapid sharing of the current with the stabilizer is impaired by the time needed to diffuse the current away from the cable. In addition, it is known that each stressing cycle decreases the RRR of the stabilizer[23] and, in fact, the stability margin computations for the CMS coil have been performed for degraded pure aluminum with an apparent RRR of 400 under field. Thus, the drive to procure superpure aluminum with an RRR of 3000 (at zero field and no stress) seems a little vain. Hence, the idea of trying to replace the pure aluminum stabilizer by an *ad hoc* alloy with better mechanical properties still having a sufficient RRR.

8.2. *Compound-reinforced conductor considerations*

To use a compound conductor for large high-field, indirectly cooled, potted solenoids seems still a good solution. It allows incorporating in an easy way the quantity of high-strength alloy required by strain considerations.

One alternative to the compound-reinforced conductor is to have a thick mandrel, but, as discussed above, such a thick mandrel seems very difficult to manufacture by welding with the necessary mechanical properties, and one loses the advantage of reacting the magnetic force where it is created, leading to dangerous shear stress inside the coil, in particular at the winding–mandrel critical interface region.

Another alternative to the compound conductor is to wind an aluminum alloy reinforcement section in parallel with the conductor proper. This alternative has been rejected by the design team, as it would render the winding process and the module-to-module coupling much more complicated, giving a winding of inferior quality with more fragile and numerous interfaces filled with fiber-glass-epoxy or even pure resin, and it is known that these are really the weak points of potted coils.

8.3. *Is it necessary to improve the mechanical properties of the alloys for the reinforcement and the mandrels?*

When one is planning to increase the field of a large thin coil, an important question arises concerning the choice of the reinforcing alloy. Do more performing grades have to be used? In fact, the alloys used for CMS for the mandrels and the reinforcement are performing sufficiently for the next generation of coils, and thus it is not necessary to look for stronger alloys. This is coming from the fact that the stored magnetic energy per unit of volume inside the thin solenoid in J/m^3, and the equivalent pressure acting inside the thin solenoid in N/m^2, can have the same expression: $B^2/2\mu_0$, in which B is the central induction in T and μ_0 is the permeability of the vacuum in H/m. First, referring to the left of Fig. 11, we can introduce α, the ratio of the structural aluminum alloy inside the coil (meaning that the pure aluminum stabilizer is considered in the plastic state transmitting only

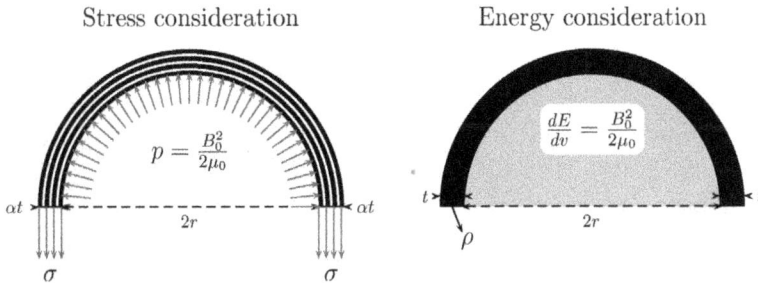

Fig. 11. The coil can be considered as a thin tube submitted to an internal magnetic pressure or to a mass recovering the magnetic energy stored inside its volume.

pressure and taking no circumferential stress); Y, its Young modulus; σ, the circumferential stress in the alloy; and ϵ, the corresponding elongation; and we can write the force equilibrium equation for a thin tube, using also $\epsilon = \sigma/Y$:

$$2r\frac{B_0^2}{2\mu_0} = 2\alpha\sigma t = 2\alpha t Y \epsilon.$$

Secondly, referring to the right of Fig. 11, we can equate the magnetic energy stored per unit length of cold mass with the energy recovered by the same unit length of cold mass (of specific mass ρ) when all the energy is dumped into it, respecting the ratio E/M:

$$\pi r^2 \frac{B_0^2}{2\mu_0} = 2\pi r t \rho \frac{E}{M}.$$

Comparing the two equations, one can relate the magnetic energy per unit mass of cold mass E/M to the hoop strain ϵ of the coil:

$$\frac{E}{M} = \alpha \frac{Y}{2\rho}\epsilon. \tag{1}$$

Thus, the hoop strain ϵ is also a measure of the stored energy per unit mass of cold mass E/M.

On one hand, it seems difficult, respecting construction codes, to exceed a hoop strain of 0.15% as, in addition, the axial stress due to the compression force is not negligible. In the case of CMS, these combined stresses correspond to a maximum von Mises stress of 140 MPa, requiring alloys with $R_{p0.2} > 210$ MPa and $R_m > 420$ MPa at 4.2 K.

On the other hand, one has to be careful about increasing the ratio E/M beyond 12 kJ/kg because, as the enthalpy of aluminum is negligible at $T = 4.5$ K, this ratio is representative of the mean temperature of the coil after a fast dump. In the case of CMS, as 50% of the energy is extracted and dissipated in the dump resistors, this corresponds to a safe mean temperature of 80 K for the cold mass. It covers also the ultimate accidental case, in which the quench detection system or an active element of the dumping system is not functioning and all the energy is dissipated in the cold mass by the passive quench-back protection scheme.

Thus, one can tentatively conclude that the selected alloys, EN AW-6082-T51 for the reinforcement and EN AW-5083-H321 for the mandrels, are perfectly suitable for a 5 T coil designed for ultimate reliability, even in case of failure of an active element of the protection system.

8.4. *Improving the compound-reinforced conductor by replacing the pure aluminum stabilizer*

The considerations on the stability margin open the door to using a less-well-performing stabilizer in terms of RRR in the virgin state, but better-performing in terms of mechanical properties.

There is a development project[26] to manufacture 200 m of an improved CMS reinforced conductor in which the pure aluminum is replaced by the EB-weldable doped alloy Al-0.1wt%Ni, developed by KEK for the Atlas thin solenoid.[24] This alloy in the cold work state reaches a yield strength $R_{p0.2}$ of 85 MPa at 4.2 K, even after a 130°–15 h curing cycle, while maintaining an RRR of 590.

This new compound conductor would be better-performing mechanically, as in Eq. (1) the factor α would be nearly 1. Another clear advantage would be of staying in an almost fully elastic state, thus exhibiting constant properties with time in terms of mechanical properties and RRR.[25]

8.5. *Attaching the reinforcement sections to the insert*

The way the reinforcement sections are attached to the insert can certainly be reviewed. The EB welding process may look like a deluxe one, although it has numerous advantages: the quality is superb and the process is easy to monitor and can be stopped and restarted at any moment during the manufacturing process. In the end the EB welding proved easy to apply.

The design team considered for a while the possibility of attaching the reinforcement sections by continuous soft soldering. This looked promising but, through lack of available effort, this study was stopped as soon as the EB solution proved able to succeed. However, techniques have progressed and it is worthwhile to re-examine the situation and explore new possibilities like laser welding or friction welding.

9. Conclusions

At the start of the CMS coil project, in the early 1990s, several problems were identified that had no known solution, and the construction proper even started with some of these problems still in the pending state. It took years to develop and validate some of the retained solutions.

Problems connected with superconductivity were quickly solved. In fact, looking back, it appears that most of the difficult problems were connected with mechanics, stress and strain distribution in the cold mass and in the insulation layer, and procuring components with the necessary mechanical properties, in the required size for mandrels, or along the full conductor lengths. With the drive to increase the field, these mechanical problems are likely to become even more dominant. New processes can be developed; however, the passing of the superconductor through all the manufacturing processes imposes severe constraints on these processes not to degrade the electrical performances of the superconductor.

Knowing what is known today the CMS Magnet Collaboration, as a group, concluded that if it had to build a new large high-field thin super-conducting coil, it would basically use what is now referred to as the CMS design, with the following possible improvements: replace the pure aluminum stabilizer by a doped alloy and look for an alternative to electron beam welding if it proves to be clearly less expensive. Otherwise the design would be maintained; the basic reason to resist changes is the fact that one cannot risk constructing a nonperforming large coil by introducing more risky solutions just because they are potentially less expensive.

The CMS design, which is clearly adapted to DC fields staying within the range of NbTi with a sufficient enthalpy stability margin at 4.5 K, can thus be extrapolated up to 5 T. More demanding solenoids, designed for higher fields for example, would necessitate a different approach, like the use of cable-in-conduit technology profiting from future technical developments coming from the ITER project.

References

1. G. Acquistapace *et al. The Magnet Project: Technical Design Report*, Vol. LHCC 97-10, CMS TDR 1 (CERN, 1997).
2. B. Levesy *et al.* CMS solenoid assembly, *IEEE Trans. App. Supercond.* **16**(2) (2006) 517–520.
3. D. Campi *et al.* Commissioning of the CMS magnet, *IEEE Trans. App. Super-cond.* **17**(2) (2007) 1185–1190.
4. F. Kircher *et al.* Magnetic tests of the CMS superconducting magnet, *IEEE Trans. App. Supercond.* **18**(2) (2008) 356–361.
5. V. Klyukhin *et al.* Measurement of the CMS magnetic field, *IEEE Trans. App. Supercond.* **18**(2) (2008) 395–398.
6. F. Kircher. Magnetic field geometries, in *Proc. 42nd Workshop on Innovative Detectors for Supercolliders* (World Scientific, Sciences and Culture, 2004), pp. 31–43.
7. J. Baze *et al.* Design construction and test of the large superconducting solenoid Aleph, *IEEE Trans. App. Superconduct.* **24** (1988) 1260–1263.

8. F. Brédy *et al.* Experimental and theoretical study of a two-phase helium high circulation loop, in *Proc. Cryogenic Engineering Conference, CEC-ICMC05* (American Institute of Physics, 2006), Vol. 51, pp. 496–503.

9. F. Juster and P. Fabbricatore. Thermal stability of large aluminium-stabilized superconducting magnets: theoretical analysis of the CMS solenoid, in *Proc. ICEC-17* (Bournemouth, England; Jul. 14–17, 1997), pp. 369–374.

10. P. Fabbricatore *et al.* Experimental study of CMS conductor stability, *IEEE Trans. App. Supercond.* **10**(1) (2000) 424–427.

11. J. Lottin *et al.* Conceptual design of the 4 T CMS solenoid, in *Advances in Cryogenics Engineering* (Plenum, New York, 1996), Vol. 41 A, pp. 819–826.

12. I. Horvath. Superconducting cable development for future high energy physics detector magnets, *Nucl. Phys. B (Proc. Suppl.)* **44** (1995) 672–676.

13. I. Horvath *et al.* Development of an aluminium stabilized reinforced superconducting conductor, *IEEE Trans. App. Supercond.* **32**(4) (1996) 2200–2202.

14. B. Blau *et al.* The CMS conductor, *IEEE Trans. App. Supercond.* **12**(1) (2002) 345–348.

15. S. Sequeira Tavares *et al.* An improved billet on billet extrusion process of continuous aluminium alloy shapes for cryogenic applications, in *Proc. AMPT, 2001*, Vol. 143–144, pp. 584–590; *J. Mater. Process. Technol.* (Dec. 2003).

16. B. Curé *et al.* Mechanical properties of the CMS conductor, *IEEE Trans. App. Supercond.* **14**(2) (2004) 530–533.

17. M. Castoldi *et al.* Possible fabrication techniques and welding specifications for the external cylinder of the CMS coil, *IEEE Trans. App. Supercond.* **10**(1) (2000) 415–418.

18. S. Sgobba *et al.* Design, construction and quality tests of the large Al-alloy mandrels for the CMS coil, *IEEE Trans. App. Supercond.* **12**(1) (2002) 428–431.

19. P. Fabbricatore *et al.* The winding line for the CMS reinforced conductor, *IEEE Trans. App. Supercond.* **12**(1) (2002) 358–361.

20. A. Desirelli *et al.* Finite element stress analysis of the CMS magnet coil, *IEEE Trans. App. Supercond.* **10**(1) (2000) 419–423.

21. P. Fabbricatore *et al.* The manufacture of modules for CMS coil, *IEEE Trans. App. Supercond.* **16**(2) (2006) 512–516.

22. L. Bottura. A practical fit for the critical surface of NbTi, *IEEE Trans. App. Supercond.* **16**(1) (2000) 1054–1057.

23. B. Seeber *et al.* Variation of the residual resistivity ratio of the aluminium stabiliser for the Compact Muon Solenoid (CMS) conductor under dynamic stress at 4.2 K, *IEEE Trans. App. Supercond.* **10**(1) (2000) 403–406.

24. K. Wada *et al.* Development of high-strength and high-RRR aluminum-stabilized superconductor for the Atlas thin solenoid, *IEEE Trans. App. supercond.* **10**(1) (2000) 373–376.

25. S. Sgobba *et al.* Toward an improved high strength, high RRR CMS conductor, *IEEE Trans. App. Supercond.* **16**(2) (2006) 521–524.

26. A. Gaddi. Considerations about an improved superconducting cable for linear collider detectors. CERN, CLIC-PH-2009-001 (Apr. 2009).

27. ILD Collaboration, in *ILD Letter of Intent for the International Linear Collider* (Mar. 2009).

28. SiD Collaboration, in *SiD Letter of Intent for the International Linear Collider* (Mar. 2009).

29. 4th Concept Collaboration, in *4th Letter of Intent for the International Linear Collider* (Mar. 2009).

30. L. Linsen *et al.* Magnetic considerations for a CLIC detector. Private communication, May 2009.

Chapter 13

THE ATLAS MUON SPECTROMETER

Giora Mikenberg

Department of Particle Physics, The Weizmann Institute,
P.O. Box 26, Rehovot 76100, Israel
Giora.Mikenberg@cern.ch

1. Introduction

The ATLAS muon spectrometer[1] was designed and constructed to trigger and accurately measure high momentum muons in a stand-alone mode over a large rapidity range that extends to 2.7. This has been achieved by embedding the spectrometer instrumentation in a set of large air-toroid superconducting magnets that provide an average field of 0.5 T in a volume of 16,500 m^3. The toroidal geometry of the field allows for equal bending of the muon tracks, independently of the production angle, and therefore a constant momentum (and not transverse momentum) resolution within the rapidity acceptance. Furthermore, the inherent open geometry of an air-toroid allows the minimization of the material encountered in the passage of muons, therefore not only reducing multiple scattering, but also — more important for high momentum muons — reducing the probability of high energy losses in interactions with materials (catastrophic loss), which are then limited to the calorimeters, where such energy loss can be measured. Such a choice for a spectrometer also has inherent problems, in particular the nonuniform magnetic field, which requires a dynamic mapping (since the magnets move inside such a large volume, mainly due to the magnetic forces) with a large number of sensors, to follow the movements of the coils to within 2 mm, and also the fact that the muons' trajectories have to be reconstructed over large distances (up to 14 m) with deviation from a straight line that can be as small as 0.5 mm. These deviations need to be measured to a precision of 0.05 mm, which requires detectors constructed with very high precision (0.02 mm) and a very complex alignment system to follow deformations and movements.

To complement the above requirements, one of the main issues at the LHC is to provide a fast and clean muon trigger that copes with the 40 MHz repetition rate of the machine, while providing a transverse momentum threshold behavior that is uniform over a rapidity range extending to 2.4 and being insensitive to the high background levels present in the ATLAS experiment.

All the above-mentioned requirements have been achieved in the ATLAS muon spectrometer, by constructing a very complex system that encompasses more than 3300 tracking and trigger chamber modules, constructed in 21 different institutions, with more than 12,000 alignment elements and 1800 magnetic probes. Indeed, the full assembly of the ATLAS muon spectrometer has been tested during a period of three months, at the end of 2008, using cosmic tracks. These long cosmic runs have allowed a preliminary evaluation of the spectrometer performance, which approaches the original goals.

Section 2 presents the overall spectrometer design; Sec. 3 deals with the construction and testing of the various detector components; Sec. 4 describes the integration of the various components into the ATLAS experiment; the last section deals with the first test results of the spectrometer performance.

2. Spectrometer Design

The muon system forms the outer part of the ATLAS experiment[2] and it is designed to track penetrating muon tracks over a pseudorapidity range $|\eta| \leq 2.7$, while providing a trigger over the range $|\eta| \leq 2.4$. The performance goal of the spectrometer is to achieve a stand-alone momentum resolution of better than 10% for muon tracks of 1 TeV. Such a requirement translates into a sagitta of 500 μm, leading to the required position resolution of 50 μm, in each of the measurement planes. The low momentum range is also important for some of the LHC physics processes, in particular the physics of the b quarks. This sets the upper limit on the magnetic field, which allows the ATLAS muon spectrometer to trigger muons down to 5 GeV/c.

The spectrometer design is shown in Figs. 1 and 2; it consists of a barrel and an end-cap part. The barrel part of the spectrometer is made up of three concentric layers of tracking detectors (MDTs), located at three different radii at the beginning (BI, $r \sim 5$ m), the middle (BM, $r \sim 7.5$ m) and the end (BL, $r \sim 10$ m) of the toroidal magnetic field, with an average value of 0.5 T in the space between coils. Due to the presence of the toroid coils, the various planes are divided into large chambers (BIL, BML, BOL) between two coils, and small chambers (BIS, BMS, BOS) for chambers in the coil planes. Two stations of trigger detectors (RPCs) sandwich the middle (BML, BMS) tracking layer and a third station is located on one side of the outer (BOL, BOS) tracking layer.

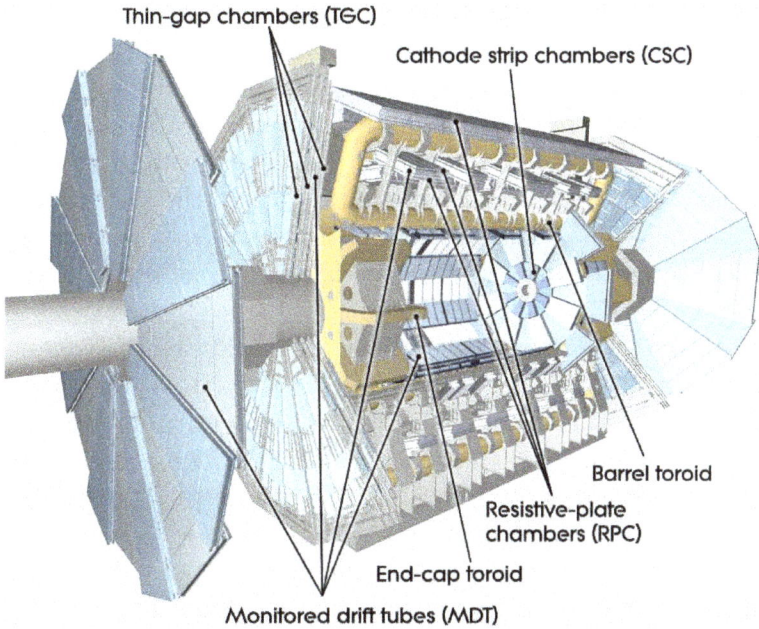

Fig. 1. General view of the ATLAS muon spectrometer, showing the position of the tracking chambers (MDT-CSC) and of the trigger chambers (RPC-TGC), as well as of the three toroidal magnets.

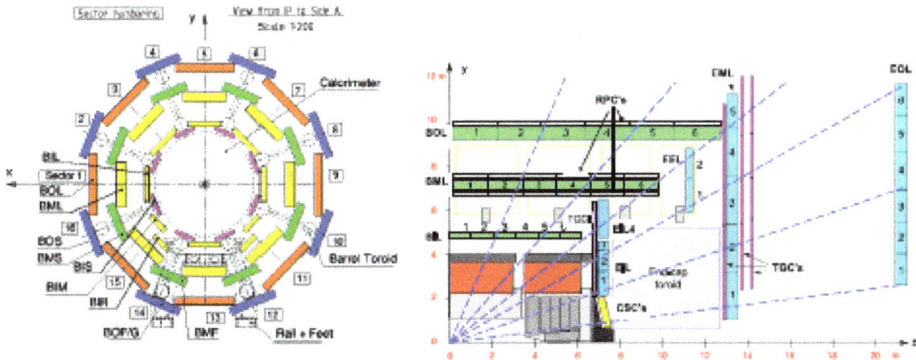

Fig. 2. Cross-sections of the ATLAS muon spectrometer perpendicular to the beam axis (showing the barrel part) and parallel to the beam axis (showing the end-cap part).

The end-cap part of the spectrometer is organized in large circular disks of tracking and trigger detectors planes, located mainly outside the magnetic field (due to the fact that the end-cap toroids are located in a closed vacuum vessel) at various positions along the beams, at $|z| \sim 7.4\,\mathrm{m}$ (EIS, EIL),

10.5 m (EEL, EES; inside the barrel magnetic field), 14 m (EML, EMS) and 21.5 m (EOL, EOS) of the interaction point. Small tracking detectors are also located parallel to the beam axis, on the outer radius of the end-cap toroids (BEE). The tracking detectors are of the same type (MDTs) as in the barrel part, except for the innermost layer, at the high rapidity ($2.0 < |\eta| < 2.7$) region, where, due to the higher detector occupancy, cathode strips chambers (CSCs) are being used. Three stations of trigger detector planes (TGCs, T1, T2 and T3) sandwich the EM tracking layer, and one trigger station plane is in front of the EI layer (TGC-EI).

The main tracking element consists of monitored drift tube chambers (MDTs), which combine high measurement accuracy ($\sim$80 μm per tube), predictability of mechanical deformations and simplicity of construction. The basic structure of an MDT chamber is shown in Fig. 3. It consists of two multilayers of three (four for the BI and EI chambers) layers of pressurized 30-mm-diameter aluminum drift tubes separated by an Al spacer. The tube length varies between 1 m and 6 m, depending on its position in the detector, while the width of such a chamber varies between 1 m and 2 m. The structure of a drift tube is shown in Fig. 4. Its main merit is the simplicity of construction; however, to obtain the needed accuracy, it requires a high quality Al extrusion, as well as a very precise wire positioning device (end plug) at the ends of the tube. This has been achieved, and every single tube, of the 400,000 that constitute the MDT system, has been measured and included in a chamber, provided that the centricity of the central wire is smaller than 10 μm. The end plugs have, furthermore, the gas input/output, as well as

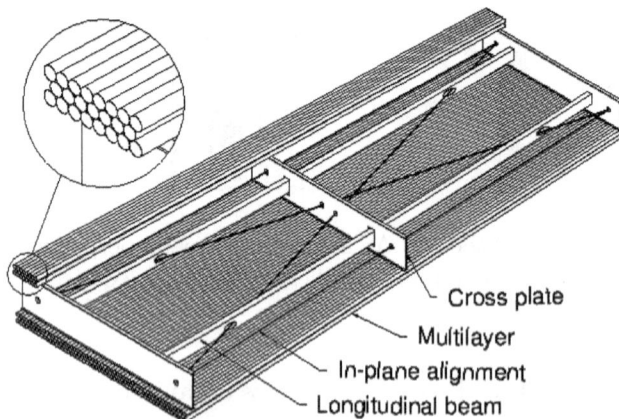

Fig. 3. Schematic view of an MDT chamber showing the two multilayers and the spacer between them, which include the in-plane alignment system.

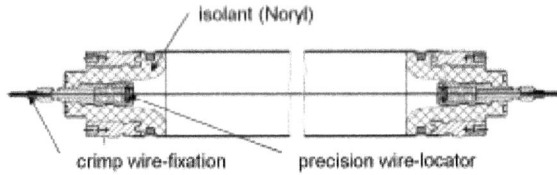

Fig. 4. Longitudinal view of an MDT tube, showing its wire positioning element.

ensuring that there are no gas leaks from the tube (operating at 3 bars), and providing the passage for the HV of the central wire. The drift tubes operate with Ar–CO$_2$ gas (93%–7%) at 3 bars. Ionization is collected at the central 50-μm-diameter gold-plated tungsten–rhenium wire, at a potential of 3080 V. The gas has been chosen due to its ageing properties, with no deposits being observed in the wires after long irradiation campaigns.[3] The disadvantage of this particular gas mixture is its nonlinear drift time relation (which can be corrected for) and its long drift time for the 15 mm maximal drift distance in the tube (or 700 ns maximal drift time). This 700 ns maximal drift time becomes in reality a dead time, since due to the circular nature of the tube one always gets drift electrons from the wall of the tube, independently of the position closest to the wire where the muon has traversed the tube.

The fully tested drift tubes were glued to the Al spacer frame, or to each other using high precision jigs ($\sim$ 3–5 μm) in temperature control clean rooms, to be able to transfer the inherent tube precision to the constructed chamber.[4–7] 15% of the constructed chambers were then measured in a high precision tomograph that allows one, through the use of the wire absorption of x-rays, to find their position with a 3 μm precision.[8] This allowed one to qualify each production site, provided that all wires were at their correct position to within 20 μm, with respect to the particular geometry corresponding to each site. Such high precision can obviously not be kept in a chamber that is made of a very light structure, but since its deformation can be predicted, a system that follows them in real time has been developed. This is done by an internal alignment system, based on four alignment rays (shown in Fig. 3), that are included in every chamber. Such an alignment system, called Rasnik, is shown schematically in Fig. 5.[9] It is based on a checkerboard pattern that is being monitored through a lens by either a CCD camera or a CMOS sensor. This allows one to obtain a transverse resolution to the image of 1 μm and a resolution in the magnification of 0.00005. Images of the checkerboard are being taken continuously and the deformations are followed and corrected in real time.

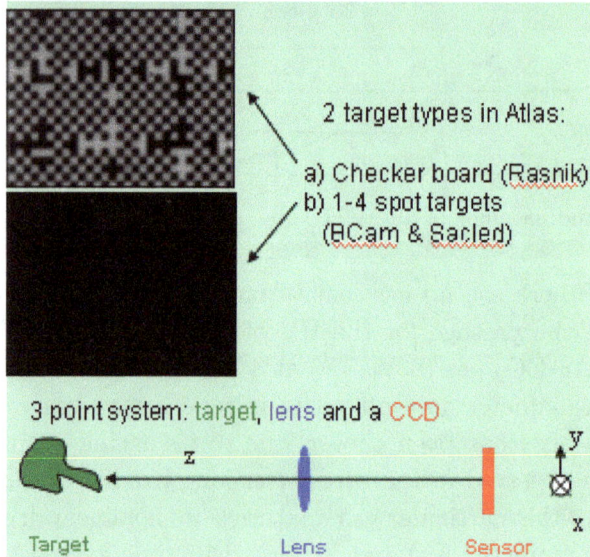

Fig. 5. Schematic of the three-point alignment system. For the checkerboard arrangement it is called Rasnik, while the 1–4 spots are BCAM or Sacled.

The drift signal from the wires is read out through a front-end board that serves 24 tubes.[10] This readout board contains four custom-made ASD (amplifier/shaper/discriminator) chips, the output of which is sent to a custom-made TDC (time-to-digital converter), as well as an ADC (analog-to-digital converter), with the digital information stored in a local buffer. The readout boards of a full MDT chamber are controlled by a local processor called the CSM (chamber service module),[11] which broadcasts the synchronization signals to all the readout boards, and collects the digital information from them when a trigger is issued. This digital information is then formatted at the CSM and sent via an optical fiber to the external readout driver.[12] A lot of emphasis has been put on the fact that 98% of the readout boards could be accessible during a long access (since each one only serves 24 channels, while all the CSM boards in the barrel region can be accessed during any short shutdown of the LHC). For the majority of the end-cap chambers, owing mainly to cost issues the CSMs are accessible only during long shutdowns.

To reconstruct muon tracks with the needed precision (~ 0.05 mm at each point along the bending direction), it is not enough to know the local deformations of each chamber, but one also needs to know where each detector is located in real space and, more important, where each of the three layers that a muon will traverse is located with respect to each other. This

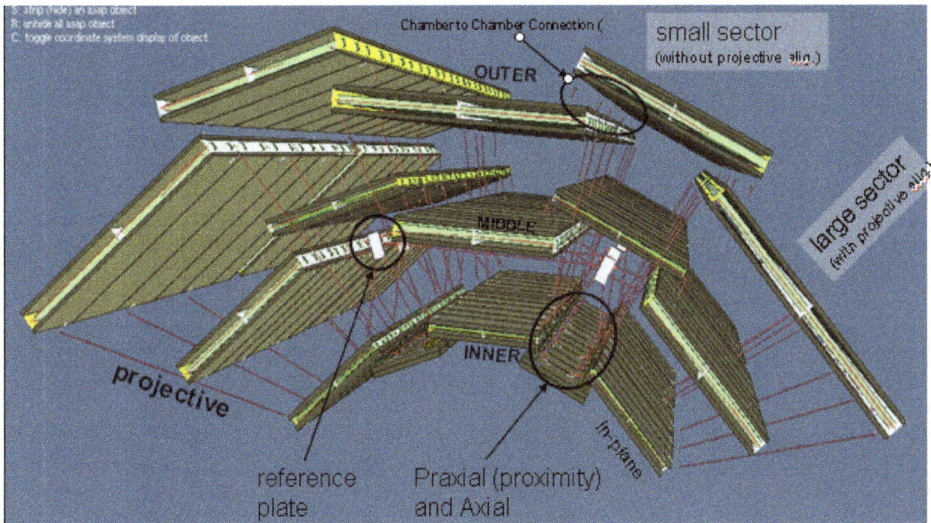

Fig. 6. Schematic of the barrel alignment system, showing the projective rays and the axial–praxial rays, as well as the interconnection between the small and large chambers. The reference system is also shown.

is done by a very complex alignment system that permits one to know within a 0.03 mm precision, where each chamber in a given layer is located relative to the others, and by using triangulation on projective alignment rays, where each layer is located with respect to the others. This process is done differently in the barrel[2] and end-cap regions,[13] due to the presence of the vacuum vessel of the end-cap toroids. The schematics of the alignment system are shown in Figs. 6 and 7, for the barrel and end-cap regions, respectively.

The barrel alignment consists of a praxial system connecting the corners of two adjacent chambers, using two short-crossed Rasniks and a praxial system, parallel to the beam direction that connects a row of six chambers with long Rasniks. Four pairs of projective lines in each half-octant of the barrel (symmetric with respect to the interaction point; see Fig. 6) allow one to know where each chamber layer is located with respect to the others by locating the Rasnik, with the innermost chambers (BIL) containing the mask, the middle layer (BML) containing the lens and the outermost layer (BOL) containing the sensor. Since only the large sectors are equipped with projective lines, there is an additional set of optical lines that connect the small and large sectors (CCCs), as well as optical elements that connect the alignment system to a reference system located in the toroid. In total, 5817 sensors are part of the barrel alignment system.

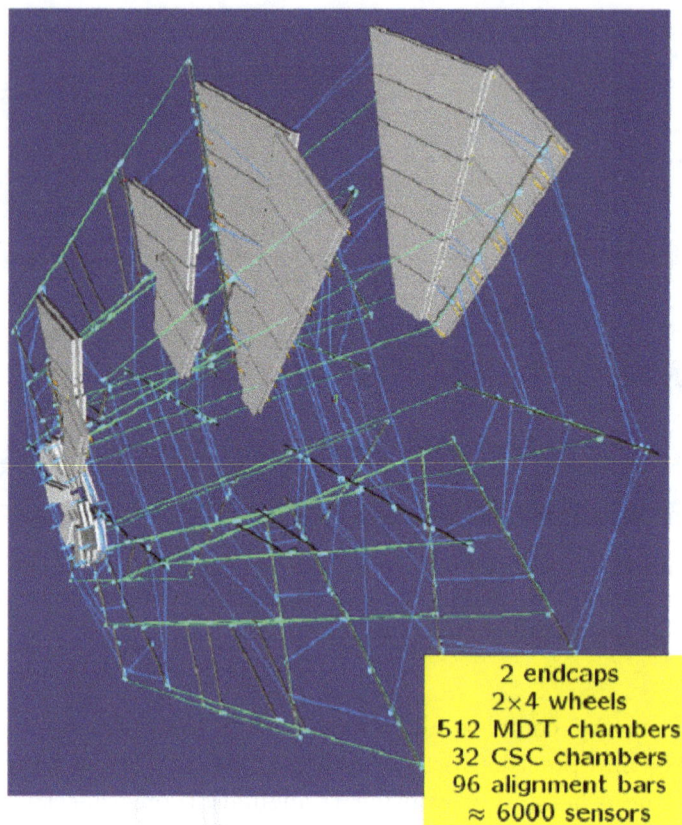

Fig. 7. Schematic view of the alignment rays for one of the end-cap systems, showing the alignment bars and the connections between them. Also shown are chamber-to-bar lines.

The end-cap alignment system is based on a network of optically instrumented Al tubes, called alignment bars.[14] These alignment bars are equipped with Rasniks, which measured their deformations to within a few microns, as well as temperature sensors to calculate their elongation. They are interconnected by a set of projective and azimuthal optical lines that consist of a CCD camera that monitors the position of two or four laser diodes through a local lens. This arrangement, called BCAM, is such that each CCD camera houses also a set of laser diodes that can be viewed by another BCAM. This kind of device can achieve an angular resolution of 5 μrad at a distance of 16 m. The set of alignment bars with their various BCAMs constitute an alignment grid to which the positions of the close-by chambers are being referenced by the use of a Rasnik system. In total, 6536 optical sensors are part of the end-cap alignment system, which also monitors the position of the CSC tracking detectors in the innermost layer of the end-caps.

The alignment system of the muon spectrometer allows one not only to have a very good starting point to improve or corroborate the absolute position of the detectors using tracks, but also to precisely know, at the 1 mm level, where the 1800 magnetic field sensors are located, so as to be able to map with great precision the highly nonuniform toroidal field. Furthermore, the very high relative accuracy of the alignment system allows one to transform high accuracy position measurements obtained with straight tracks (magnetic field off) into the situation where the toroids are turned on, and the various detectors move a few mm.

The tracking system of the ATLAS muon spectrometer is complemented in the innermost layer of the end-cap by a set of 16 four-layer packages of CSCs on each side of the spectrometer. The CSCs are multiwire proportional chambers operating at a gas gain of 60,000 in an Ar/CO_2 (80%/20%) gas mixture with two planes of cathode strips to measure the bending coordinate (5.08 mm pitch) and the orthogonal coordinate (pitch ranging from 12.5 to 21.0 mm). The above-mentioned pitch allows one, from the charge distribution in neighboring strips, to measure the bending coordinate with a precision of 65 μm at a background rate of 1000 Hz/cm^2. A description of the CSC electronics is given in Ref. 2. Their accessibility is such that they can be reached only during long shutdowns of the LHC.

While the majority of the tracking elements of the ATLAS muon spectrometer (MDTs) measure the bending coordinate with very high precision, they provide no information on the azimuthal coordinate of the muon tracks. Furthermore, their long drift time (700 ns) provides no information on the bunch crossing that has produced the muon. This information is provided by the trigger detectors, which have three complementary roles: (1) to provide a trigger to the spectrometer for muons with a transverse momentum higher than a given threshold, (2) to identify the bunch crossing at which the muon was produced, and (3) to measure the azimuthal coordinate with moderate precision (8–10 mm). The ATLAS muon trigger is provided by two types of detectors, RPCs in the barrel region and TGCs in the end-cap region, due to the higher background rates.[15] It is, however, not only a change of technology but also of architecture going from the barrel to the end-cap, since in the barrel the trigger detectors are imbedded in the magnetic field, while in the end-caps they are located outside the magnetic field, due to the presence of the cryostats of the end-cap toroid. The schematics of the muon trigger system are shown in Fig. 8. The basic principle is that if an

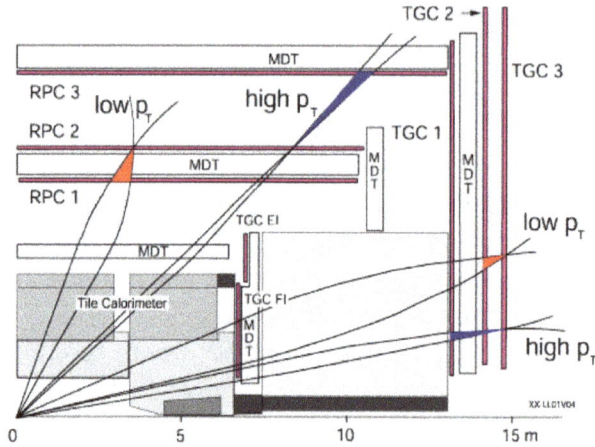

Fig. 8.   Schematic of the muon trigger system in the barrel and in the end-caps.

infinite momentum muon hits the so-called pivot plane (RPC2 for the barrel; TGC3 for the end-cap), which consists of a doublet of either RPCs or TGCs, a final momentum muon will hit the RPC1 or TGC2 doublet layers within a given cone, whose opening angle will be inversely proportional to the muon momenta, for which a coincidence is formed. In the case of a very high momentum muon (10–30 GeV/c), a further-away plane is required to perform the coincidence, so as to have the needed resolution (RPC3 doublet or TGC1 triplet). For the case of the end-cap, a triplet of detectors is used for the higher momentum measurement due to the higher background in the end-cap region. Furthermore, for the end-cap region, an additional doublet of TGCs in the innermost layer can be used in the coincidence to improve the background rejection, as well as to provide the azimuthal coordinate to the precision measurements of the MDTs. This particular layer was very important in changing the trigger chamber technology from the barrel to the end-cap, since possible background rates can reach in this region up to $1\,\mathrm{kHz/cm^2}$.

Figure 9 shows the internal structure of an RPC doublet, where two units are interleaved to avoid dead areas. Each unit consists of two independent gas volumes of RPCs, each equipped with a set of two orthogonal readout coordinates. The gas gaps, with their readout, are sandwiched with paper honeycomb structures, to avoid any change of the internal RPC gap, in case of internal gas overpressure. The ATLAS RPCs are operated with a gas mixture of $C_2H_2F_4/Iso–C_4H_{10}/SF_6$ (94.7%/5%/0.3%) at a nominal operating voltage of 9.7 kV. This condition allows one to operate in an avalanche mode

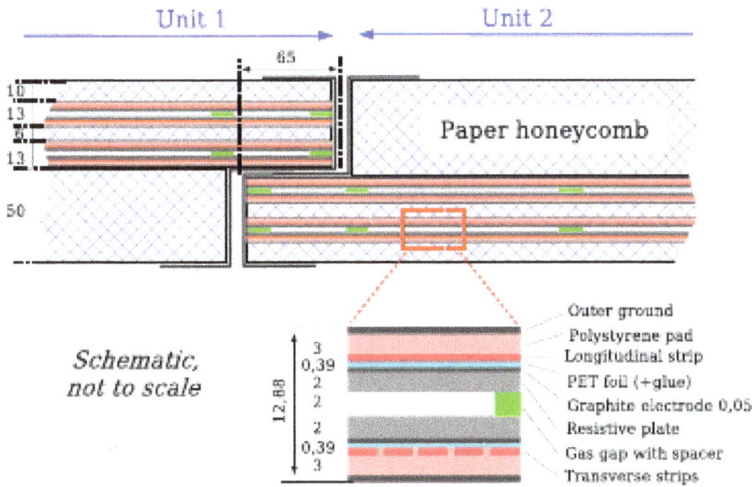

Fig. 9. Schematics of two doublets of RPCs.

with less than 1% of streamers. This last point has been a driving element in optimizing the ATLAS RPC operation, since the presence of streamers in the high background LHC environment could lead to a shortened detector lifetime. In order to make use of the excellent time resolution properties of the RPCs (1.5 ns time jitter), the pickup strips have been implemented as a transmission line, with a Cu ground reference, and the 25–35 mm strips separated by a 2 mm gap with a 0.3-mm-thick ground strip. The strips are coupled directly to a GaAs ASIC, which provides an amplifier–discriminator, via a transformer integrated in the printed circuit. The output of the GaAs ASIC is single-ended, to reduce the power consumption, since they are part of the RPC unit, which is sensitive to temperature increases. The single-ended signals are then combined and transformed into differential input signals to the trigger logic units, outside the RPC doublet modules, located in positions that are accessible during any short shutdown.[16]

Figure 10 shows the internal structure of the triplets and doublets of the TGC units. The units are installed in such a way that there is always full overlap between any two (or four in the corners) units, to avoid any dead area. The TGCs are MWPCs with the characteristic that the wire-to-wire distance (1.8 mm) is larger than the wire-to-cathode distance (1.4 mm), while the cathodes are made of high resistivity (0.5–1.0 MΩ/mm) graphite coating on FR4. The azimuthal pickup strips are located behind the graphite, which is transparent to the fast signals, while the bending coordinate is measured by the wires, which are grouped according to the requirements

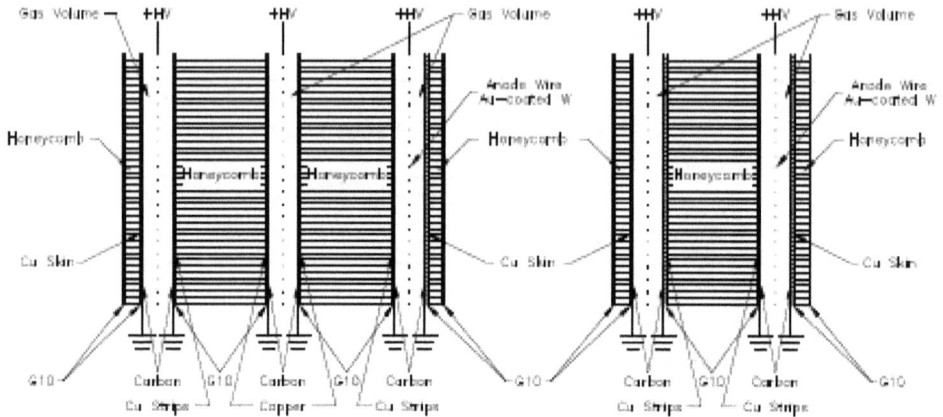

Fig. 10. Cross-section of a TGC triplet and doublet. The triplet has three wire planes but only two strip layers.

of the corresponding rapidity region. The detectors are operated with a highly quenching flammable gas mixture of CO_2/n-pentane (55%/45%) which totally avoids the production of streamers. Unlike the RPCs, the front-end amplifiers–discriminators[16] are located outside the detector units and can be reached during any long shutdown. Their output is sent to the trigger logic units, which are accessible during any short shutdown.

One of the main emphases in the design of the ATLAS muon spectrometer has been the reliability of operation, taking into account that failures do occur during the operation of such a complex system, during its lifetime. For this reason every layer (or multilayer in the case of MDTs) of each of the four detector types has its own gas supply line that can be switched on/off remotely, and for each chamber it can be switched off/on during a short access. Furthermore, every layer (or multilayer in the case of MDTs) has its independent HV power supply, for each chamber in the case of MDTs, TGCs and CSCs, and a group of chambers in the case of the RPCs, while the power supplies themselves can be accessed during any short access. But in any large and complex system it is very hard to predict what can go wrong during such a long experiment. For this reason, a full network of access platforms has been constructed inside the barrel toroid and in front of the end-cap wheels which allow safe access even during a short shutdown of the collider to the barrel CSM modules of the MDTs, the various chamber gas connections, as well as to all the trigger logic units of both RPCs and TGCs. Access platforms of this kind also provide access during long shutdown periods to the front-end electronics of the MDTs, as well as to each individual tube connection to gas and HV. Such access possibilities are crucial for keeping the spectrometer running reliably for many years to come.

3. Testing and Quality Control

Constructing such a complex object as the ATLAS muon spectrometer, with detectors or parts of detectors being constructed in 21 different institutions around the world, requires a large number of quality control tests, to ensure that each detector will work with minimal failure and that the full assembly that constitutes the spectrometer will work coherently, including its alignment and magnetic field measuring devices. Therefore a system of strict testing and quality control has been imposed on all production sites (different for each of the technologies), complemented by common tests at CERN before, during and after the installation in the ATLAS cavern. This has led to a total failure rate that is at the % level in the detectors themselves, with most of the problems already corrected, and a much larger number of problems related to connectivity of the various services, which have all been practically solved.

The largest and most complex subsystem of the ATLAS muon spectrometer is the MDT. With the large number of construction sites involved in its production,[15] a common set of quality control steps was imposed on each site. These quality control steps included tests at the individual drift tube, where gas tightness, HV stability and wire position were measured and rejection imposed, as well as mechanical measurements of the wire position using a central x-ray tomograph at CERN for every 1/8 chamber produced at each site, to control the reproducibility of their position at the level of $20\,\mu$m (see Figs. 11 and 12).

But mechanical precision is not the only important element in such a large system, where failure rates have to be below the 1% level and chambers have to be gas-tight so as to avoid gas impurities, since the gas is being recycled. For this reason, every chamber was tested for gas leaks at the production laboratories and then after its arrival at CERN. Furthermore every chamber, with its final electronics, was exposed to cosmic rays at the individual production laboratories, and again after their arrival at CERN.[18] Partial results of one such cosmic test of one chamber are shown in Fig. 13. All these activities required major infrastructures and manpower contribution from the various production laboratories, but the end result was that only a few detectors required important intervention after the installation, while most of the problems encountered during the commission were due to connectivity. In the case of the end-cap MDT chambers, the situation is slightly

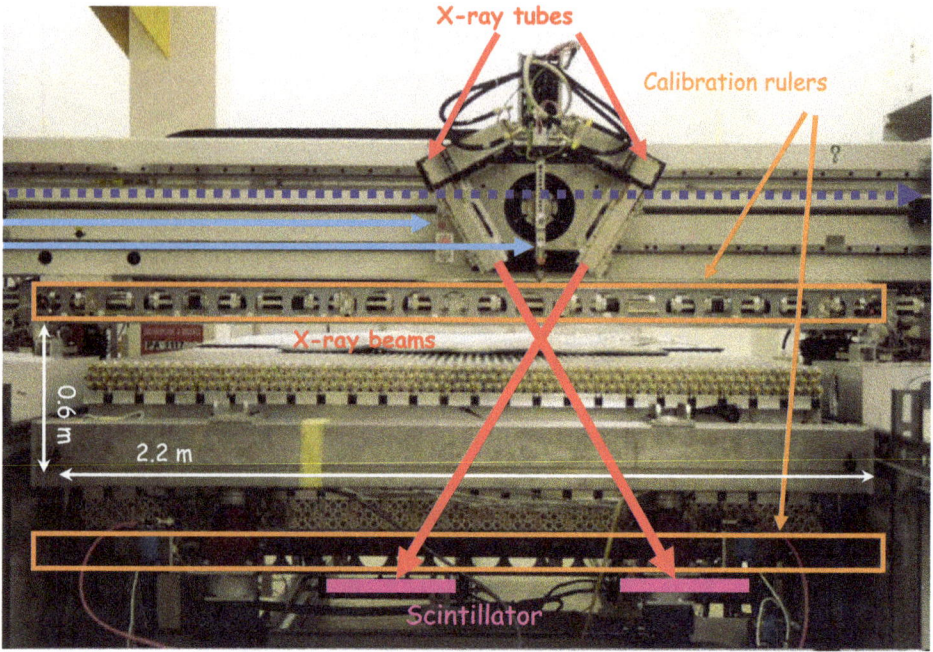

Fig. 11.   Picture of the x-ray tomography facility, showing the two x-ray tubes as well as an MDT chamber being measured.

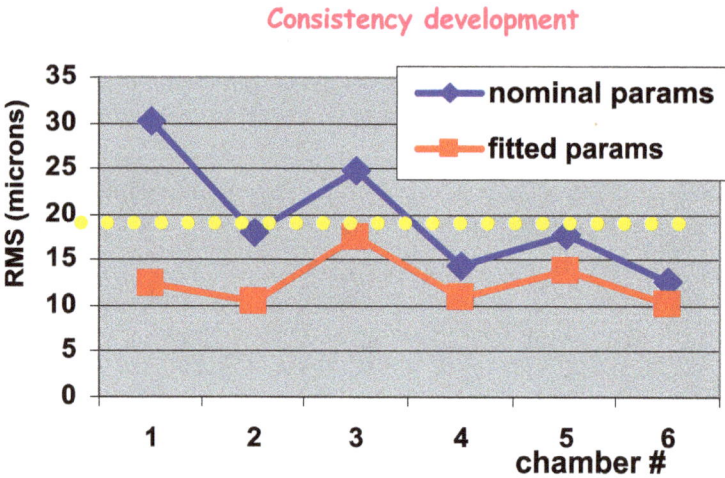

Fig. 12.   Typical reproducibility of the MDT wire positions in the tomography for six chambers of the same type.

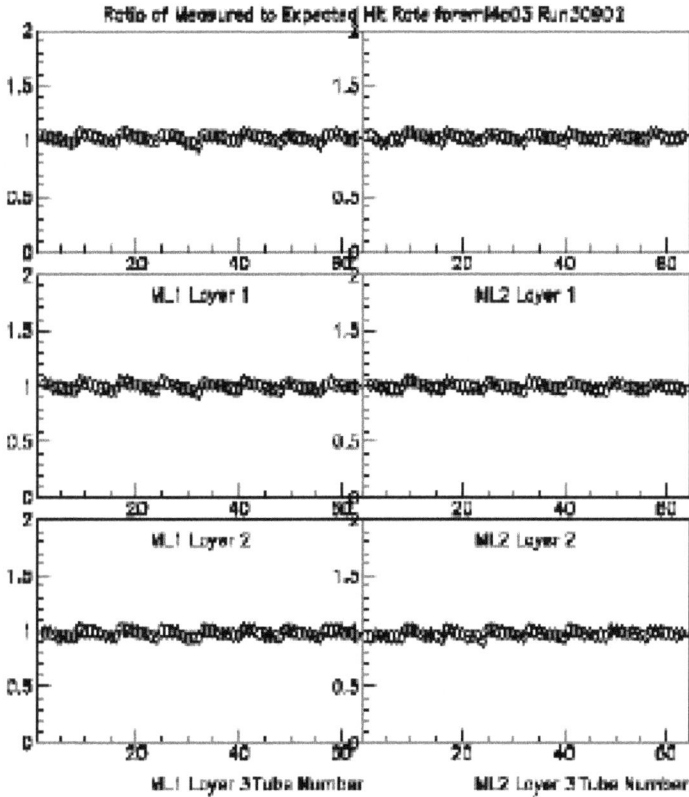

Fig. 13. Typical ratio of observed to expected rate in a given tube for each multi-layer in an MDT chamber.

different, since instead of individual chambers, full sectors of 5 chambers or 32 chambers (small wheel) could be tested with cosmics before installation in the ATLAS experiment, reducing even further the failure rate.

The trigger detectors constitute a crucial element of the ATLAS muon spectrometer. Also here a number of extensive tests were performed at the construction laboratories and then at CERN, before the installation into the ATLAS experiment. In the case of RPCs, single doublet modules, which included the built-in front-end electronics, were scanned using cosmics at the three production laboratories, and gas gaps were rejected if they did not have a uniform efficiency.[19] Also, gaps showing high current were rejected. Finally, fully assembled trigger stations, together with their corresponding MDT detector, were scanned at CERN using cosmic rays, and problematic detectors were replaced. One of the main problems that continued to appear in a small percentage, even after their installation, was the fragile gas connections to the gas volumes, which in most cases were reinforced during the testing period.

The TGCs, which provide the trigger to the ATLAS muon spectrometer in the end-cap region, have also been subjected to a number of stringent tests, before their installation into the wheel sectors of the spectrometer. Each detector went through a cosmic ray scan[20] at three testing laboratories, and if any of its planes (two or three for doublets or triplets) had a region of more than 5% with less than 99% efficiency, the detector was rejected (see Fig. 14 for a typical efficiency map). Then 2/3 of the produced detectors were irradiated for 1/2 h in a 3 MCu Co(60) source, leading to observed rates of 300 Hz/cm^2. The source was then switched off and the detectors had to go back to their initial current to within 1 μA. Finally, all detectors were operated continuously at CERN for a period of three weeks, to ensure that there was no infant mortality. This stringent set of tests resulted in a small failure rate (less than 1‰ replacements and 0.1‰ nonoperational readout channels) during the ATLAS installation.

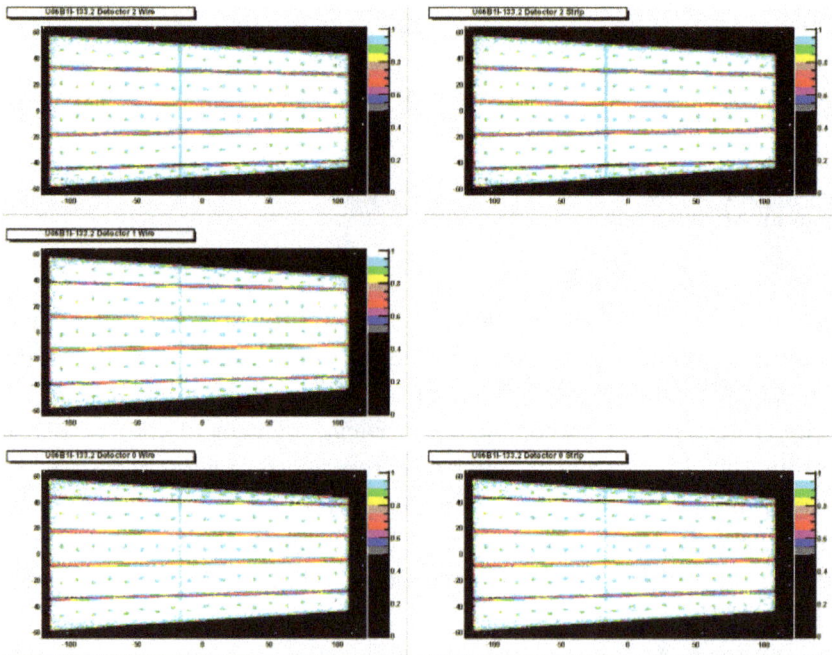

Fig. 14. Typical efficiency scan using cosmics of a TGC triplet. The left side corresponds to the wires, and the right side to the strips (only two layers of strips on each triplet). One can clearly see the support structures, while the rest has an efficiency of over 95%.

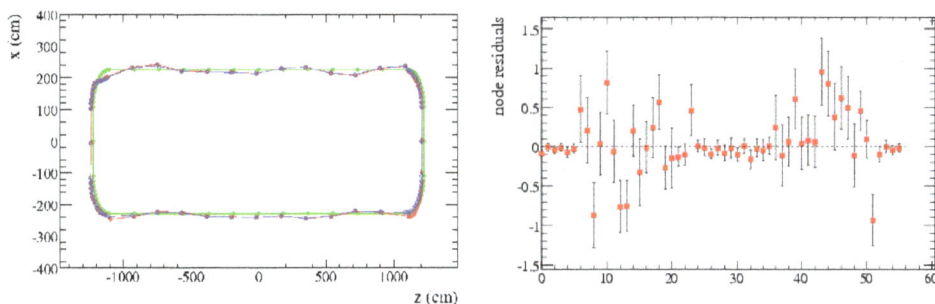

Fig. 15. Measured deformations of one toroidal coil and the respective residuals.

Since the ATLAS muon spectrometer is a magnetic spectrometer with a highly nonuniform toroidal field, it was extremely important to perform measurements of the magnetic field in order to reproduce the field parametrization.[21] This was already done during the first test of powering the ATLAS barrel toroid, with more than 50% of the barrel chambers installed, as well as their magnetic field probes, which allowed reconstruction of the magnetic field, as well as the magnet deformation to within the millimeter level. This is shown in Fig. 15, where the deformations of one of the coils are shown, as well as the residuals, after implementation of the reconstructed geometry.

4. Initial Performance of the ATLAS Muon Spectrometer

In September 2008, the first beams were injected into the LHC, which allowed a first timing exercise. This first injection was followed a few days later by a major incident in the LHC collider, which restricted the bulk of the commissioning of the ATLAS muon spectrometer to be performed using cosmics. The results of this commissioning exercise are being prepared for publication and therefore quantitative results will not be included in the present review; however, qualitative results, to be finalized in the future publication, will clearly show that the main goals of the spectrometer are close to being achieved.

In terms of the operation of the detector, the tracking MDT chambers were all operational, with a total channel failure rate of less than 1.5%; the majority of them have been repaired, leading to a future start LHC run with less than 1% of nonworking elements. Similarly, the efficiencies of each drift tube exceed 99%. This combined with the working alignment system has allowed one to track cosmic rays through the spectrometer and obtain first results on its resolution. Indeed, by comparing track intersection in the middle layer with the expected chamber position after alignment, one can check the quality of the alignment; for the end-caps, one finds that the distributions are centered at 40 μm, while for the large barrel planes, which are covered by the projective alignment system, this value is better than 150 μm and still at the millimeter level for the small barrel chambers, where the alignment is indirect. The achieved position resolution in each plane is approaching, to within factors of two, the expected resolution, which is partly to be expected for cosmic tracks. The CSCs were not fully operational during the early cosmic ray running, due to readout problems that are presently being solved. They were used in a small number of runs, with 98.5% of working channels, but due to their small solid angle for cosmic rays, only first indications of their resolution could be obtained.

The barrel trigger detectors (RPCs) were at the commissioning stage during the 2008 cosmic run, and therefore only 60% of them were used, while a major effort to complete their commissioning took place later, and presently more than 95.5% of the channels are fully operational. The efficiency of individual layers, for the cosmic run, exceeds 90%, including support structures for the large majority of individual planes. A clear correlation between the hits in the tracking and trigger chambers is observed during all the commissioning runs (see Fig. 16 for an example of such a correlation) and their position resolution is consistent with the expectations.

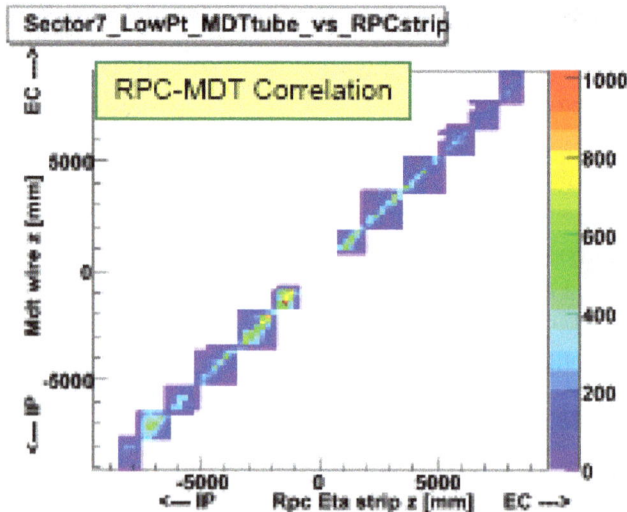

Fig. 16.   Position correlation between an MDT hit and its corresponding RPC for a particular sector, during the cosmic commissioning.

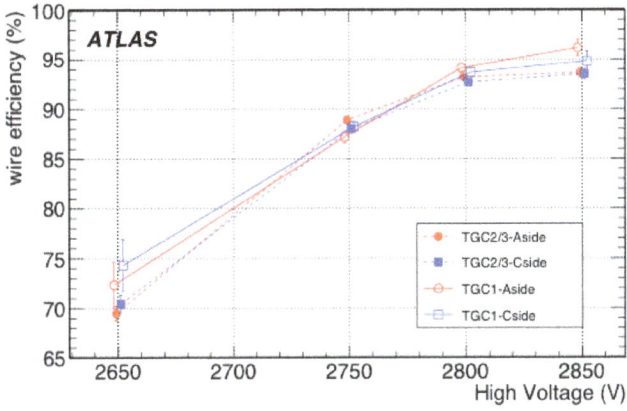

Fig. 17.   Operational plateau for 99% of the TGC detectors.

Fig. 18.   View of the trigger end-cap wheels before their closure.

Fig. 19. View of the barrel muon spectrometer after the installation of its chambers.

The end-cap trigger detectors (TGCs) were fully operational during the 2008 cosmic run, with some minor problems in 0.4% of the channels, which have been repaired in the meantime, leaving 0.8‰ nonworking channels. The efficiency of individual layers, obtained by tracking cosmics through the spectrometer, as a function of the operating HV, is shown in Fig. 17. One can see that it exceeds 93%, including the support structures, with the majority of the detectors exceeding 90%. Also, the position resolution has been checked with respect to cosmic tracks being obtained from the tracking chambers, which are consistent within factors of two (before alignment of the TGCs) with the expected position resolution. The timing of the TGCs has been adjusted — using test pulses — to within one bunch crossing and should be close to the starting point of the LHC running.

To conclude, it is incredible that such a complex system (see photos in Figs. 18 and 19) is so close to its design goals so shortly after its completion. This has only been possible by the combined effort of almost 500 physicists and over 3000 man-years of engineers and technicians that through their strong dedication permitted its finalization. It is clear that nobody in particular can take credit for this achievement, and certainly not the author of this article, but it is this combined effort of very motivated people, with some of them putting more than 15 years of their scientific life into this project, that has enabled the completion of one of the most complex scientific instruments ever made.

References

1. The ATLAS Muon Spectrometer Technical Design Report. CERN/LHCC/97-22 (1997).
2. G. Aad *et al.* The ATLAS experiment at the CERN Large Hadron Collider, *JINST* **3** (2008) S08003.
3. S. Horvat *et al.* Operation of the ATLAS muon drift-tube chambers at high background rates and in magnetic fields, *IEEE Trans. Nucl. Sci.* **53** (2006) 562–566.
4. K. Bachas *et al.* The construction and the quality assurance — quality control of the 112 MDT-barrel inner small chambers of the ATLAS muon spectrometer, *Nucl. Instrum. Methods Phys. Res. A* **581** (2007) 198–201.
5. M. Livan *et al.* Construction of the inner layer barrel drift chambers of the ATLAS muon spectrometer at the LHC, *Nucl. Instrum. Methods Phys. Res. A* **546** (2005) 481–497.
6. F. Bauer *et al.* Large-scale production of monitored drift tube chambers for the ATLAS muon spectrometer, *Nucl. Instrum. Methods Phys. Res. A* **518** (2004) 69–72.
7. A. Borisov *et al.* ATLAS monitored drift tube assembly and test at IHEP (Protvino), *Nucl. Instrum. Methods Phys. Res. A* **494** (2002) 214–217.
8. S. Schuh *et al.* A high-precision x-ray tomograph for quality control of the ATLAS muon monitored drift tube chambers, *Nucl. Instrum. Methods Phys. Res. A* **518** (2004) 73–75.
9. H. v. d. Graaf *et al.* RASNIK technical system description for ATLAS. NIKHEF Note ETR 2000-04 (2000).
10. Y. Arai *et al.* ATLAS muon drift tube electronics, *J. Instrum.* **3** (2008) P09001.
11. J. Chapman *et al.* On-chamber readout system for the ATLAS MDT muon spectrometer, *IEEE Trans. Nucl. Sci.* **51** (2004) 2196–2200.
12. H. Boterenbrood *et al.* The read-out driver for the ATLAS MDT muon precision chambers, *IEEE Trans. Nucl. Sci.* **53** (2006) 741–748.
13. S. Aefsky *et al.* The optical alignment system of the ATLAS muon spectrometer endcaps, *J. Instrum.* **3** (2008) P11005.
14. C. Amelung *et al.* Reference bars for the alignment of the ATLAS muon spectrometer, *Nucl. Instrum. Methods Phys. Res. A* **555** (2006) 36–47.
15. A. Aloisio *et al.* The trigger chambers of the ATLAS muon spectrometer: production and tests, *Nucl. Instrum. Methods Phys. Res. A* **535** (2004) 265–271.
16. G. Aielli, *et al.* The RPC first level muon trigger in the barrel of the ATLAS experiment, *Nucl. Phys. B, Proc. Suppl.* **158** (2006) 11–15.
17. O. Sasaki and M. Yoshida. ASD-IC for the thin gap chambers in the LHC ATLAS experiment, *IEEE Trans. Nucl. Sci.* **46** (1999) 1871–1875.

18. G. Avolio *et al.* Test of the first BIL tracking chamber for the ATLAS muon spectrometer, *Nucl. Phys. B, Proc. Suppl.* **133** (2004) 137–143.
19. M. G. Alviggi *et al.* First results of the cosmic rays test of the RPC of the ATLAS muon spectrometer at LHC, *Nucl. Instrum. Methods Phys. Res. A* **518** (2004) 79–81.
20. E. Etzion *et al.* The cosmic ray hodoscopes for testing thin gap chambers at the Technion and Tel Aviv University, *IEEE Trans. Nucl. Sci.* **50** (2003) 3744–3749.
21. F. Bauer *et al.* The control on the deformation of the ATLAS barrel toroid warm structure, *Nucl. Instrum. Methods Phys. Res. A* **572** (2007) 145–148.

Chapter 14

THE CMS MUON DETECTOR: FROM THE FIRST THOUGHTS TO THE FINAL DESIGN

Fabrizio Gasparini

Dipartimento di Fisica "Galileo Galilei"
Universita di Padova, Via Marzolo 8
I-35100 Padova, Italy
fabrizio.gasparini@pd.infn.it

The project presented in this chapter, the CMS muon detector, is fully described in Ref. 1. It was designed from 1991 to 1994[5] by a community of physicists and engineers from universities and institutes of several countries. The details of the design were fixed in 1997 with the submission of the CMS Muon Technical Design Report.[4] This chapter reflects the views, the memoirs and the understanding of the author, who is responsible for the choice of the subjects and for any omission or error.

1. Introduction

The presence of muons is an important signature of rare and interesting processes generated by the close interactions of quarks and gluons, the basic constituents of the high energy protons supplied by the Large Hadron Collider at CERN.

Muons are heavy leptons, and final leptonic, states can be detected by looking for muons or electrons or a mixture of the two. Hence a good muon detector allows one to look at the same process in different channels, with different leptonic signatures enriching the final sample. The efficiency in the detection may depend on the presence of different backgrounds and/or on the performance of the full detector, and muons are often favored by their easier identification in a much-less-crowded environment. The production of Z and W bosons will be copious at the LHC and their multiple production will be possible up to very high energies. The first process is well understood and will be essential for the calibration and monitoring of the performance of the detectors; the presence of anomalies in the multiple production at high energies would be an important sign of New Physics. The final state, consisting of four charged leptons or the presence of two charged and two neutral leptons, detected by looking at the missing Et in the calorimeters, comprises "gold signatures" of the Higgs particle. Supersymmetric (SUSY) Higgs bosons decaying in $\tau\tau$ and detected by the presence of a couple of charged leptons and missing Et would be an important discovery. Very heavy gauge bosons with masses in the TeV region are predicted by SUSY, grand unification and multidimensional theories, and can be detected studying the dielectron or dimuon mass distribution. The identification of the charge of the final leptons is important for studying the backward–forward asymmetries, which should allow to distinguish among the different models. A good measurement of the position and an independent measurement of the momentum are important for identifying the muon inside jets in the crowded inner detectors and this is relevant to many Top and SUSY searches. The detection of low momentum muons is essential for reconstructing the Y, Y′, Y″ decay in muons in heavy ions collisions and for measuring their relative production rates for various nuclei to investigate the formation of quark–gluon plasma.

This short summary shows that the final performance of an omnipurpose detector at the LHC will depend on the integrated performance of all its parts, and that a powerful and reliable muon detector is an important ingredient for reaching the ambitious goals in the study of the still poorly known region around 1 TeV. At the same time it identifies the requirements for the muon system:

- Safe identification of a muon within the full solid angle;
- Good measurement of the transverse momentum up to and above 1 TeV;
- Unambiguous charge identification up to and above 1 TeV;
- The capability of detecting and measuring very low p_T muons.

A critical task of an LHC detector is related to the very low probability of the kind of events described above. At the nominal LHC luminosity of $10^{34}/\text{cm}^2\,\text{s}$, two bunches of 10^{11} protons cross each other at a rate of 40 MHz. Because the total inelastic cross section is around 80 mb, 10^9 inelastic events will be produced per second and each of the rare, interesting events will be accompanied by 20 inelastic events in the same crossing. The rate of hard collisions is expected to be 10^{12} times lower — one every 100 crossings or every few microseconds. The amount of information associated with each interaction crossing and the bandwidth limitations of the most advanced technology allow recording them on final tapes at a rate of a few 100 Hz. To get rid of the largest possible fraction of inelastic events, the data from the detectors are sampled at 40 MHz, pipelined and selected in different consecutive steps, called "levels of trigger." As usual, the first of them is the most critical: it must cut the rate by a factor of around 10^4 in a couple of microseconds. Close interactions of the basic constituents are characterized by the presence in the final state of particles of high transverse momentum, so the first level selection stands on suitable combinations of the fast recognition by the calorimeters of large deposition of energy, energetic electrons or photons, and/or by the presence of a high momentum muon or several low momentum muons. An unavoidable requirement is that the detectors involved in the first level selection are capable of tagging unambiguously the beam crossing that generated their first level data.

A second important requirement is a prompt measurement of the muon momentum. Muons are heavy and long-lived particles insensitive to the nuclear force, so they are identified by their capability of crossing a very large amount of matter without being stopped.

The material thickness in front of the muon system, measured in collision and radiation lengths, is the same in ATLAS[2] and CMS: 11 collision and about 110 radiation lengths. This value is large enough to reduce the rate of hadron or electron punch-through to below the irreducible rate of prompt and decay muons.

So what comes out of the calorimeters should only be muons. These are prompt muons from decay of heavy quarks, from Z and W boson decay, and muons from kaons and pions decaying in the inner detectors. The most important component is from b and c quark decay. The components have different spectra but their superposition shows that for a p_T larger than several GeV the spectrum in the central η region is quite steep, roughly as $d\sigma/dp_T \sim p_T^{-4}$ [Fig. 1(a)].[3]

The expected number of single muons above a given value of p_T decreases by four orders of magnitude from a p_T of a few GeV to a p_T of a few tens of GeV [Fig. 1(b)].[4] By design the CMS first level output trigger rate should not exceed 30 kHz, shared between muon and calorimeter triggers, and about 5 kHz for the single muon trigger. To reach this value a threshold in momentum on single muons should be put around 10–15 GeV at the first level, but large uncertainties on cross section and background at the LHC require a sound safety margin. Higher thresholds should be foreseen up to and possibly above 50 GeV to further cut the rate by at least one order of magnitude. Any threshold has some width of $\pm\sigma_{P_T}$ around the chosen P_T value, and a fraction of the lower $-\sigma_{P_T}$ bin will pass through and a fraction of the upper bin will be cut. The steepness of the p_T distribution demands sharp thresholds in order to minimize the promotion of lower momentum particles to a higher

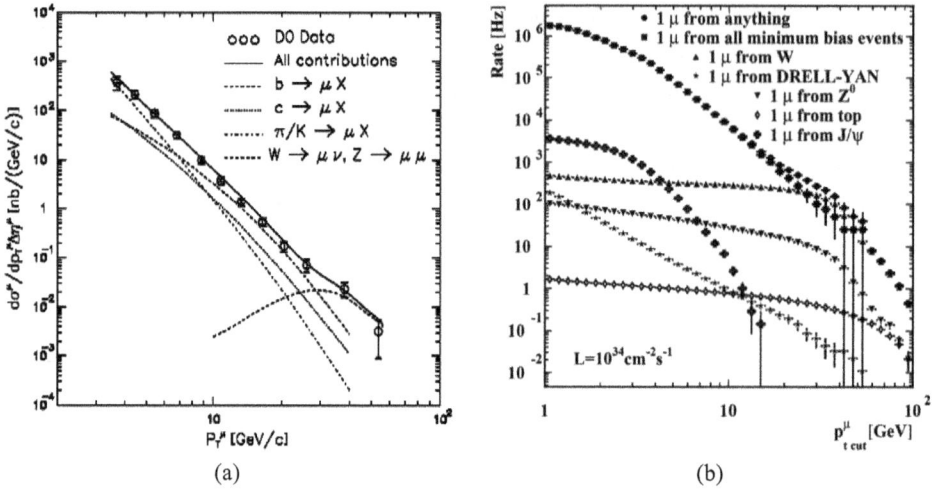

Fig. 1.   (a) Inclusive muon spectrum at the Tevatron[3]; (b) single muon expected rate in CMS.[4]

momentum. To cope with this, the muon trigger should be capable of providing suitable thresholds in a range from 10 to 50 GeV with a resolution in p_T of 20–30%. This is quite demanding at the first level of selection, because it is only two times worse than the ultimate momentum resolution of the CMS muon system alone in the same range (see Fig. 3). The required value of 20–30% translates into a similar amount for the measurement of the bending angle or in 1 mm position resolution.

In summary, and for trigger purposes, the tasks of the muon detector are to tag the crossing that generates the particle and supply a prompt and as-accurate-as-possible information about its position and momentum in a time of about 100 proton bunch crossings.

The issues of time resolution and trigger had the decisive and more important impact on the design of the CMS muon system and of its detectors.

2. The MUON Detectors

The two big experiments, at LHC, ATLAS and CMS are fully described in Refs. 1 and 2.

They made different design choices for the muon system, related to the choice made for the central magnetic coil (Fig. 2).

CMS made the choice in favor of a 4 tesla field, and because the magnetic pressure is proportional to B^2 the coil could not be thin so as not to degrade the performance of the calorimeters. To house the calorimeters inside the

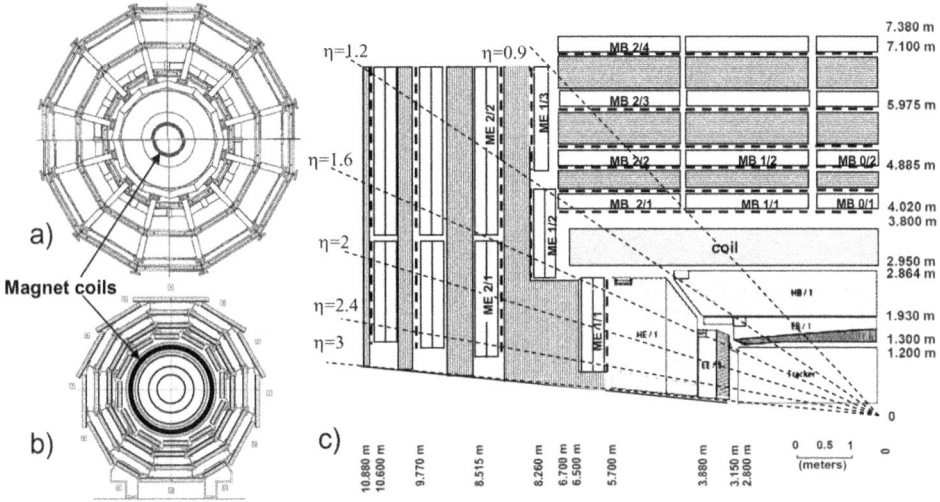

Fig. 2. ATLAS (a) and CMS (b) at the same scale: visible are the different size of the magnet coils and the same size of the block of the central detectors. (c) A section of CMS with the chamber names: for the Barrel MB (wheel/station) and, for the endcap ME (disk = station/ring). Dotted lines represent RPCs.

coil its radius is 3.4 m and its length 12.5 m; its thickness was set at four radiation lengths.

The outer radius of the central part of CMS, taking into account the coil and its vacuum tank, is the same as in ATLAS, i.e. 4 m.

The size of the return yoke is very large, in order to capture the huge magnetic flux generated by the central coil. In CMS the muons have to cross another 10 absorption and 100 radiation lengths in the iron yoke before reaching the outermost station. The muon detector of ATLAS consists of three stations of 4–8 detector layers each in the air, and that of CMS consists of four stations of 6–12 layers each around and inside the iron yoke. The first station is just after the calorimeters (and the coil in the barrel), two are embedded in the iron yoke and a fourth is outside the yoke. The muon p_T is measured in ATLAS with the help of a second magnet which generates a toroidal field in the air, while in CMS the magnetic field is the return field inside the magnetized iron.

To judge the performance one has to take into account the complete detectors — tracker and muons. Figure 3 compared to Ref. 2 shows that the design dp_T/p_T muon resolution of ATLAS and CMS is expected to be similar, around a few percent up to 300 GeV and approaching 5–8%, depending on η at 1 TeV.

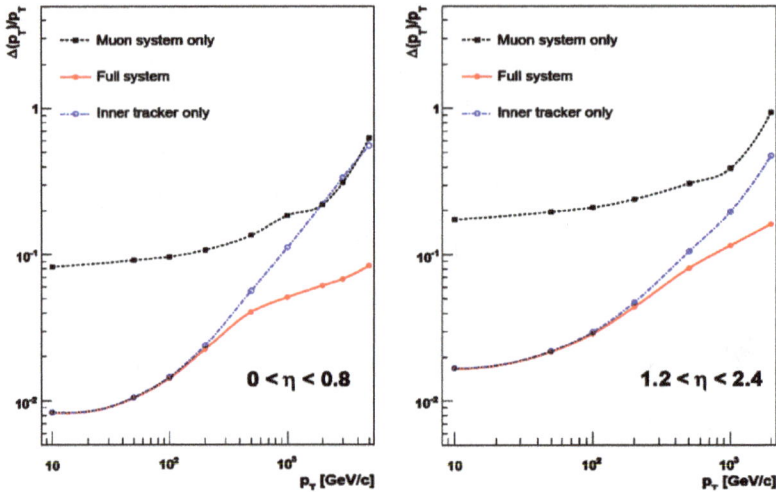

Fig. 3. CMS muon transverse momentum resolution as a function of p_T.[1]

This result is determined in ATLAS mainly by the muon system alone, while in CMS it is obtained from the tracker up to 300 GeV and by combining muon and tracker data above this value. The CMS curves, both for the inner tracker and the muon system, show a constant dp_T/p_T resolution at low momentum followed by a linear increase with p_T.

The value of the momentum, $P(\text{GeV})$, can be obtained from the measurement of the angle of bending $\phi = 0.3\ BS/P$ and $dp_T/p_T = d\phi/\phi = d\phi P/(0.3BS)$, where B = magnetic field (T), S = arclength (m).

The measure of ϕ is affected by several uncertainties coming from the multiple Coulomb scattering (MCS), the detector resolution (σ_d), the alignment uncertainties (σ_a), etc.

$$\sigma_\phi = \sqrt{\text{MCS}^2 + \sigma_d^2 + \sigma_a^2 + \cdots}, \quad \text{MCS} = 13.6\sqrt{X/X_0}/P(\text{MeV})$$

$(X/X_0 = \text{number of radiation lengths})$,

$$dp/p = \sqrt{\text{const}^2 + b^2\sigma_d^2 P^2 + b^2\sigma_a^2 P^2 + \cdots}, \quad \text{const} = 13.6\sqrt{X/X_0}b,$$

$$b = 1/0.3BS.$$

At low p_T the detector resolutions are masked by the constant term due to the MCS, while at high p_T the dispersion is a function of p_T and is dominated by the intrinsic resolution of the detectors. The extent of the region dominated by multiple scattering depends on the number of radiation lengths; in the central region $1 > \eta > -1$ it is about 0.5 for the tracker and more than 100 for the muons. The 10–15% resolution of the muons up to 300 GeV does not limit the global performance, because it is enough to guarantee an unambiguous matching between muons and tracker to exploit its full potentiality.

The resolution of the detectors is then dictated by the requirements at very high p_T. The heaviest objects might be the Z' bosons, whose mass is supposed to be around 1 TeV or above, with a typical width of around 4%: this demands a comparable value of dp_T/p_T that can be obtained only combining the information from tracker and muon.

The difference between the approaches of the two experiments is reflected in the requirements and the design of the detectors.

The requirement of precision in the CMS Muons is two to three times less stringent than in ATLAS (e.g. dp_T/p_T at 300 GeV is 10% in CMS and 3% in ATLAS).

The number of stations and layers is 30–50% larger in CMS, which then deploys a larger number of electronics channels (Table 1).

The first point is related to the radius of the CMS tracker, ~ 1 m, which is one third the radius of the coil and then profits from only 30% of the full bending power. Once the momentum is high enough to disregard the MCS in the calorimeters the muon system does not need a challenging resolution to allow CMS to exploit the full power of its $BS = 12$ Tm bending. This clearly demands a very good understanding of both detectors and a perfect matching between them. The second point is related to the decision of CMS to exploit the negligible contribution of the chambers to the MCS, as opposed to the contribution of the iron, to design multilayer chambers for a very effective bunch crossing identification and to set sharp momentum thresholds. This short summary shows that the differences between the two systems are not in the expected final performance but in the way the performance is obtained. The final performance will depend on the way the two experiments are able to understand the behavior of the full detector, dealing with different backgrounds, different calibration and intercalibration problems, and on their capability of meeting the alignment requirements.

Table 1.

	Volume (m^3)	Stations	Resolution (μm)	Layers/station	Chambers	Channels
ATLAS	10.000	3	35–40	4–8	1182	385 K
CMS	2.500	4	55–120	6–12	790	580 K

3. A Few Further Remarks

The proposers of a multilayer chamber design for CMS, capable of providing the bunch crossing identification and the information for the first level trigger, had several arguments to support their choice. Some of these are as follows:

(1) The existence of many layers makes the requirement of resolution per layer much less stringent.
(2) Trigger information from precise chambers allows precise and efficient cuts in P_T at the first level of selection.
(3) A self-triggering chamber allows the checking of its performance (overall and per layer efficiency, resolution, etc.) at any moment and without any external reference during construction and operation, outside or during data taking.
(4) A self-triggering chamber allows modulation of the trigger parameters station by station and chamber by chamber for good overall trigger tuning.
(5) A self-triggering chamber is capable of triggering and reconstructing random time tracks as cosmic rays or low β particles from the interaction vertex.
(6) During data taking a time resolution of a few ns per station and the use of several precise stations allow quick recognition not only of the position but also of the direction of flight of the particle. This is important for protecting the calorimeters against large energy deposition from a cosmic ray.
(7) A very precise trigger on cosmic rays would be of invaluable help in the synchronization of the full CMS detector prior to the LHC startup.
(8) A drift chamber with a time resolution of few ns allows one to measure and monitor in an unbiased way the local drift velocity and Lorentz angle in the presence of a magnetic field.

The way these goals were achieved is the main subject of this chapter.

4. The CMS Iron Yoke and the Chamber Type

The CMS yoke has a dodecagonal shape so as to be as symmetric as possible around the coil to prevent uncompensated-for stresses. The barrel yoke is segmented in five wheels along the beam line, and in three layers along its radius, each endcap yoke is segmented in three disks perpendicular to the beam line. A picture of the magnetic field lines and of the expected particle fluxes in CMS can be found in Fig. 4.[5] A rough estimate of the size of the yoke

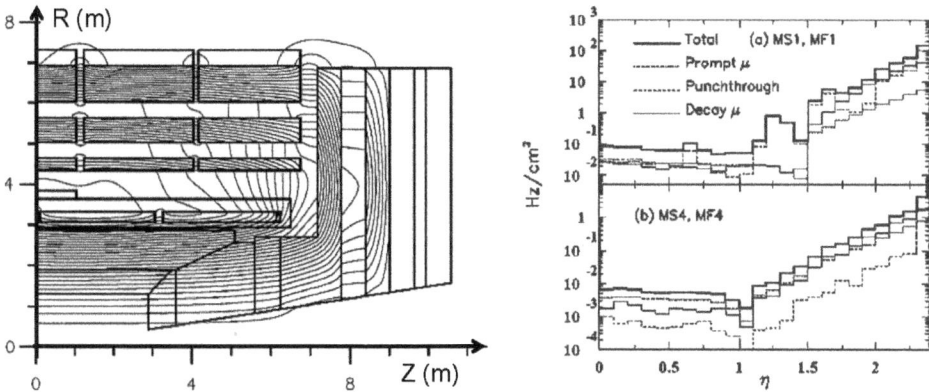

Fig. 4. Left: Typical magnetic field line distribution in CMS.[5] Right: Expected particle fluencies in the CMS muon detector as a function of η at the LHC nominal luminosity.[4]

can be obtained by requiring that the total flux through the central section of the CMS coil ($\sim 32\,\text{m}^2$) match the total flux across the central barrel yoke. To minimize the value of the magnetic field in the drift chambers volume, the barrel iron should not be saturated and B should not exceed 1.8–2 T. This gives an outer radius of the central yoke of 7 m and a yoke cross-sectional area of $\sim 59\,\text{m}^2$. The first iron layer of the barrel is fed by the B lines coming out across the coil, while the second and third layers are fed via the endcap disks. This scheme demands the first barrel layer to be quite near to the coil. At the exit of the coil the B-field lines open radially in their path to the Barrel yoke and the field intensity in the iron decreases along the radius. The larger intensity in the barrel will be in the first iron shell at a radius of about 4.5 m.

If there the B field is $\sim 2\,\text{T}$ it will be $\sim 1.6\,\text{T}$ and $\sim 1.4\,\text{T}$ in the second and the third layer with a total bending power of $\sim 2.5\,\text{Tm}$. Because of that the muon trajectory will be different from an arc of a circle and this will be worsened by the energy losses of the muon across the iron.

The above values of the Magnetic Field in the yoke tell that a fraction of the flux returns outside the yoke. Furthermore some changes in the size of the air gaps between the iron elements were introduced later on, once the design of the detector services was finalized.

The increased reluctance of the magnetic circuit results in a final field map slightly different from the old plot in Fig. 4. An intensive campaign is going on to check and remap carefully the field in the yoke profiting from the cosmic data collected in fall 2008, when the complete CMS was operated for the first time (see also the end of Sec. 9).

The thickness of the iron depends on the layer; the first iron disk and the first barrel ring are only 30 cm thick — about two absorption lengths.

Due to the energy losses in the iron the measurement of momenta as low as 3 GeV can be done only by the first two barrel stations and this demands the first layer to be thick enough to absorb the tails of the hadronic showers from the calorimeters and thin enough to allow low momentum particles to reach the second station.

The choice of the type of detectors was driven by two factors: the behavior and intensity of the magnetic field in the chamber area and the expected particle fluxes (Fig. 4). The B field lines are radial in the endcap disks and parallel to the beam line in the barrel. Hence the most important coordinate for the measurement of the momentum both in the barrel and in the endcap is along circumferences centered on the beam line (R–Φ, where Φ is the CMS azimuth angle and R is the radius of the circumference). Figure 4 shows clearly that the CMS magnetic field has a very large component perpendicular to the chamber planes in the full endcap, where the field is bent to feed the barrel yoke, and that sizeable field components enter the barrel chamber volume in the gaps between all the barrel wheels and in the more external chambers of stations 1 and 2 in the outer wheels. The figure shows also that the expected muon rate from prompt and punchthrough muons is small in the central barrel but increases significantly by three orders of magnitude in moving to the inner endcap. The shape of the magnetic field and the expected particle rates led quickly to the decision to equip the barrel with drift chambers, more sensitive to the magnetic field and with limited rate capability because of the long drift time and the long wires, and with multiwire proportional chambers with cathode strips read out (cathode strip chambers, CSCs) the full endcap.

CSCs are much better suited for this area because their short drift path (half the wire pitch) makes them much less sensitive to the presence of a magnetic field and gives them an excellent rate capability. The important R–Φ coordinate is well measured in the barrel by drift chambers with wires parallel to the beam line, and is very easily measured in the endcap by CSCs whose cathodes can be segmented in thin trapezoidal strips running along the CMS radius. In spite of the fact that strips of variable width and angle with the wires make the spatial resolution dependent on the position of the track, a nontrivial advantage is that the CSC chambers can take the more natural trapezoidal shape to fit the 12 sectors of the iron disks. The full barrel is covered by 250 drift tube chambers (DT) and the endcaps by 540 CSCs. A few typical parameters are summarized in Table 2.

Table 2.

Chamber	Size (m²)	Weight (kg)	Gas gain	Landau peak (fC)	Integr. time (ns)	Range	Least count
Barrel	5–10	780–1500	$\sim 10^5$	125	25		0.8 ns/380 ns
Endcap	0.5–4	60–276	$\sim 7 \times 10^4$	112^{cathode} 142^{anode}	150 30	12	noise 2.2 counts

Chamber	Number of read channels	Number of wires	Cathode strips (not read)	Field-shaping strips (not read)
Barrel	180 K	180 K	360 K	360 K
Endcap	180 K wires/220 K strips	~2.000 K	360 K	

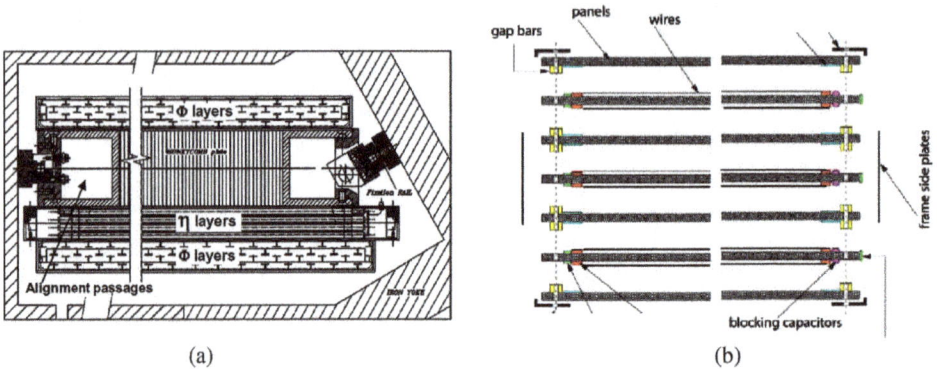

Fig. 5.   (a) Cross section of a drift tube chamber in the CMS bending plane.[5] Indicated are the two groups of Φ and the group of η layers. (b) Cross section of a CSC[4]: three panels support the wires and four panels the cathode strips.

The CMS drift chambers [Fig. 5 (a)] that equip the first three stations of the barrel yoke are made up of three independent units of four layers each of rectangular tubes with a cross section of $13 \times 42 \, \mathrm{mm}^2$. Each group is shaped by four arrays of aluminum I beams sandwiched between five 1.5-mm-thick aluminum plates. The three groups are individually tested before being glued to a thick aluminum honeycomb plate.

The direction of the wires of the two outer groups (Φ groups) is along the Z axis, thus measuring the R–Φ coordinate, and they all have the same length (2.5 m). The third group measures the Z, or η, coordinate and the wire length depends on the R position of the station. The chambers of the fourth and outermost station have only the two Φ groups, the Z one being substituted by an equivalent additional honeycomb panel. The decision was taken under the pressure of funding limitations and also to avoid wires longer than 4.5 m.

The total thickness of a chamber is larger than 0.25 radiation lengths (about 25 mm of aluminum). The endcaps are equipped with CSCs [Fig. 5(b)]. They have six gaps 9.5 mm wide (or 7 mm wide in ME1/1), and each of them provides both the Φ and the R coordinates. The R is obtained from the position of the hit wires, and the Φ by an interpolation of the amplitudes of the signals induced on the cathode strips. Each chamber is made up of seven honeycomb panels clad by fiberglass and sheets of copper: three of them support on opposite sides two planes of wires and four of them the etched strips for the cathode readout. The panels have a density of $8 \, \mathrm{kg/m}^3$ and fiberglass and copper are 1.6 mm and 34 microns thick (0.13 radiation lengths).

Each muon station is backed by one or more layers of resistive plate chambers (RPCs) [Fig. 2(c)]. They are trigger-dedicated detectors, which constitute a second and independent trigger system, the primary one being covered by the tracking chambers.[1,4]

The overlap of the barrel chambers in the same station is prevented by the 12 iron ribs that separate the iron layers. The endcap chambers of the same station can overlap because of the lack of connection between the iron disks in the chamber area. To reduce the chamber size the CSCs of stations 2, 3 and 4 are split into rings two along R without overlap and, for technical reasons, into two (inner rings but ME1/1) or three (outer rings) overlapping parts in Φ in the same sector. The dead regions in R are not pointing to the interaction vertex. The first endcap muon station is split into three rings, instead of two, to allow the first ring of chambers to stay just at the mouth of the coil.

The requirement of a precise measurement of the position and direction of a particle track can be fulfilled only if the position and the orientation of each chamber with respect to the central tracker are known within a value comparable to its internal resolution. This is a serious difficulty because of the size of CMS and the large number of units, supported by the 11 independent and movable elements of the yoke.

5. The Requirements of Spatial Resolution

The muon system of CMS is optimized for the best performance, especially for the Level-1 trigger, for muon momenta between 20 and 100 GeV, the important range for the Higgs search. The space resolution is determined by the overall performance at high p_T. An approximate, but realistic, estimate can be obtained in a simple way from the required value of a dp_T/p_T around 5% at 1TeV for $\eta < 0.8$ and around 10% for η between 1.2 and 2.4 (Fig. 3).

The momentum can be measured by the radius of curvature of the track, **r**, or by the bending angle, ϕ, or by the sagitta, **s**. The sagitta of an arc of circumference of radius r is given by $s = r(1 - \cos(S/2r))$, where S is the length of the arc. If, as is the case, $r \gg S$, then $s = S^2/8r$. Since $r = P/(0.3B)$, one has $s = S^2 0.3B/8P$ and $ds/s = dp_T/p_T$. Let us disregard the MCS for a particle with a p_T of, or above, 1 TeV and approximate the arclength S to the length of the cord of the arc.

If the sagitta of the track is measured inside the tracker, S is in good approximation the tracker radius R and, assuming for the tracker a typical resolution $\sigma_\tau = 20\,\mu m$, one gets $dp_T/p_T = \sigma_\tau/s = 13\%$. The limit is due to the tracker radius (~ 1 m), which does not exploit the full bending power of the 3 m radius of the coil.

If the tracker could use the full bending (i.e. it had a 3 m radius), the goal of $dp_T/p_T = 5\%$ could be reached by measuring the sagitta at $R = 1.5$ m (half the coil radius) with a resolution as comfortable as 68 μm. This goal can be achieved including in the measurement of the sagitta the information of the first muon station. The last tracker layer is at $R = 1$ m and the first barrel station (MBm/1) is at ~ 4.2 m. To fulfil the 68 μm resolution at 1.5 m one must weight the contribution of the respective resolution by the ratio R (at the sagitta)$/R$ (of the detector position).

$$\sigma^2 = (1.5/4.2\sigma_\mu)^2 + (1.5\sigma_\tau)^2 = 68\,\mu\mathrm{m}^2$$

where σ_μ and σ_τ are the muon and tracker resolution.

With $\sigma_\tau = 20\,\mu$m, the 5% resolution at 1 TeV is reached if the spatial resolution of MBm/1 is $\sigma_\mu = 168\mu$m.

Let us repeat the exercise in the FW region: the outer radius of ME1/1 is 2.5 m. This is still larger than the tracker radius but smaller than the radius of the coil. ME1/1 does not exploit the full bending. The maximum sagitta is at 1.25 m and one has to take into account that the resolution of the outer layers of the Tracker Endcap (TEC) is around 50 μm. To reach $dp_T/p_T = 0.1 = ds/s = ds \times 8 \times P/(2.5^2 \times 0.3 \times B)$ and ds must be $ds = 93\,\mu$m. The requirement that $(1.25/2.5 \times \sigma_\mu)^2 + (1.25 \times \sigma_\tau)^2 = 93\,\mu\mathrm{m}^2$ gives $\sigma_\mu = 130\,\mu$m for ME1/1. The same procedure for ME1/2 gives a comparable figure.

The values above include the effect of the "disregarded" MCS in the calorimeters and the magnet coil. Taking it into account the intrinsic chamber resolutions must be around 150 and 80 μm. The conclusion is that the desired dp_T/p_T resolution at 1 TeV is attainable by combining the information from tracker and muons if the resolution of the Muon Chambers is 150 microns in the barrel and in a large area of the endcaps, and 80 microns in ME1/2 and ME1/1. Because the chambers can be made up of many layers of drift tubes or CSCs, the requirement per layer can easily be more than two times larger, a comfortable value for this type of detector even when taking into account the presence of a difficult magnetic environment.

Those rough estimates were carefully checked by a complete simulation of the detector, taking into account the expected magnetic field, the uncertainties on the alignment internal to the muon and toward the tracker, and the energy losses of the muons due to ionization and radiative processes (bremmstrahlung and pair production) in the calorimeters and in the iron of the CMS yoke. While ionization losses are gently distributed along the track and increase logarithmically with energy, the radiative losses are due to small cross section processes and increase linearly with energy. Because they are localized and have large energy fluctuations, they can seriously affect the energy or momentum measurement. In the iron the radiative losses for a muon become important around 100–200 GeV and comparable to the ionization losses around 300–400 GeV. The lost energy is released in energetic electromagnetic showers that can "blind" a muon station. To be inside the iron is a disadvantage that can only be mitigated by the presence of multiple stations decoupled by a suitable amount of absorber and by a station design with many sensitive layers to cope as well as possible with the many tracks of a shower. It was shown that the resolution can be improved making a fit that includes the tracker and all the four muon stations, confirming anyway that the performance of the first station has a dominant role. For the best performance the relative position of the muon chambers, and namely of the first station, with respect to the tracker must be known in position and angle with a precision better than or equal to the chamber resolution and small with respect to the bending angle. This requires that the Φ position of the first muon station versus the tracker is known with a precision better than 40 μrad. The requirements are less stringent for the outer stations.

6. The Trigger Issue and the Number of Layers

Having a clear idea of the requirement for the position resolution, one can deal with the question of the number of layers.

The idea was that the number of layers had to be driven not only by the space resolution but also by the more difficult goal of reaching the time resolution needed for the bunch crossing identification. The time taken by a relativistic muon to cross and leave the full detector ranges from a minimum of about 20 ns to about 36 ns, larger than the 25 ns time between two consecutive bunch crossings. To allow the time synchronization of the trigger detectors and of the data sampling in the full CMS, the LHC clock is distributed all across the detector delayed by the time taken by a relativistic particle to fly from the interaction vertex to each detector or element of it.

A relativistic particle then crosses all the different parts of CMS "at the same virtual time." An "interesting crossing" will be quickly recognized by suitable multiple time coincidences of signals from selected parts of CMS.

However, the condition that each triggering element of CMS should have a time resolution much better than 25 ns cannot be dropped.

Drift and CSCs are detectors designed for spatial and not for time precision. The gas amplification is done in a small volume around the wires and this generates an intrinsic variable delay due to the time taken by the electrons to migrate to it. In the most common gas mixtures at atmospheric pressure, the drift velocity of the electrons is typically of the order of 40–60 μm/ns, so already the typical 1.5 mm path in a standard MWPC introduces a time dispersion that can be as large as about 40–60 ns. To overcome the problem the usual choice is to put together a hybrid system: the precision in the track position is obtained from drift or proportional chambers, and the precise timing from a set of dedicated devices like RPCs, or plastic scintillators.

At the time the design started, the RPCs were in a tumultuous phase of study and improvements, and the muon designers agreed to exploit the freedom on the number of layers per station to obtain the desired time and momentum resolution from the "slow" tracking chambers.

The goal was reached in each single station thanks to a fast process of the signals from a suitable number of detecting layers.

The p_T assignment requires the measurement of the radius of the bent trajectory and the knowledge of its position in at least three stations. The endcap chambers can overlap in Φ in the same station but overlapping is prevented in the barrel by 12 iron structures that generate 12 dead areas in Φ in each station.

The dead regions combined with the possible presence of radiative cascades generate a sizeable inefficiency when requiring at least three out of four stations.

The difficulty was bypassed by asking each station in the barrel to provide a segment instead of a point; in this way two stations is sufficient to measure the bending of the trajectory.

6.1. *The drift tubes case*

A maximum drift path of 21 mm, or a tube width of 42 mm, was chosen taking into account the low particle rate and the wire length. The number of layers per station and the design of the internal field-shaping electrodes were determined by the requirement of the time resolution for the bunch crossing identification.

The starting point was to apply to a drift tube device[5,6] the well-known "mean time" technology, widely used to have a constant time delayed coincidence of the signals from two photomultipliers looking from opposite

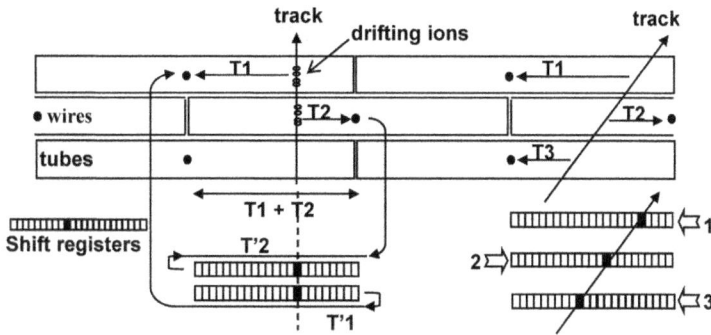

Fig. 6. Playing with two and three layers of drift tubes and delay lines. With three or more layers the track pattern is reproduced in the digital delay lines after a fixed time equal to the maximum drift time.

sides to the light generated by a charged particle in the same plastic scintillator. In a drift chamber this process can be applied to a couple of successive layers of the same station, staggered in position by half the wire pitch (Fig. 6).

For a track perpendicular to the wire planes and if the drift velocity does not depend on the position all along the drift cell, the sum of the times of drift in the two tubes is equal to the maximum possible drift time, M_t, in a tube.

This is not true for an inclined track, because the two lines are not fed, as in a scintillator, by the same original signal but by two independent signals. The resulting error is proportional to the angle of the track. The problem can be solved using three or more layers. The signals from each wire feed each a digital delay line switching on a bit that is moved along it at the same average speed as the drift velocity in the gas.

If the delay lines are represented by a matrix the hits align in the matrix after a time equal to the maximum drift time. The pattern gives the position of the track segment with respect to the hit wires and its angle to the chamber plane. The lines are clocked at 80 MHz, synchronously with the LHC local clock, and about 31 steps of 12.5 ns are needed to cover the maximum drift time of 385 ns in a 85/15% gas mixture of argon/CO_2. The number of steps is programmable, in order to fit possible changes in the drift velocity due to changes in the gas mixture or to the presence of a magnetic field. Energetic electrons (delta rays) generated in the gas or in the tube walls and passing nearer to the wire than the parent muon, mask its pulse, making its apparent drift time shorter. The quantification of this effect was quite easy by looking at the number of faulty layers after the track was reconstructed. Studies done on muon beams between 50 and 300 GeV with a four-layer prototype showed that the probability of this kind of events is around 25%, but that in 80% of the cases only one layer is affected, because the delta ray is stopped by the thick aluminum plate that separates two layers. The use of four layers and the requirement of an alignment of at least three out of four bits gave the correct time of passage of the particle in 95% of the cases. All these basic tests on the Drift Tubes (and CSC, see later) were possible thanks to the large RD5 experiment and collaboration at CERN.[7]

The uncertainty in the time of alignment is one step (12.5 ns, 700 μm at the nominal drift velocity) and the time resolution is roughly $12.5/\sqrt{12} \sim$ 3.6 ns. Taking into account that the maximum time spent by the signal to walk along the 2.5-m-long wires is 11 ns, the combined uncertainty is less than 5 ns, enough to identify the parent crossing.

The drift time is accurately measured by separate TDCs with a time bin of 0.78 ns. Once the track has been reconstructed one can make the "mean time" operation off-line.

For three layers the maximum drift time is $M_t = (t_1 + t_3)/2 + t_2$; in the case of four layers M_t can be obtained from any combination of t_1, t_2, t_3 and t_4. M_t is a fixed time signal that allows one to compute the drift velocity. Under the reasonable assumption that the single layer time resolution is layer-independent, the single layer time and space resolutions can be obtained

$$\sigma_t = \sqrt{2/3}\,\sigma_{Mt}, \quad \sigma_x = \sigma_t V_d,$$

where V_d is the drift velocity.

Figure 7(a) shows the dispersion of M_t in a test beam; assuming a uniform drift velocity of 54.5 μm/ns the value of 3.8 ns gives a resolution per layer of 170 μm. Figure 7(b) show the dispersion of the drift velocity along all the wires of all the chambers as measured during the CMS cosmic runs

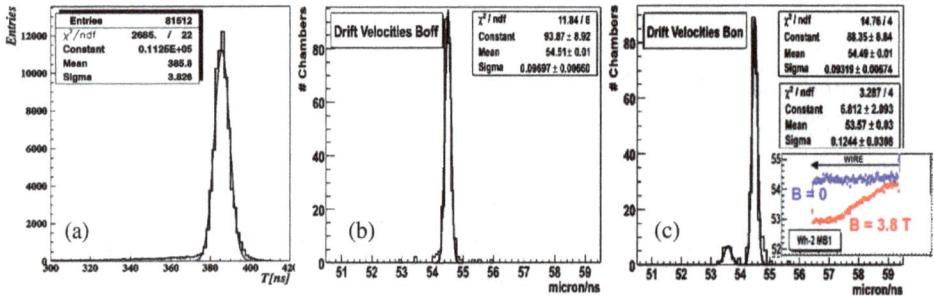

Fig. 7.   See text below. (a) is from Ref. 1; (b) and (c) are from Ref. 9.

in fall 2008, with the complete barrel operational. The dispersion of the drift velocity is below 0.096 μm/ns, and the deviations from a uniform value are below 0.2%. Figure 7(c) shows what happens when the CMS magnet is switched on at 3.8 T.[8] The small peak is the drift velocity in the MB 2/1 chambers, where the B field in the chamber volume increases along the wires up to about 0.6 T. The behavior of the drift velocity along the wires of those chambers is shown in the small insert.

The delay lines and the pattern recognizer are housed in one electronic chip.[4] Its complexity is related to the intrinsic left–right ambiguity in each tube and increases with the number of wires to be looked at in the function of the angle of the track. In the final design each chip looks at nine neighbor wires with a superposition of four wires between two neighbor chips. The limit in the angular acceptance depends mainly on the ability in finding a nonambiguous alignment of the bits in the shift registers when the angle goes above a given value. The only, and fruitful, way was to exploit the possibility of having a 12-layer station, with two groups of four layers reading the critical Φ coordinate and one group of four layers reading the less critical Z.

The trigger signal is generated by the simultaneous presence of an alignment signal from at least two of the three groups of four layers.

As said above, in the barrel the goal was to provide a segment to the first level trigger. The two Φ groups sandwich the Z one and a thick honeycomb plate providing a distance in R between their midplanes of 20 cm. A fast track correlator and sorter combines the information of the two Φ groups for a resolution in angle of 3 mrad. This value complies with the desired 20% resolution in the bending angle on a 10 GeV muon crossing the first iron layer (29 cm) in a field of 1.8 T.

Simulations showed that the resolution is degraded below 10 GeV by the MCS and above 50 GeV by angular resolution.

The design estimates about position and angular resolution at trigger level were fully confirmed in a test at CERN of a couple of final chambers on a muon bunched beam of 120 GeV. They are better than 1 mm and 3 mrad in an angular range of $\pm 30°$.[10]

The bunch crossing identification efficiency was tested on a muon beam[11] with momentum from 15 up to 300 GeV. It was proven to be better than 98% per chamber and to be degraded by 3% at 300 GeV because of the presence of hard electrons or particle showers generated by a 40-cm-thick iron block in front of the chamber. It is worth recalling that 300 GeV is the momentum value where the energy losses due to ionization and radiative processes become comparable.

To summarize, two groups of four staggered layers of tubes provide, after a fixed time of a little more than 380 ns, the time of passage of the track with a resolution better than 5 ns and its position and angle within 1 mm and 3 mrad. This nice result is achieved with excellent efficiency and noise immunity.

6.2. *The CSC case*

A CSC layer delivers two coordinates: one from the wires and one from the cathode strips. The signal from the wires, which gives a coarse radial resolution depending on the wire pitch, is treated for timing purposes as having a fast rise time (30 ns). The signal induced on the cathode has a typical bell shape that can be sampled by measuring the charge induced in a few neighbor strips to obtain the position of its peak, i.e. the Φ coordinate of the track. This coordinate must be precise. Depending on the position of the track, 2–4 strips are fired, because of their variable width. The amplitudes are measured by 12-bit ADC with an integration time of 150 ns. The signals from the strips are "slow signals." The delay coming from the different shaping time is appropriately compensated for to match the wire and strip information of the same track. Tests in the beams showed that the dispersion in time of the signals from a single CSC layer is 40–66 μsec, but that, with a small staggering of the wires in a multilayer chamber, it is well inside 20 μsec in the fastest layers (Fig. 8(a)[4,12]). It was shown that in a CSC with a wire pitch of 3.2 mm a 99% efficiency in the BX tagging can be reached with at least 6 layers through a fast process of the signals delivered by groups of 5–16 ganged wires per plane in a small radial interval. This is achieved by requiring the presence of 4 out of 6 planes in a time window of 2 crossing (50 ns), and assigning to the track the time, and the bunch crossing, of the second- or third-fastest layer.

The probability of local generation of delta rays is very similar to that of the drift tube case and their chance of affecting more than one layer is very small due to the thickness of the panels. Their presence is harmless in the CSC time tagging because of the much shorter drift time. However, they generate distortions in the width, position and symmetry of the image induced on the cathode strips. This effect is limited by the redundancy in the number of layers.[7]

The signal from the strips allows one to know the position of the track in one station and to measure its momentum looking at two or three stations. The procedure of making a fit to the amplitudes of the strip signals is too slow for triggering purposes. A much faster way is to make only a prompt comparison between the signals of the strips concerned [Fig. 8(b)]. A suitable

(a)

(b)

(c)

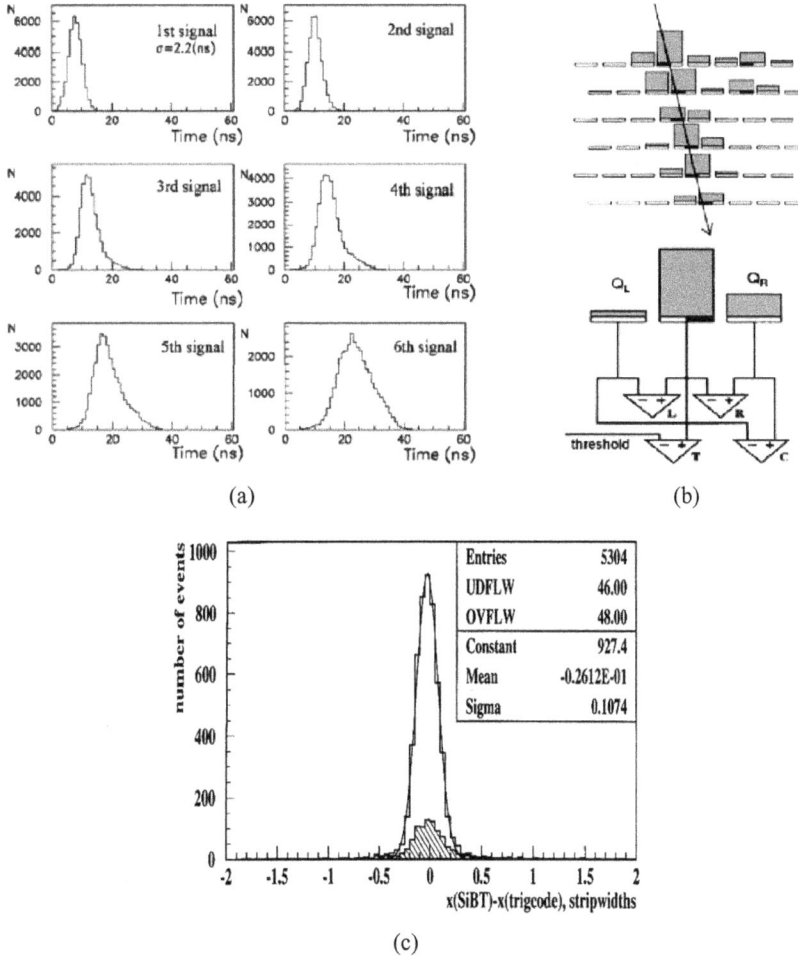

Fig. 8. (a) The time dispersion in the slowest and fastest layers in the ME1/1 chambers: its value in the second- and third-fastest layers is around 4 ns and well inside a 25 ns window. (b) How the fast comparator machinery works (see text). (c) Resolution of the comparator in units of strip width. The white area is the distribution from "clean" events and the shaded area from nonisolated tracks.

threshold is set on the amplitude of the signal of each strip. The signal from the strips is sampled at 40 MHz and once a signal is over the threshold a gate of 150 ns is opened to allow the signals to reach their peak value. Then three comparisons are made with the two neighbor strips and between them. The highest strip is taken, and the track is assigned to its left or right half following the result of the comparison between the two neighbor ones. What is needed is a circuit with four comparators per strip, one of them in common with the next strip.

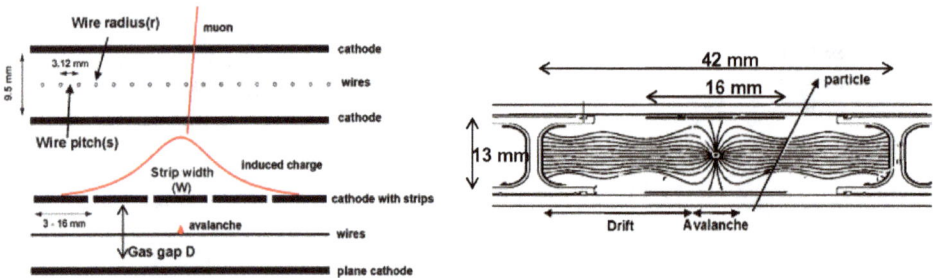

Fig. 9. Relevant parameters for a CSC and a drift cell design (see text).

The position is assigned to the center of the higher strip when only two strips are fired. The procedure provides a quick pattern of half and full hit strips, one per detecting layer. If the strips of the six planes are staggered by half a strip, comparison with a look-up table in which a number of possible patterns of half and full strips are loaded in the function of the position and angle of the track allows one to find the position of the track within a 0.11 strip width.[13] The variation of Φ between the first and the sixth layer of the chamber, whose distance is 15 cm, is also recorded in a fraction of the strip width. Because the strip width depends on the radial coordinate, the information from the strips must be combined with that coming from the wires that are processed in parallel in a much shorter time. The width of the strips ranges (with the exception of ME1/1) from 6.6 to 16 mm, and the position of the segment is located with a resolution between 0.7 and 1.8 mm. In a six-layer chamber the efficiency of finding the segment is 99.8% for clean events but still as high as 95% in the presence of electromagnetic showers generated by a 40 cm iron block in front of the chamber.[13] The momentum measurement is obtained by comparing the variation of the Φ coordinate between two or three chambers. Its resolution spans between 20% at 10 GeV and 40% at 50 GeV, the average being around 29%, in agreement with the trigger requirements.[14]

7. The Choice of the Detector Parameters

The choice of the wire length does not usually appear among the important parameters for the design of a detector, but it had some relevance in the CMS case. In a multiwire chamber a wire becomes mechanically unstable under the electrostatic forces if its mechanical tension T is not above a critical value determined by its length and by other chamber parameters.[13]

$$T > (CVL/s)^2/(4\pi\varepsilon_0), \quad L = \text{wire length}; \ V = \text{operating voltage};$$

$$s = \text{wire pitch}; \ C = \text{capacity}.$$

The wire length is limited by the maximum affordable wire tension and by the capability of the chamber frame to withstand the force of many hundred wires stretched up to several hundred grams each.

Taking advantage of the possibility of overlapping the chambers in Φ, which had allowed the use of particle tracks for their relative alignment in the same iron disk, the chambers of the same sector were split into two or three in Φ. This limited the wire length in the CSC to 1.3 m and its tension to 250 g. The limited area of the chamber also allowed one to solve the problem of the planarity of the electrode panels (see below).

The relevant parameters for the chamber design are shown in Fig. 10. For a DT they are the ratio between the extension of the avalanche region, in which the electric field is highly inhomogeneous and grows as $1/r$, and the drift region, in which the electric field is uniform. The "thickness" of the drift region generates a dependence of the shortest drift path on the track angle.

In the CMS drift tubes the three parameters must be tuned to meet the goal of a uniform drift velocity along the full cell, and this requires a careful design of the field-shaping electrodes.

In a CSC the important parameters are the wire pitch s, the gas gap D, the wire radius r and the strip width W. The swarm of positive ions leaving the amplification region around an anode wire induces a charge on the cathodes whose density, projected in a direction perpendicular to the wires

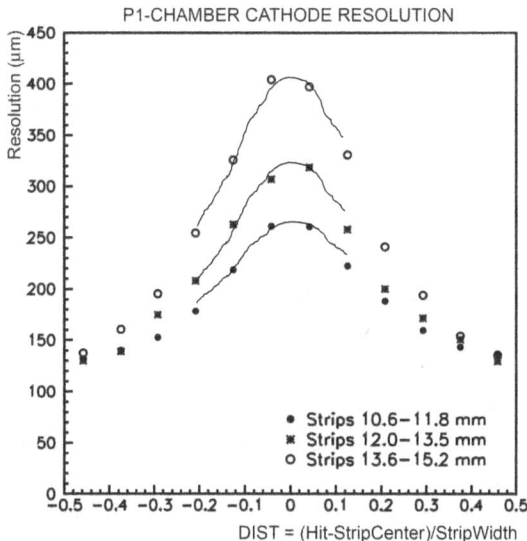

Fig. 10. Dependence of the CSC resolution on the position of the track with respect to the center of the strip for different strip widths solid lines are to guide the eye).

(the strip direction), is bell shaped and can be precisely computed by electrostatics. For a practical design the distribution must be described by a formula containing explicitly the important parameters of the detector. It was shown[16] that its shape normalized to the anode charge can be best fitted (within 0.1%) by

$$\Gamma(\lambda) = K_1(1 - \tanh^2 K_2\lambda)/(1 + K_3 \tanh^2 K_2\lambda)$$

where $\lambda = D/2x$, (x is the position of the cluster peak), and K_1, K_2 and K_3 are three empirical parameters whose value depends on s/D.

A further small approximation allows one to express K_1 and K_2 in the function of K_3 only. The best value for K_3 can be computed once D/s and r/s are fixed.[17] Once this is done the fit depends on two free parameters: the value of x and of the total charge Q.

From the shape of the charge distribution one can extract useful hints about the best width of the strips. Let us assume that the position of the track is at the center of one strip; one can find the charge deposited on the neighbor strip integrating Γ between $-W/2$ and $+W/2$, between $+W/2$ and $+3W/2$ etc. Half of the signal is left to the outer strip if $W = D/2$, one-eighth if $W = D$ and very few if $W = 2D$. (see also Ref. 12).

This is relevant because the error in the position comes in a fraction of W and depends on the resolution in the measurement of the charge.

$$\sigma_x = \sqrt{(n-1)}W\mathrm{ENC}/\Sigma i Q_i, \quad n = \text{number of strips},$$

$$\mathrm{ENC} = \text{electronics noise}.$$

The precision on the charge is limited by the electronics noise and the error is larger if Q is small, so the geometry must be arranged in order to have a small number of hit strips, typically between three and five, and the usual choice is to have $W \sim D$. As mentioned above, in CMS the best strip geometry to measure the Φ coordinate is to have them running along the radius and this implies that their width increases with the radius and that the angle between the strips and the wires depends on Φ, and so on the wire length. Both conditions violate the golden rule that for the best performance the strip width must be uniform and the strip-to-wire angle a constant. Taking advantage of the large number of layers, the effort was concentrated on the ambitious goal of reaching a charge resolution per strip of $dQ_i/\sum_i Q_i/ \sim 1\%$ of 1% in the full endcap area and taking the best compromise for the average strip width. It was set to 1.3 D for the chambers expected to have a 150 μm resolution and to 0.8 D for the most demanding ones. The critical area is at the higher radius border of the chambers, where the strip width reaches

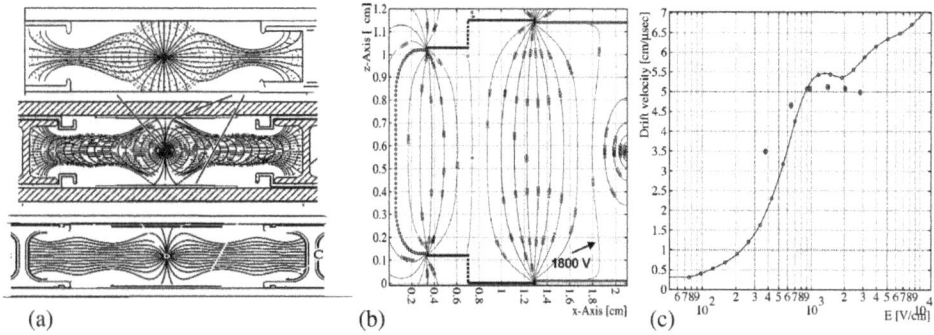

Fig. 11. (a) From top to bottom: development of the design of the CMS drift tube cell. (b) Cell equipotentials. (c) Drift velocity as a function of the electric field in the Ar/CO$_2$ gas mixture. The dots show, for comparison, the behavior of a typical (70/30) Ar/isobutane mixture.[14]

1.8 times the gap width. The problem was solved by staggering the strips of the different layers by half a strip: in this case the critical track passing at the center of a strip in one layer will cross the subsequent in between two strips. The dependence of the resolution on the strip width is shown in Fig. 10[4] in the function of the position of the track.

The resolution for a track between two strips is around 140 μm, independent of the strip width. Let us take the example of ME2/1, whose strips range from 7 mm to the larger value of 16 mm at the outer border of the chamber. With no staggering the resolution of a 6 layers chamber for a track in the center of the 16 mm strips would be $\sigma \sim 500/\sqrt{6} \sim 200\,\mu$m. With staggered strips it can be obtained from $1/\sigma^2 \sim 3/\sigma_1^2 + 3/\sigma_2^2$, and with $\sigma_1 \sim 140\,\mu$m and $\sigma_2 \sim 500\,\mu$m one obtains $\sigma \sim 77\,\mu$m. The magnetic field and the angle of the track with respect to the chamber plane will degrade this figure, but the design value of 150 μm is clearly at hand.

The choice of r and D was guided by practical constraints. The choice of a 50 μm wire was driven by the requirement of robust wires stretched at a comfortable tension due to their relatively short length (the only exception was made for ME1/1, which is equipped with 30 μm wires). The choice of D followed the unavoidable decision of using for the 3500 panels standard panels of industrial production. The problem was to fix the constraint on the panel planarity. The reference was the size of the collected charge, which depends on the spread induced by the gas gain variations on the wire and by the unavoidable Landau fluctuations of the ionization.

For a fixed voltage the gain, G, depends exponentially on the variation of the field, E, on the wire $dG/G \sim \exp(20\,dE/E)$.

Simulations showed that the two effects become comparable if the gain variation is not much larger than 2, and this requires that the local variation of dE/E should not exceed the 3% and an obvious similar value for the planarity of the panels. Commercial panels of the typical area of $3 \times 1\,\text{m}^2$ were available with planarity around 200–300 μm, larger panels were not satisfying such a strict specification, and this was an important argument in favor of the chamber splitting in Φ. The choice was made for a gap D of 9.5 mm and a wire pitch of 3 mm for all the CSCs with the exception of ME1/1. The definition of the wire pitch is not completely independent of the choice of the gas gap, because the current induced by the movement of the swarm of positive ions is shared between the cathodes and the neighbor wires following their relative capacitance. This effect was taken into account in the overall simulation of the electrostatic properties of the chambers.

The best value for K_3 is 0.33 for all chambers and 0.45 for ME1/1.

A positive outcome of the multiplication of the number of chambers due to the segmentation in Φ was that the panels were also robust enough to withstand the wire tension and that a robust and self-supporting CSC was feasible by simply bolting together the seven panels.

Figure 11 shows, from top to bottom, the development of the design of the DT drift cell from the first prototypes to the final design. The design was determined by the crucial requirement of a uniform drift velocity in the full cell for trigger purposes.

The first model was inspired by the L3 experiment at LEP and the improvements by the tubes designed for the muon system of the SDC experiment at the planned US supercollider. What makes the difference is the size of the CMS design, which foresees a tube cross section 15–20 times smaller.

The tubes are rectangular, to minimize the thickness of a multilayer chamber. They are 42 mm wide and 13 mm high. The height determines the number of primary electrons generated by an ionizing track and could not be further reduced without loss of efficiency. The thickness of the drift region is about 6 mm, half of the height of the tube, and about 50 primary electrons, out of the 100 generated in the full gap, reach the anode. The first design had only three electrodes: the wire, the I beams (insulated from the plates by thick plastic extruded profiles) and the grounded aluminum plates. It showed an unsatisfactory behavior in the presence of a magnetic field. To cope with that in the second version, two electrodes (made of thin Mylar-insulated aluminum strips) were inserted in front of the wires and set at positive HV. They allowed better control of the drift field and of its homogeneity along the full cell. The last change was to make the cathodes with the aluminum-insulated strip glued to I beams that were set to the ground. This improved the field in the cathode region and the chamber robustness, removing the weak plastic-to-metal gluing and halving the number of layers of structural gluing. As shown in Fig. 11(b),[1,18] the intensity of the electric field in the drift region is controlled by the voltage of the strips in front of the wire (the 1800 V equipotential in the figure), and by the voltage of the cathode (−1200 V).

The distance between the two is about 1.9 cm and the resulting field is ∼1.5 kV/cm, which corresponds to the center of the V_d plateau of the Ar/CO_2 gas mixture.

The difficulty of the design can be appreciated by looking at the short plateau shown in Fig. 11(c).

Both the DT and the CSC are sensitive to the presence of a magnetic field: if uniform over the detector area one can correct for it; if not uniform it will contribute to the degradation of the resolution. The only region in the muons where B is uniform enough to allow a correction is the small area covered by the ME1/1 chambers discussed at the end of the next section.

8. The Constraints from the Magnetic Field

An electron traveling in a gas in the presence of an electric field E has a stop-and-go motion under the force $F = eE$.

A simple expression for its average velocity, i.e. for the drift velocity, is

$$V_d = e/2mE\tau, \quad e, m = \text{electron charge and mass},$$

$$\tau = \text{mean time between collisions}.$$

The time between collisions has been found experimentally to be largely independent of E, and at high-enough values of E all the practical gas mixtures show a more or less wide region in which V_d is independent of E. The reason is that the cross section and the fraction of energy lost per collision with the gas molecules depend on the local energy gained by the drifting electrons between two collisions. At high fields there is a compensation between the two effects that makes V_d independent of the strength of E.

In the presence of a magnetic field a new force appears: the Lorentz force, F_L. If V_d and B are perpendicular to each other, $F_L = eV_dB$. If V_d is saturated its modulus does not change but its direction takes an average angle tang $\Psi = V_dB/E$. Substituting V_d with $V_d = e/2mE\tau$, tang Ψ is often expressed in the function of the cyclotron frequency, ω, of the electron in the local magnetic field $\omega = e/2mB$: tang $\Psi = \omega\tau$.

This expresses the fact that the electron path between two collisions keeps the same length but it is no longer a straight segment but an arc of circumference, and so it moves a bit along the direction of the Lorentz force and less along the electric field E.

In the region where V_d is saturated, the size of the Lorentz angle can be tuned by playing with the value of E, but the drift velocity in the direction perpendicular to the wire is in any case reduced by a factor $\cos \Psi$, $V_d' = V_d \cos \Psi$.[19] This shows the two relevant things for a drift chamber: if B is not constant along the wire length the drift time will depend on the position of the track; if the saturation region is wide enough the resulting time dispersion can be limited, reducing the Lorentz angle Ψ playing with the value of E.

In the MB1 and MB2 chambers of the outer wheels, the magnetic field is inhomogeneous and its intensity perpendicular to the wires plane increases along the wire from 0.2 to 0.4 T in MB2 and from 0.4 to 0.8 T in MB1. Tests in a magnetic field showed that below 0.5 T the Lorentz angle in one layer was below 10°, generating a maximum increase of the drift time smaller than 5 ns (just a bit more than 1%). But the increase of the magnetic field from 0.3 to 0.8 Tesla along the same wire in MB1 had introduced a delay as large as 12 μsec between the signals from simultaneous tracks crossing the chamber at the two ends of the wire.

Because the LHC clock is distributed following a local delay a local shift of its time of 6 ns had reduced this uncertainy to $+-6$ ns, well below the 12.5 ns of the delay lines step and sufficient for the bunch crossing identification. Actually, the results reported in Ref. 8 and in Fig. 7(c) show that the B field in that region is lower than expected and that the variation of the maximum drift time is about 9 ns instead of the 12 reported above.

In presence of 1 T field the increase of the drift field from 1.5 to 1.8 kV/cm was measured to reduce the Lorentz angle from 26.4° to 22.4°, in agreement with the above formula.[4]

The case is different for a CSC. For perpendicular tracks crossing the wire plane in the same position x along a wire the region concerned with the multiplication avalanche is roughly a point if $B = 0$, but in the presence of B it becomes a segment with a length $L = s$ tang Ψ, where s is the distance of the track from the wire (Fig. 12). The segment extends to the right or to the left of the point at $B = 0$ according to whether the track is above or below the wire. This introduces a dispersion of $(s/2$ tang $\Psi/\sqrt{12})$. With the exception of ME1/1 and ME1/2 the field B is below 0.5 T, and tang Ψ was measured to be below 0.1 so that $\sigma_x < 40\,\mu m$, which is a negligible degradation for the large fraction of chambers from which the expected resolution is $150\,\mu m$.

The exception is the chambers in the two inner rings of the first disk, where B takes values as high as 1 or 3 T. In ME1/2 the axial field ranges from 1 T at the inner radius to 0 T at the outer border. The best possible precision in P_T requires a chamber space resolution of $80\,\mu m$, which corresponds roughly to $160\,\mu m$ per layer. The value of tang Ψ was measured for several magnetic field intensities, and is 0.15 at 1 T. The induced dispersion of $65\,\mu m$ per layer does not dramatically compromise the chamber performance. The target resolution of 160 microns per layer was restored, or even improved, reducing the strip width, whose average value in ME1/2 is 8 mm instead of the typical 12 mm in the other chambers.

From Fig. 12 one can see that the distortions induced by a uniform magnetic field can be canceled by tilting the wires with respect to the strips by an appropriate angle. This was not possible in ME1/2 because of the large inhomogeneity of the magnetic field, but was possible and mandatory in ME1/1.

The ME1/1 region is critical also because, due to the higher momentum, the rate of punch-through is increasing and it is expected to reach a value around $300\,Hz/cm^2$, while the total random rate of single hits due to the background is around $600\,Hz/cm^2$. The effect of the random hits is largely mitigated by the multilayer structure of the stations.

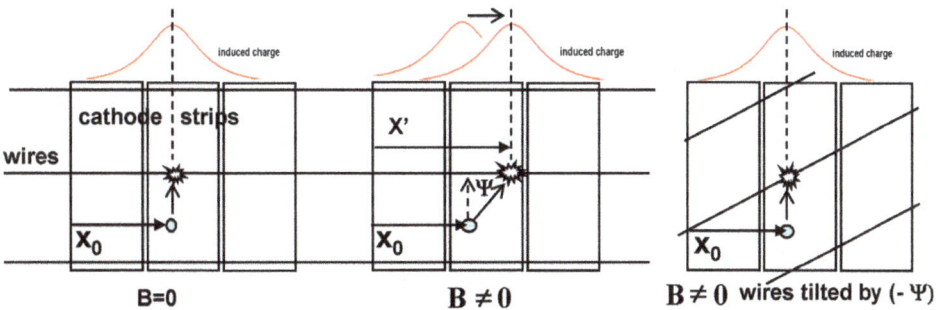

Fig. 12. At $B = 0$ the path of the electrons generated by a particle at $x = x_0$ is perpendicular to the wires and the induced charge distribution is centered at x_0. In the presence of a magnetic field the paths are tilted by the Lorentz angle α and the charge distribution is moved from x_0 to x'. The correct position is restored by tilting the wires by $-\Psi$. If B is not uniform, α depends on the position of the particle, and the correction is impossible.

The only way to face a higher rate is to reduce the drift time inside the chambers and the size of the charge image on the cathode. The drift time depends on the wire pitch, and the size of the image on the gas gap. In the first ring the best result was obtained by reducing the wire pitch from 3.2 to 2.5 mm and the gas gap from 9.5 to 7 mm; the wire diameter was reduced to 30 μm. The average strip width was reduced by a factor of 2 with respect to the other CSC chambers. Those changes demanded an improvement in the mechanical assembly of the chambers and a reduced tolerance in the manufacture of some critical elements, namely for the panel flatness, whose requirement was moved from 200 to 50 μm. The panels had to be done at home in the participating institutes.

The ME1/1 chambers are positioned just at the mouth of the coil in the presence of a very high axial magnetic fields perpendicular to the drift paths. The result of the careful adjustment of the chamber parameters discussed above had been largely masked by the presence of the Lorentz force if the magnetic field at the mouth of the coil was not regular enough to permit a correction by tilting the wires. This correction cannot be perfect, because the angle between the wires and the radial strips of the 168-cm-long chambers is 90° only at the center and reaches ±5° at the borders. Thanks to the three-fold segmentation per sector, this angle is still small compared to the 25° of the Lorentz angle in a 3 T axial field.

Figure 13 shows that the nominal resolution in the presence of an intense magnetic field could be almost completely recovered.

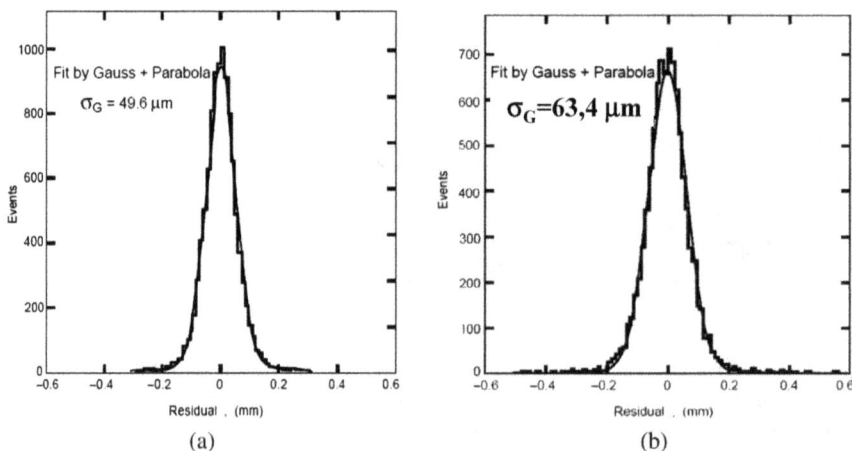

(a) (b)

Fig. 13. The resolution of two ME1/1 prototype chambers: (a) without tilted wires in the absence of a magnetic field and (b) with tilted wires in the presence of a magnetic field.[4]

9. Few Final Comments

It is worth noting that CSCs and drift tube chambers are independent autotriggering devices: each of them is an independent unit capable of tagging the time of passage, and measuring the position and angle of any track. In spite of the fact that their electronics is clocked at 40 MHz, to be synchronous with respect to the expected time of passage of the tracks in the LHC, they can detect with a small reduction of efficiency or resolution, due to the asynchronous sampling of the signals, even particles at random time as cosmic rays or low beta massive particles generated in the LHC. Cosmic muons come randomly within the 25 ns windows while the clocked electronics generates the signal trigger with a time pitch of 25 μsec introducing a time error of $25/\sqrt{12}$ ns $=\sim 7$ ns.

The time resolution of a drift tube is ~ 4 ns and the combined dispersion reaches 8 ns, which corresponds to a spatial resolution of about 500 μm. However, the TDC of the four hit wires in a four-layer group will all be stopped at a time that is delayed with respect to the time of passage of the track by the same amount of time. A fit to reach the best alignment of the hits determines quickly the common correction to be applied to each wire restoring the nominal resolution and giving the absolute time of passage of the random track with the usual 4 ns resolution.

For the CSC this time represents the "error" in the sampling of the signals for wires and strips, and it is reasonably small compared to the 150 ns shaping time of the cathode strips; so the degradation of the resolution is very small. The muon detector is playing a decisive role in the preparation of CMS for the LHC startup. The capability of the CSCs and drift tubes to give the absolute time and position of the cosmic rays permits a careful synchronization between all the chambers and of the muon versus the tracker and the calorimeters in the absence of a beam. The variation of the drift velocity from chamber to chamber or inside the same chamber gives an independent measure of the intensity of the local Magnetic field.

After the tracker has been put into operation, giving a good measurement of the muon momenta, the data from the muon are allowing an intensive campaign to check and correct the CMS B field map in the yoke area, and to improve significantly the methods and performance of the detectors' alignment with tracks.

10. The RPC

In the early stages of CMS three different designs were submitted for the muon chambers, all based on CSCs and drift tubes. Two of them were hybrid systems, in which the time tagging was based on RPC layers coupled to the tracking chambers. When the choice was made in favor of the third one, described in this chapter, CMS decided to maintain an RPC system, fully decoupled from the tracking chambers, so as to have an independent way to check and monitor the trigger efficiency of the detector.[1,4]

The two systems are treated in an independent way by the overall trigger of CMS. The RPC is a parallel plates gaseous detector made up of two 2-mm-thick highly resistive Bakelite plates, with a bulk resistivity of $\sim 10^{10}\,\Omega/\mathrm{cm}$, and with a typical gap of 2 mm. Because of the high resistivity the plates are "transparent" and swarms of ions moving across the gap induce an electric signal on a plane of insulated strips put behind the cathode plate.

The planar electrodes generate a uniform electric field across the gap. It must be intense enough to allow the avalanche multiplication to start with the first mean free path of the primary ionization electrons. The electric field felt by the electrons across the 2 mm gap is then comparable to the field they feel in the last $100\,\mu\mathrm{m}$ of their path toward the wire in a proportional chamber, typically $5 \times 10^6\,\mathrm{V/m}$. To exploit efficiently the small number of primary electrons (typically 10), the gas gain must be as high as 10^7 and the fluctuations of the collected charge are as large as 10^3, ranging from 1 fC to 1 pC. One weak point of the RPC is the rate of random signals or discharges that can be generated by the large number of photons from the ion recombination or by local spontaneous discharges due to the unavoidable local surface defects of the Bakelite panels. Because of the very high field the energy dissipated in the gas is large: at a rate of a few $100\,\mathrm{Hz/cm^2}$ it may reach a value of $1\,\mathrm{W/m^2}$.

The above observations explain the strength and the weakness of an RPC. The avalanche multiplication is local; there is no delay due to any drift time, and the only delay is due to the fixed time of 40–50 ns of the signal growth determined by the electrical parameters of the detector and of the front end electronics. The time fluctuations in the generation of a useful signal are a fraction of the time taken by one electron to run across the 2 mm gap — 1 ns or less.

A difficult issue remains: the very coarse space resolution. The very large signal fluctuations and the large extension of the multiplication region (2 mm, to be compared to the "pointlike" size in a DT or CSC) translate into unpredictable fluctuations of the size and shape of the charge image on the cathodes, preventing signal processing similar to that used in a CSC. The strip width must be set large enough to have only one or two fired electrodes — typically a couple of cm. The resolution cannot be improved by a local multilayer structure. A reasonably precise tracking or momentum measurement can be obtained only by multiple RPC planes well separated in space. CMS is "compact" by definition, and a muon trigger based on the RPC only was judged to be inadequate and risky.

RPCs have, however, some important advantages: the extremely precise time tagging, the lack of drift time and of any type of local fast or slow signal processing makes an RPC trigger twice as fast as the DT/CSC part. Their resolution is poor enough to be substantially insensitive to the presence of a magnetic field and of electromagnetic showers, but the sharpness of the momentum threshold is two times worse than the DT/CSC value for P_T larger than 10–15 GeV. These properties are so clearly complementary to those of the more powerful trigger based on the tracking chambers that CMS decided to keep alive an option in favor of a simple RPC-based trigger to be used as monitor and backup of the main trigger line.

In CMS two RPC layers sandwich the first two barrel stations and one layer is present in the two outer stations. Four layers are present in the end-cap and a plan exists to raise the number to six.[1] The trigger is based on a quick comparison between the simple pattern of six, or four, fired strips and a number of possible patterns stored in large look-up tables. A fast and simple process that could only be made critical by the presence of random hits would cause the number of patterns to be looked at to diverge. This was the main risk, and the final green light to set up an RPC trigger was not given until a convincing demonstration showed that the critical requirement of a random signal rate well below $10\,\mathrm{Hz/cm^2}$ was achievable. Currently this rate is below $1\,\mathrm{Hz/cm^2}$, a nontrivial result obtained after long research on very effective quenching gas mixtures and on careful surface treatments.

The promotion of the RPC from a simple local time tagging machine to a detector suitable for a trigger system at the LHC is the result of ten years of passionate work of a few people who were able to improve the RPC's basic performance by more than two orders of magnitude.

At the start of this long run the RPC was a detector operated in the so-called "saturated streamer mode," generating signals in the range of 100 pC, with a noise rate of around $100\,\mathrm{Hz/cm^2}$ and a comparable rate capability.

Those figures must be compared to the actual charge smaller than 1 pC, to a noise rate below $1\,\mathrm{Hz/cm^2}$ and to a rate capability exceeding $1\,\mathrm{kHz/cm^2}$.

11. Alignment

The knowledge of the relative positions of the tracking detectors has been always a critical issue for any high energy physics detector.

The detectors of the CMS muon system are supported by the eleven heavy elements of the iron yoke.

With the exception of the central wheel, that is fixed and contains the coil and the central detectors, the other elements are movable along the beam line to allow one to open the yoke for the installation and maintenance of the detectors and the services. Gravitational distortions lead to static deformations of the yoke elements that generate displacements of the detectors with respect to their nominal position, up to several mm. The displacements can be measured within few hundred mm by photogrammetry when the detector is open. The repositioning of the large elements of the yoke after a CMS opening is very precise but, taking into account their size and weight, can not be better than few mm.

The switch-on of the magnetic field induces deformations and movements that may be as large as few cm. The eleven elements are compressed and possibly slightly tilted, the endcap iron disks are bent, the central part of the first disk is pushed toward the interaction point by a couple of cm and the chambers move with respect to their nominal positions. During the operation the thermal equilibrium of the yoke is reached after several months, the effects are expected to be in the sub-millimetre range.

All those displacements are either partially or totally non repeatable, and their typical value is an order of magnitude larger than the target figures of accuracy.

This picture shows the series of constraints on the possible designs of the alignment system which led, after several steps, to a solution based on a number of precise rigid structures supported by each yoke element and moving with it. Those structures contain the optical elements that look at the relative positions of the chambers in the same element. The connection among the structures sitting on the various elements is possible only when CMS is closed, and is obtained through a network of laser beams and a number of local distance sensors and digital cameras.

The CMS alignment is described with all details in Ref. 1. It is made up of three independent parts: the internal alignment of the tracker, of the barrel and of the endcap muon. A fourth element, the LINK, locks them together and monitors the displacements of the heavy structures of CMS during the critical phase of the closing and during the normal operation.

A sketch of the full system is shown in Fig. 14. It consists of many optical paths lying on three "planes" all across CMS and staggered in Φ by 60° to comply with its 12-fold symmetry. The figure shows the layout of one of the three optical planes. Each endcap station is monitored through radial

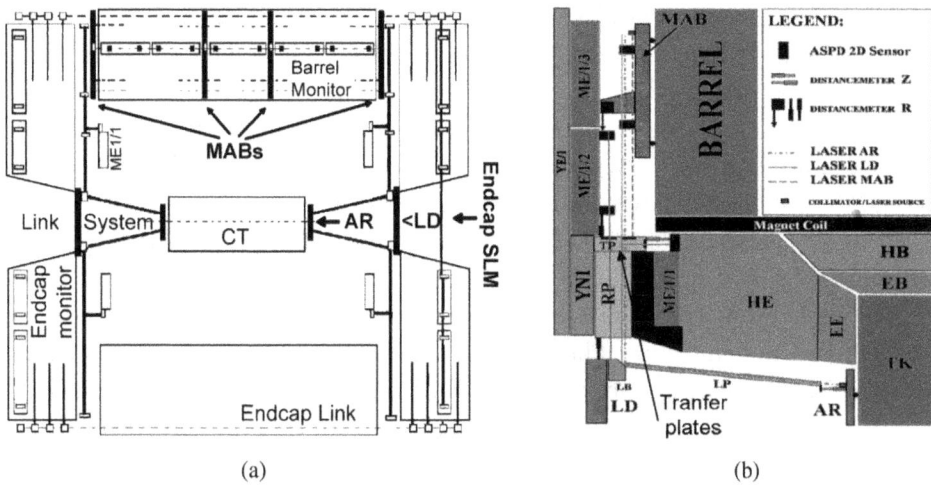

Fig. 14. (a) One of the six alignment planes of CMS, showing the position of the endcap SLM, the endcap link lines, the LNK system and the MABs inserted in the iron gaps of the barrel yoke. (b) Shows the details of a quarter of a plane of the LINK.

straight line monitors (SLMs) running along the full diameter of the supporting disks. Each SLM has two laser beams, sent out at the two opposite ends of the SLM line from a couple of rigid transfer plates and detected by a couple of sensors in each of the four crossed chambers. The rest of the chambers are aligned with respect to them, detecting tracks that pass through their overlapping regions. The lines monitor the position of the chambers in R–Φ and also their displacement along Z, due to the disk deformation in the presence of the magnetic field.

The positions of the 250 barrel chambers are monitored by a floating network of 36 carbon fiber structures (MABs) optically connected together. Two groups of six MABs are on the outer faces of the external wheels and six inside each gap between the wheels; the two groups in the gaps between the movable wheels are not shown in the figure because they are displaced by 30° in Φ with respect to the alignment planes. Each MAB holds eight digital cameras that look at LED sources precisely installed in the barrel chambers. MABs are placed in Φ along the separation between two barrel sectors and look at the neighbor sides of the eight chambers of the two adjacent sectors.

The two endcaps are connected together by endcap link laser transfer lines, two per alignment plane, measuring their relative movements in Φ. Those lines are also detected by sensors sitting on top of four of the six arrays of the barrel MABs (shown in the figure).

The more peculiar part of the CMS alignment is the LINK system, whose main issue is to link together the three CMS tracking detectors: the barrel endcap muon and tracker.

The LINK is the part of the alignment whose final design was changed several times and fixed very late, in 2003–2004, when the final details of the design of the CMS tracker and of its services were fully defined.

The key elements of the LINK are two carbon fiber rings (link disks, LD) supported by the first endcap iron yoke at the two opposite sides of CMS. Their diameter is 1300 mm and they are positioned at $\eta = 3$, centered with respect to the beam line. They are supported by three long aluminum bars attached to three of the six reference plates [TP; see Fig. 14(b)] attached to the iron at the outer border of the nose. The six TPs lie inside the alignment planes at the border between ME1/1 and ME1/2. The LD generates a number of laser lines along the CMS radius that lie on the three alignment planes. The lines are detected by ASPD sensors with a nominal precision in the range of 5 μm. Two of them are in each MAB, two in the ME1/2 chambers and two in the TP. The detection of those lines links together the three alignment planes and gives the relative Φ position of the barrel MABs with respect to the ME1/2 chambers and to the first iron disks. The position of the barrel chambers is provided by the MABs and that of the endcap chambers by the endcap alignment in which the ME1/2 chambers are integrated by six SLMs running on the surface of the inner face of the first endcap disks (not shown in the figure). The relative position in the transverse plane of CMS of those structures between themselves and with respect to the LD is obtained through a number of different types of distance sensors, whose precision has been tested to be in the range of 30 μm. Some of them are visible in the figure. Distance sensors connect also in R and Φ the ME1/1 chambers to the TP. The full system is floating and its absolute position in the CMS reference system depends on the knowledge of the position of the fiducial references (e.g. the TP) that are measured by photogrammetry in the range of 300 mm. But the relative position of the different monitored structures depends only on the performance of the ASPD and of the distance sensors, which is in the range of 20–30 μm.

The first and crucial issue of the LINK is to measure and monitor the relative position of the central tracker with respect to the muon, and mainly with respect to their first stations. Two carbon fiber rings with a diameter of 730 mm (alignment ring, AR) are on the outer external faces of the tracker, attached to the support plane of the outer silicon layer of the tracker endcap. They are equipped with six laser collimators and three distance sensors. The laser lines reach the LD and are deflected radially and detected in the MABs by the same sensors that look at the lines generated by the LD itself. The relative displacements of the AR planes with respect to the LD are monitored by the three distance sensors that are touched by the long aluminum tubes, supported by the LD. One of them (LP) is visible in Fig. 14(b). The axial range of those sensors is 5 cm, to allow one to monitor closely the critical movements of the first iron endcap disks during the final steps of the CMS closing and the deformations of the disks when the magnet is switched on and off. The switching-on in sequence of all the laser lines allows one to monitor any relative displacement of the three main detectors with a precision that should be dominated by the high resolution of the different types of sensors. The different participating structures (MAB, LD, AR, etc.) are equipped with many ancillary devices, like tilt meters and clinometers, which measure their absolute angle with respect to the gravity, and temperature and humidity sensors to correct for possible thermal dilatations.

In terms of final performance, it is worth noting that the connection of the AR to the TK volume is weak. It is just done by the precise mechanical attachment of the AR to the most external tracker flange. Thus the two sides of the LINK system, forward and backward, are weakly linked among them, as well as to the internal tracker alignment.

After huge calibration work on sensors, MABs, LD and AR done on a special full size bench in a large hall at CERN, the first partial but realistic test was eventually possible only during the first test of the CMS magnet in the CMS surface hall in August 2006.

Only one-quarter of two adjacent planes was equipped on the positive side of CMS. Thanks to the quite long range (5 cm) of the potentiometer attached to the AR, the system allowed one to monitor the movements during the critical phase of the endcap closure, the deformations of the endcap disks and the compression along Z of the barrel wheels during the ramp-up and ramp-down of the CMS magnet. Their size agrees well with the FEA results.

Because the tracker was not yet installed in CMS, the ARs were supported by a carefully positioned mock-up. The reported conclusion[20] was that the relative spatial locations and angular orientations between the muon chambers and the AR were measured with a resolution better than 150 μm for distances and about 40 μrad for angles — not far from the requirements reported in the closing lines of Sec. 5 and without exploiting all the redundancies of the complete system.

A more complete test, whose results are still under study, was possible during the cosmic runs in fall 2008 after CMS was lowered in its cavern.

The displacements of the CSC chambers in the presence of a 3.8 T field with respect to their position at zero field were measured by the endcap alignment and by the LINK. The two independent measurements on the ME1/2 chambers of the first iron disk agree within a few tens of μm.

12. Conclusion

In fall 2008 the full CMS was operated for the first time and 300 million cosmic tracks were recorded. Data analysis shows that the muon detector works safely and that its performance is as good as its design. The 15 years that separate the beginning of this chapter from this final and first full operation were spent to prototype, improve, understand, build, test, install, tune and understand the detector.

The author dedicates this chapter to all the friends, colleagues and others in the CMS muon community who shared with him this long and fascinating venture.

References

1. CMS Collaboration (S. Chatrchyan *et al.*), The CMS experiment at the CERN LHC, *JINST* **3** (2008) S08004.
2. ATLAS experiment at the CERN Large Hadron Collider, ATLAS Collaboration (G. Aad *et al.*), *JINST* **3** (2008) S08003.
3. Stapelberg *et al.* Measurement of the inclusive muon production cross section. D0 note 4325 (2004-01-20).
4. CMS Muon Technical Design Report. CERN/LHCC 97–32 (1997).
5. CMS Technical Proposal. CERN/LHCC 94–38 LHCC/P1 (1994).
6. CMS Letter of Intent. CERN/LHCC 92-3 (1992).
7. RD5 Collaboration (C. Albajar *et al.*), Electromagnetic secondaries in the detection of high-energy muons, *Nucl. Instr. and Meth. A* **364** (1995) 473–487.
8. M. C. Fouz *et al.* Measurement of drift velocity in the CMS barrel muon chambers at the CMS Magnet Test Cosmic Challenge. CMS Note 2008/003.
9. CMS DP-2009/002 and CMS Collaboration, Calibration of the CMS Drift-Tubes System and drift velocity measurements with cosmic muons. To be submitted to *JINST*.
10. P. Arce *et al.* Bunched beam test of the CMS drift tubes local muon trigger, *Nucl. Instrum. Methods Meth. A* **534** (2004) 441–485.
11. M. Aldaya *et al.* Results of the first integration test of the CMS drift tubes muon trigger, *Nucl. Instrum. Methods A* **579** (2007) 951.
12. M. M. Baarmand *et al.* Tests of cathode strip chambers prototypes. CMS Note 1997/078.

13. M. M. Baarmand *et al.* Spatial resolution attainable with cathode strip chambers at trigger level, *Nucl. Instrum. Methods A* **425** (1999) 92–105.

14. The CMS Level-1 Trigger Technical Design Report. CERN/LHCC 2000-038 (15 Dec. 2000).

15. F. Sauli. Principles of operation of multiwire proportional chambers and drift chambers. CERN Yellow Report 77–09.

16. E. Gatti *et al.* Optimum geometry for strip cathode or grids in MWPC for avalanche localization along the anode wires, *Nucl. Instrum. Methods* **163** (1979) 83–92.

17. E. Mathieson *et al.* Cathode charge distributions in multiwire chambers, *Nucl. Instrum. Methods* **227** (1984) 277–282.

18. A. Benvenuti *et al.* Simulations in the development of the barrel muon chambers for the CMS detector at LHC. *Nucl. Instrum. Methods A* **405** (1998) 20. (See also T. Rovelli CMS Note 1998/051.)

19. W. Blum and L. Rolandi. *Particle Detection with Drift Chambers* (Springer-Verlag) 1993.

20. A. Calderon *et al.* CMS Note 2009/04.

Chapter 15

THE WHY AND HOW OF THE ATLAS DATA ACQUISITION SYSTEM

Livio Mapelli[*] and Giuseppe Mornacchi[†]

Department of Physics, CERN
CH-1211 Genève 23, Switzerland
[*] livio.mapelli@cern.ch
[†] giuseppe.mornacchi@cern.ch

1. Introduction

The selection of interesting physics channels and the acquisition of their data requires a trigger and data acquisition system (TDAQ) based on innovative architectural solutions providing a multilevel selection process and a hierarchical acquisition tree, expected to provide an online selection power of 10^5 and a total throughput in the range of terabits/s. Its implementation consists of a combination of custom electronics and commercial products from the computing and telecommunications industry.

The concept and design of the ATLAS TDAQ[1,2] have been developed to take maximum advantage of the physics nature of very-high-energy hadron interactions, i.e. exploit the high level of rejection against QCD background provided by the analysis of single-particle signatures before acquiring the bulk of the full-event data.

The final system will consist of tens of thousands of processing units, for the majority of commercial PCs, interconnected by a multilayer Gbit Ethernet network, whose central core comprises multihundred-port switches. The selection and data acquisition software has been designed in-house and production releases are regularly made and extensively used for the operations at the ATLAS experimental site[3] and in ATLAS laboratories and institutions worldwide.

This chapter describes the realization of the ATLAS data acquisition system, with emphasis on the reasons for the architectural and technological choices and the way they have been implemented. The full construction cycle is analyzed, from the conceptual design, driven by the physics and the experimental environment, to the final implementation, through a long phase of R&D and prototyping, during which emphasis was put on the utilization of the prototypes of increased completeness and complexity in real life, mainly at test beams for single-detector and multidetector combined data taking.[4]

Disclaimer. Although in ATLAS the trigger and data acquisition systems are normally referred to as a unique system (TDAQ), the first-level trigger and the algorithmic aspects of the higher levels are beyond the scope of this chapter. However, the infrastructure and system aspects, both hardware and software, of the high-level trigger are so highly integrated with the data acquisition that they cannot be separated. In the following they are referred to as DAQ or DAQ-HLT.

2. Problem Description

The challenges created by an experiment of the size and complexity of ATLAS[5] on the data acquisition are unprecedented in high-energy physics. They span all the aspects of data movement, event selection and controls.

The first challenge is due to the *size of the detector* itself (Fig. 1). With its 44 m in length and 22 m in diameter, the outer elements of the detector (muon endcap) are as far as ∼25 m from the interaction point, i.e. particles will take ∼75 ns, or the equivalent of three bunch crossings, to reach them. Furthermore, the length of cables connecting the detector electronics to the

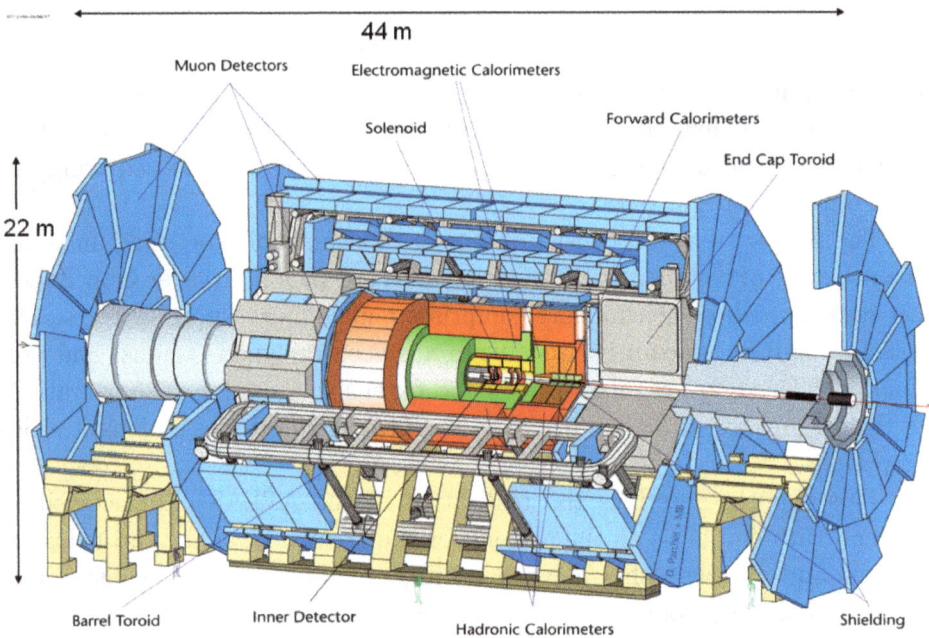

Fig. 1.   The ATLAS experiments at the CERN Large Hadron Collider.

first level of selection (level-1 Trigger, LVL1[6–8]) is ~100 m, i.e. the signals produced in the detectors reach the LVL1 in 0.5 μs. Adding all the components of the processing for the first-level decision (time of flight, signal propagation, signal processing and computation and, finally, redistribution of the LVL1 decision), the total latency of the LVL1 trigger amounts to 2.5 μs, i.e. the equivalent of 100 bunch crossings.

The *LVL1 latency*, combined with the LHC bunch crossing frequency of 40 MHz, constitutes the second big challenge for the detector readout and the data acquisition. In order to avoid a dead time for data taking that would completely spoil all the advantages of the high luminosity of the LHC, the data from 100 consecutive events must be stored during the entire LVL1 latency of 2.5 μs, and this at a pace of 40 MHz. This is achieved by equipping each of the tens of millions of electronic channels of the ATLAS detectors with a buffer, i.e. the front-end electronics is based on a pipelined architecture. The LVL1 trigger also had to be designed as a systolic pipelined system clocked at 40 MHz, synchronous with the readout pipelines, so that, in case of an accepted event, at the end of the 2.5 μs processing, a signal is sent to the pipelines and the corresponding data are moved out and sent via fiber links to the next stage of the readout chain.

The next challenge that has to be faced for the design of the trigger and data acquisition system at the LHC comes from the *selection power* required in order to extract the tiny physics signals from the dominating QCD background characteristic of high-energy hadron interactions. As an illustration, let us consider the case of the search for the Higgs boson — the main motivation for the experimentation of the LHC. Figure 3 of the introductory chapter shows that there are 11 orders of magnitude between the ~100 mb of the total p–p inelastic cross-section at 14 TeV and the ~1 pb of production of an ~500 GeV Higgs. In order to reach the lepton signature of the Higgs decay, one has to add 2 more orders of magnitude to the rejection power. In total, there is a rejection power of up to 10^{13}, i.e. 10,000 times higher than for the detection of Z^0's at the CERN and FNAL p–pbar colliders. Clearly, this rejection power is not fully required in real-time, i.e. from the trigger and data acquisition system. It would not be possible and, in any case, other physics channels of interest are less rare. The estimate is that, for the 40 MHz interaction rate, all the physics of interest is contained in ~200–300 Hz, which also corresponds to the amount of data that could be reasonably stored and analyzed with today's technology. The goal is, therefore, a reduction of $\sim 10^5$, which is achievable only with an architecture based on a multistage selection process and a hierarchical data acquisition.

The last of the major challenges for the data acquisition of the LHC experiments comes from the *large and highly distributed system*, namely for the system description and configuration and its control. There are two main issues: coherence of system behavior and scalability.

System coherence. The configuration, control and monitoring of data taking operations of various types (debugging, commissioning, calibration and physics runs) has to deal with a very large number of heterogeneous components, ranging from detector-specific front-end electronics to embedded processors in the readout chains, from high-level computing units (PCs) to switched networks. Everything must operate in a coordinated fashion and the control system must be capable of guaranteeing a coherent behavior and maintaining the overall TDAQ in a coherent state at every phase of operation. The organization of ATLAS in detectors, subdetectors and modular elements suggests, here again, a hierarchical, modular architecture.

Scalability. At the moment of the ATLAS DAQ and High-Level Triggers Technical Design Report (June 2003),[1] the total amount of computing power for the high-level triggers, by far the dominant part of the overall computing system in ATLAS, had been estimated at ∼2300 dual-processor boxes clocked at 8 GHz. The overall requirement for the control system was, therefore, to be able to scale up to ∼5000 processes to be simultaneously configured and controlled. The advent of the multicore technology for computers boosted the number of processing units, and consequently of processes running simultaneously, by at least one order of magnitude, changing dramatically the requirements for the control system. It will be shown later in this chapter that thanks to the modular and scalable design of the initial system, the overall structure of the original software and its main design guidelines could be preserved, so that the substantial change of performance requirements did not invalidate the entire communication model, thus avoiding a major disruption experiment-wide during the commissioning phase.

3. Conceptual Data Acquisition for a Typical LHC Experiment

Expressed in the simplest possible way, the trigger and data acquisition system of an HEP experiment has the task to read the data produced in digitized form by the detector and transport them to permanent storage. This process happens, for all practical purposes, in real time. Offline data handling will then take over from the permanent storage, to analyze the data produced online; this latter process does not necessarily (and not at all for most of the analysis) happen in real time.

The above definition obviously hides a lot of complexity, due to the rates of production and the amount of data associated with physics events. The *detector dead time* ought to be minimized, considering that there may be a difference of six or seven orders of magnitude between the event interarrival times and the speed at which data can be recorded in permanent storage. The reconstruction and the analysis of events is a very time-consuming process, with a difference of about nine orders of magnitude between event production and event analysis times. The TDAQ must drastically reduce the rate of events such that (1) data recording introduces practically 0 dead time and (2) the events selected for analysis are in a manageable amount and are "good" events for physics (and not background of no interest to physics results).

The *transport of large amounts of data* (GBytes per second) is the second task of the trigger and data acquisition system. Events have to be moved from the detector front-end electronics to mass storage, if detected as "good" events by the trigger system, or eliminated from the stream as soon as the trigger decides that the event is not interesting. Events, or fractions of them, must be dispatched to processing elements for triggering or online monitoring purposes. These processes have to be efficient (i.e. not introduce dead time) and reliable, data arrive at their destination and their integrity is guaranteed.

The *controls* are the third main element of the data acquisition system: to insure that all the parts (several thousands in an LHC-like system) of the trigger and data acquisition system are initialized as required and operate as expected.

Several levels of trigger are needed in order to minimize dead time and to include a complex algorithm in the decision processes. At each level additional information can be made available, thanks to the rate reduced by the previous levels, to perform better and more sophisticated event filtering. The boundary conditions are the event rate at the accelerator and the event-recording rate.

Different trigger levels need different amounts of data: a small number of channels for the faster trigger, a selection of the event data for the second level and the full event for the third level. Trigger decision latencies go, so to speak, as the inverse of the amount of data required.

Processes with different latencies coexist in the system. The rate of trigger decisions is a statistical process. For these reasons the data acquisition system includes buffer memories at different levels in the system. Buffers have the following functions: decouple the stages in the trigger and data acquisition system, absorb the latencies of the trigger decisions and smooth the statistical distribution of the event interarrival times at different levels in the system. Buffers are of course memories, of the proper size to absorb the latency of the decision steps, with processing capabilities, in order to handle and receive, handle and distribute event data (fragments or full events).

Performance by the system comes from sophistication of event selection algorithms and the processing power to run them, but also, to a large extent, from the exploitation of parallelism. Parallelism is natural at the level of the detector channel, but it can be maintained throughout most of the data acquisition system by means of multiplexing channels in a large number ($\sim$1000) of relatively small [O(1 kB)] fragments. Such fragments can be moved in parallel over parallel links, stored in parallel memories, distributed, when necessary, to trigger processes for the purpose of selecting the event they belong to, until the moment the full event is needed. Since the trigger selectivity has become so high, further rate reduction can only be achieved by an analysis of the complete event, or the event ought to be assembled it has to be finally written out to permanent mass storage. The data acquisition stage (called event building) where all fragments, belonging to the same collision, are merged into a complete event, can also be parallelized: multiple events can be built concurrently into different memories.

Taking into account the previous general considerations, a trigger and data acquisition system is made up of the following major building blocks:

- *Buffer memories*: to absorb latencies and smooth the statistical nature of the event arrival times. Buffer memories decouple different stages (e.g. detector readout or event building stages) in the data acquisition system. Logically the buffer memory has one (or more) input port(s), to accept data from upstream sources (e.g. a detector readout channel), one or more output ports, to deliver data to downstream clients (a processor performing event selection, a downstream buffer memory, input to a further system stage), and of course a memory large enough for instance to absorb latencies typical of that particular part of the system (e.g. a trigger level). If necessary the buffer memory may also contain additional processing capability, for instance to perform data conversion or to run signal processing algorithms.
- *Communication links*: the paths over which data are transferred. The links may be tuned to a particular problem (e.g. a detector data path) or may be more generic, such as those on which data are moved toward processing

units. Logically they provide one-to-one connectivity between data sources and destinations; in reality they can be point-to-point but also shared data paths (e.g. via switching units which for instance reduce an n^2 connectivity problem to one involving $2n$ connections).

- *Processors*: to provide local processing power to tasks such as event selection, monitoring and controls. Processors can be custom-made, for dedicated tasks such as data reduction or low latency event selection, or commodity processors to run high-level, offline-like algorithms.
- *Data flow supervisors*: to control and manage the flow of data fragments between processors and memories. For example, the event-building process needs to assign a place (buffer memory) where the event will be built; the assignment must be known by all the source buffer memories participating in the building of the event. Supervisors may be embedded in a buffer memory or be implemented by separate processor units.

It should be pointed out that each building block has, in addition to its "proper" function (such as providing an event selection decision), a control part for initialization, operation and monitoring. The control process needs to exchange low-bandwidth and low-rate (at least compared to the main event data path) information between all the elements in the system; logically speaking, over a second network interconnecting all the system elements.

4. Driving Principles of the ATLAS DAQ Design

The main features that have driven the design and implementation of the ATLAS data acquisition and high-level triggers system can be classified as:

- *Factorization* into main functions and partitioning;
- *Minimization* of data movement;
- *Maximization* of system uniformity and minimization of developments;
- *Staging* of selection rates and data volumes.

4.1. *Factorization and partitioning*

Very early in the design phase, the necessity of being able to operate the system as a number of subsystems running independently and concurrently has been recognized as a fundamental feature to support the various operations of the ATLAS detectors prior to and beyond the actual data taking of proton–proton collisions for physics. Furthermore, the capability of operating only part of the data flow chain, with limited functionality but full operability, was deemed to be equally important, in order to support the debugging and commissioning of the detector readout during the first installation and for problem-fixing in later stages.

To this extent, the data flow functionality has been factorized into three functions, corresponding to three steps in the data flow chain: (1) a detector-specific part, including the front-end readout electronics up to a detector-specific module for the completion of the signal processing and the event formatting (readout driver, ROD); (2) a local readout system (ROS, assembling a number of readout buffers, ROBs), based on independent but identical units for each detector (or elements thereof); and (3) full detector readout via an event-building stage. Data acquisition functionality is provided at all three levels for maximum flexibility in the use of the system, thanks to a decoupling between ROD and ROB via point-to-point communication links,[9] and between ROS and event builder via network links. This architecture guarantees independence of operations for each detector and also between detectors and the central DAQ, a feature of the utmost importance during initial commissioning of the detectors and of the trigger and DAQ systems, and for detector calibration and repairs throughout the life of the experiment.

Partitioning refers to the capability of a system to provide the functionality of the complete system to a number of subdivisions of the overall system running independently and simultaneously. For the TDAQ it is the ability to read out all or part of a specific detector with a dedicated local trigger in one partition, while the other detectors take data in one or more other independent partitions. Such a feature is an absolute must for an experiment like ATLAS in all the operation phases: during setup and commissioning, when tens of people need to work at the same time on different pieces; during physics runs, in case of problems on a part of the detector, in order to isolate it for debugging and fixing while the rest of the experiments continues to take data; during interrun periods, when all detectors need to run independent calibration procedures. The TDAQ system, therefore, must support a level of partitioning of a suitable granularity, so that multiple copies of DAQ can be run concurrently on subsets based on the modularity of detectors and subdetectors.

4.2. *Minimization of data movement*

Given the high requirement of data throughput, two mechanisms have been adopted at the moment of the conceptual design to minimize the data movement and therefore the complexity and cost of the system: (1) sequential processing on a subset of data of the LVL1-accepted events (LVL2 trigger); (2) selective readout for event building of the LVL2-accepted events. The first mechanism is based on the region of interest (RoI)[10] principle (i.e. the geographical addresses of the energy clusters in the calorimeter and the muon candidates identified by the LVL1 trigger are transferred to the LVL2 trigger to seed the selection processing). The second uses the result of the second-level trigger to transfer to the event builder only the data requested by the next stages of analysis (event filter and offline).

The RoI mechanism, and the flexibility of the data flow architecture, also make possible the realization of a scheme for fast calibration of muon chambers during physics data taking. A dedicated path from the level-2 farm takes all muon candidate tracks before LVL2 selection directly to a remote CPU farm, bypassing the event builder and event filter, thus providing the high statistics required for the determination of muon chamber calibration parameters.[11]

4.3. *Maximization of system uniformity*

All differentiation between detectors ends at the level of the ROD. Already the environment of integration — both hardware and software — of the ROD itself is common to all detectors. Furthermore, the common software environment of the ROD is common to that of the next layer up, the ROS, thus providing a complete and stand-alone data acquisition functionality (ROD crate DAQ).[12] The uniformity is also helped by the maximum adoption of commodity products — both hardware and software — with maximum exploitation of technology advances, in communication buses and links, networking and processing units, as well as operating systems, programming languages, databases and interprocess communication packages.

4.4. *Staging of selection rates and data volumes*

Lastly, given the funding profile and the uncertainty on the progression of machine luminosity, the data acquisition and high-level trigger system had to be designed to be staged, with the size and performance evolving as resources became available and as required by the luminosity performance, yet providing full functionality at every stage. The baseline architecture was to support the staging of the readout following the staging of detector installation and — the big part — the staging of the DAQ-HLT system and the corresponding event builder and networking, following the luminosity performance and availability of resources.

5. The ATLAS Data Flow Challenge

The generic trigger and data acquisition system outlined in Sec. 3 becomes specific to ATLAS when subjected to the following boundary constraints:

- A bunch crossing rate of 40 MHz; a bunch crossing results in a trigger to the detector; this is an input constraint as given by the LHC accelerator.
- On average 1.5 MB of data are produced in the detector as the result of a bunch crossing. This unit of data is defined as an "event," and is recorded on permanent storage at the end of the data acquisition process. This constraint is set by the physics of collisions at LHC energy and by the design of the ATLAS detector.
- A permanent storage-recording rate of ∼200 events per second, hence about ∼300 MB/s of data in permanent storage (magnetic disks). The combination of today's mass storage technology and the exploitation of parallelism would allow in principle, and at a cost, a higher recording rate. Already today a transfer rate from the ATLAS site to the CERN computer center can sustain up to 600–700 MB/s and it could even be scaled up. The ∼200 Hz limit is, however, set by the projected capability of the offline analysis system, whose cost scales with the number of events to be analyzed.

It has to be noted that the above parameters were set in the early 1990s, when the accelerator and the detector were designed. They were, and are, the constraints to be applied to the design of the ATLAS data acquisition system. These constraints generate the challenges that were faced by the designers and the implementers of the system.

In this section we look into those challenges that are related to the rates and the flow of the data; in a succeeding section the challenges created by the complexity and size of the implementation will be addressed.

Since the beginning, the ATLAS rate reduction scheme has been based on three levels of triggering (or event selection) (Fig. 2):

- The first level, implemented in dedicated custom hardware, based on local calorimeter and muon detector information, designed to reduce the rate to 100 kHz at nominal luminosity and with a fixed latency of 2.5 μs.
- An intermediate level, implemented in high-level hardware capable of running (selected parts of) the offline analysis software. The second level reduces the rate by a factor of about 30, with a latency of a few tens of ms.

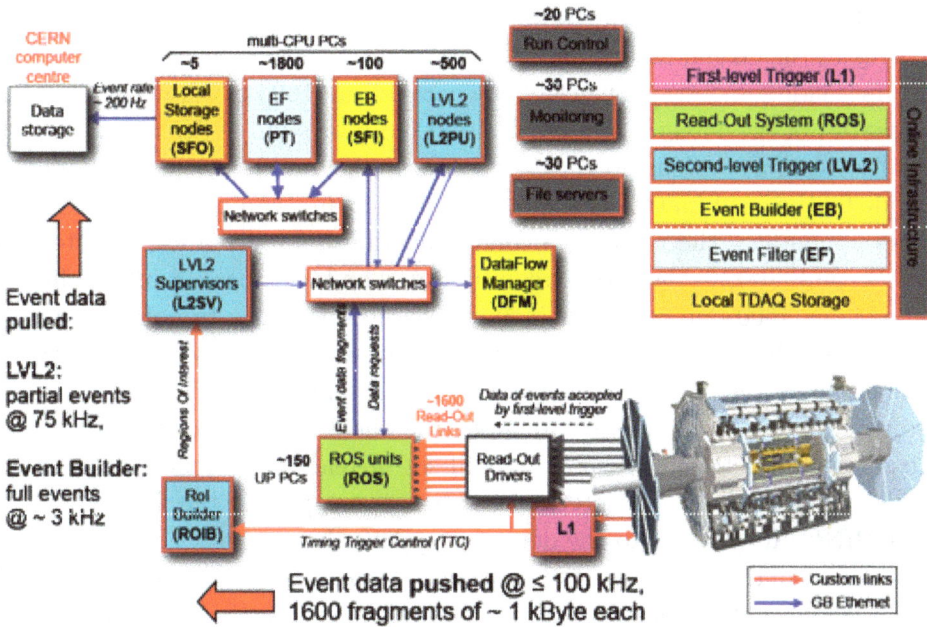

Fig. 2.   The ATLAS trigger and data acquisition architecture.

- A final level which further reduces the rate to the required 200 Hz, with a latency of seconds. Since the beginning of the design it has been clear that to achieve this final rate reduction, resulting in a global selection power of 1 in 200,000 events, complex and sophisticated algorithms (the same as those used offline) have to be applied to the complete event.

Two major challenges that the ATLAS TDAQ system had to face are beyond the scope of this chapter and will not be addressed: the whole of the first-level trigger and the algorithmic performance required to actually select the one "good" event out of 200,000. We concentrate instead on the challenges related to the movements of the data that are needed to make the above principles a reality.

The combination of the detector and the first-level trigger produces 100 kHz[a] 1.5 MB events: 150 GB/s enter the first buffer memory of the ATLAS data acquisition system. This buffer memory multiplexes three tasks: (1) to read out and buffer detector data, (2) to provide (part of) the event

---

[a]The 100 kHz includes a safety factor of 2 to take into account the fact that physics at LHC energy is not experimentally known. Monte Carlo estimates give indeed a first-level trigger rate of 40 kHz.

data to the second-level trigger, and (3) to provide the data for the accepted events to the downstream stage of the data acquisition system. The problem is first made somehow manageable by parallelism: 1600 parallel data streams come out of the detector and are multiplexed into ~150 buffer memory units (called readout systems in the ATLAS jargon). Each buffer memory unit is made by the combination of a central processor unit and peripheral I/O processors handling the three data communication links (from the detector, to the second level, to the downstream DAQ). The I/O processor interfacing to one of the 1600 detector links includes also the memory to hold the detector data. These design principles were used since the very early stages of the design of the ATLAS DAQ system. Technology evolution (hardware and software) has made it possible to implement the design mostly with commodity hardware [Fig. 3(a)] shows the full set of readout systems (commercial PCs) for the LAr calorimeter]; the only custom part is the I/O processor (ROBIN[13]) interfacing to the detector readout [Fig. 3(b)].

The second-level trigger operates on and selects from the events stored in the readout system buffer memory: the challenge is, again, represented by a rate of 100 kHz and a data volume of 150 GB/s. The rate is dictated by the first-level trigger, and is tackled by parallelism: many computers (500 in the TDR configuration for LHC nominal luminosity) run parallel event selection processes. The volume of data is strongly reduced by exploiting information produced by the first-level trigger: which regions of the detector are considered "interesting" for further analysis of the event. These regions [regions of interest (RoIs), in the ATLAS jargon] represent typically only 1–2% of the event data, which correspond to only 20–30 (out of 1600), readout fragments. This results in a second-level trigger handling "only" about 2 GB/s of data. Figure 4 is an illustration of the four RoIs (two muons and two em-calorimeter clusters) used by the LVL2 trigger for the identification of a Higgs decaying into two Z^0's, which in turn decay into a pair of muons and a pair of electrons.

The geographical information of the "interesting objects" from the calorimeter and muon LVL1 triggers is collected and passed to the LVL2 by the RoI builder, a VME-based custom device, shown in Fig. 5.

A supervisory element (actually a few tens of them), implemented by processors linked to the same data network as used for transporting the data to the second-level trigger processors, is used to translate the information received by the first-level trigger into the address of the ROBINs corresponding to the RoI, to decide which LVL2 processor will deal with that particular event and to maintain the bookkeeping of the selection process. Once more,

(a)

(b)

Fig. 3.

the mechanism of the RoI, the ATLAS way to tackle the second-level trigger challenge, was devised at the very early stage of the design. Technology today allows one to implement practically the whole second-level trigger system with commodity hardware.

At the time the second-level trigger has accepted an event, it is still in the form of 1600 parallel fragments, held in 150 readout systems. The 1600 fragments have to be merged into a single event data block; the challenge is now that of merging together (logically a sequential operation) 1600

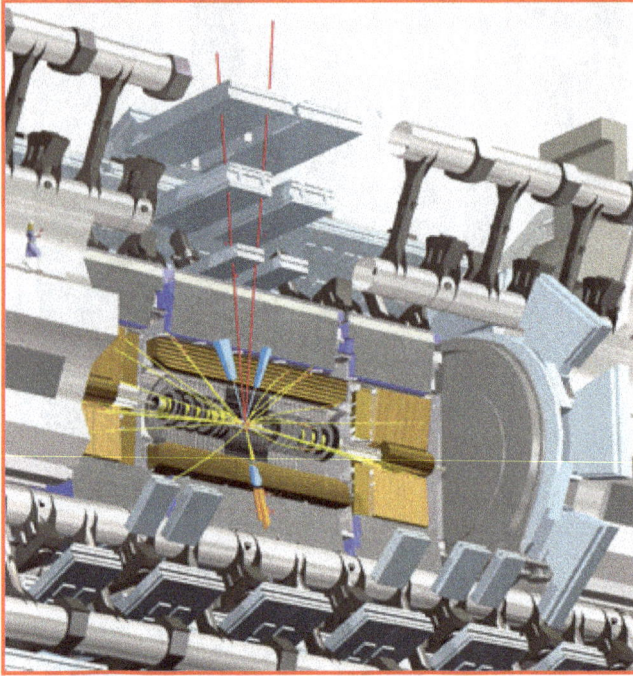

Fig. 4.　A Monte Carlo simulation of a Higgs decaying into two Z^0's, which in turn decay into a pair of muons and a pair of electrons.

Fig. 5.　The VME-based region-of-interest builder.

fragments at a rate of 2–3 kHz. Another form of parallelism is the key to tackling this challenge: we have 150 sources of data (the readout systems), if we have, say, 100 destination buffer memories, each to hold complete events, and we fully interconnect sources with destinations, we can envisage 100

Fig. 6. The three core switches, data flow (the two on the left), controls and back-end.

parallel event merging processes. This is the solution adopted by ATLAS (and the other LHC experiments): Gbit-Ethernet and the network switch provide the technology for the implementation.

The event builder consists of buffer memories, actually commodity PCs, where the event is built, a Gbit Ethernet network based on switches (Fig. 6) and a supervisory element, another standard PC, to control the event-building process (for instance to assign which output buffer memory is used to build a particular event).

The third-level trigger stage (event filter, EF) is delimited upstream by the event builder buffers and downstream by another buffer system which is in front of the permanent storage devices. In the middle there is a large amount of processing power, and eventually some 1800 PCs, each handling events in parallel and running offline-like reconstruction and decision algorithms. Gbit Ethernet and switches provide the communication paths between input buffers, processors and output buffers. Parallelism, exploiting the independence of different events, is the key to the challenge; a second aspect is that of running offline software in real time — with its issues of robustness and performance.

Figure 7 shows part of the HLT farm presently installed (about one-third of the final).

The 500 LVL2 and the 1800 EF PCs are of multicore technology, i.e. each box contains 8 or more processing units, with consequences for the scalability of the configuration and the control of the farm, as will be discussed later.

Fig. 7. Part of the ATLAS HLT farm. Each of the one-unit box is a dual-process quad-core computer. There are 31 PCs per rack.

An interesting feature of the architecture of the high-level trigger farms is the possibility of assigning about one-third of the computing nodes to either LVL2 or EF, adding flexibility to the selection process to adapt to the nature of the physics events and the maturity of the selection algorithms.

6. Global Features

Two adjectives, "scalable" and "distributed," characterize globally the ATLAS DAQ–HLT system. Scalability is a system requirement that has been present since the very early stages of its design. Distribution, of functions, is at the same time a design feature and a design and implementation challenge.

6.1. *Scalability*

The ATLAS TDAQ system has been designed to meet its nominal performance requirements as they were described in Sec. 5. Such requirements are, in the end, dictated by the nominal LHC luminosity.

The LHC accelerator will not start with, and will not reach for the first 3–4 years, its nominal luminosity. The ATLAS TDAQ system is scalable in the sense that it is capable of adapting its performance to the planned delivered luminosity. In other words, it is designed to be capable of tracking the evolution of the LHC performance. The design and functionality are unique, but the performance is modulated to that of the accelerator.

Scalability is required also for a different purpose: to match the staging of the detector itself and, in the end, the availability of funding.

Scalability is achieved first via modularity of design; the functional building blocks outlined in Sec. 4 remain present at all performance stages. Additional performance is achieved, as required, by multiplying the functional building blocks (performance by parallelism) and increasing the size of the communication networks to match the level of parallelism. For example, scaling up the performance of the event builder means (1) to add more destination buffer memories, where full events are assembled, and (2) to increase the size of the network interconnecting readout systems to event builder destination memories. This results in a higher rate of events assembled concurrently. The last triggering stage ought also to be scaled up in performance, in order to maintain the rate to permanent mass storage to such a level that no dead time is introduced, and/or the performance of the permanent mass storage must be increased (scaled up) accordingly. Once more, this is easily done with more trigger processing elements and additional network connectivity and capacity.

Scalability with respect to the detector is realized following the same principles: this time also the number of readout systems increases to meet the requirements of the detector (e.g. more readout channels).

Each of the functional blocks indicated in Secs. 4 and 5 can be scaled up, within the available financial envelope, by additional hardware fitting the system modular architecture.

In addition the ancillary functions, such as controls and databases, must scale — following the principles outlined above — with the increased size of the system.

6.2. *Distributed system*

The main system functions — detector readout, triggering, controls, etc. — are distributed over many thousands of components, instances of the building blocks outlined in Sec. 4. Each component, e.g. a level-2 processor, acts independently but is also part of the global system and must act coherently with all the others. In other words, there is a potentially high multiplicity of concurrent states, the states of the individual components, which make up for the global, coherent state.

Given the large number of concurrent actors, maintaining a global coherent state is a challenge that has been addressed by the run control system, described later in this chapter. The challenge also comes from considerations of reliability: the interplay of so many hardware and software parts, commercial and developed in-house, is prone to errors and failures. Coherence ought to be maintained also in the face of parts failing to function. For instance a trigger processor, a readout buffer, may stop functioning for any reason (hardware or software); the overall system is expected to tolerate local faults and continue to operate, possibly with a degraded performance. This is achieved first through a hierarchical description of the system and by the implementation of features which allow one to automatically isolate, and "remove" from the running system, the failing part.

6.3. *The system in operation*

When ATLAS saw the first beam in September 2008, the TDAQ system was ready and operational. The configuration of the system was tailored to provide a global performance about 30–40% of the final one.

When taking into consideration the September 2008 TDAQ system (detailed in Table 1), we observe that:

- It matches the final specifications for the detector readout: both in terms of number of channels and in terms of final performance, where each readout link can transport 160 MB/s.
- The high-level triggers and event builder were sized to meet about 30–40% of the required final performance. Extensive performance tests have of course been carried out. Figure 8 gives an indication of the performance scalability of the event builder: how performance (expressed in MB/s of the aggregate bandwidth) scales with the number of event builder buffers (SFI, which stands for "subfarm input"). The number of SFIs is an indicator of how many events can be built concurrently.
- The mass storage system (including buffer, large disks and powerful networking capabilities) matches, and can be easily scaled up to exceed, the required final performance.

7. Run Control

The DAQ control[14] occurs over a dedicated Gbit Ethernet network. The system encompasses all the software required to configure and control the data

Table 1.   Outlines the physical layout of the data flow system.

Subsystem	Component	Size	Remarks	Performance
Readout system	Link to detector readout	1577	Readout of complete detector in required final performance	Capable of standing the final performance
	Interface to detector readout (ROBIN)	551		
	Readout system core	149		
Event builder	Network core	M × N	Network infrastructure fully in place; will be scaled incrementally as more EF slices are added	4 kHz EB rate, 3.3 GB/s; infrastructure meets final specification; performance adapted to initial luminosity
	Event builder buffers (SFIs)	31	As needed to meet specifications of initial luminosity	
High-level trigger	RoI builder	1	Supports 8 level-2 supervisors	35% of final system
	Level supervisors	8		
	L2 processing units	120 8-core PCs		
	Event filter processing units	465 8-core PCs		
Mass storage	Mass storage stations (SFOs)	5 units, each with 4 high-speed network connections, 10 TB disk space	550 MB/s sustained; 700 MB/s peak	Exceeds final performance requirements

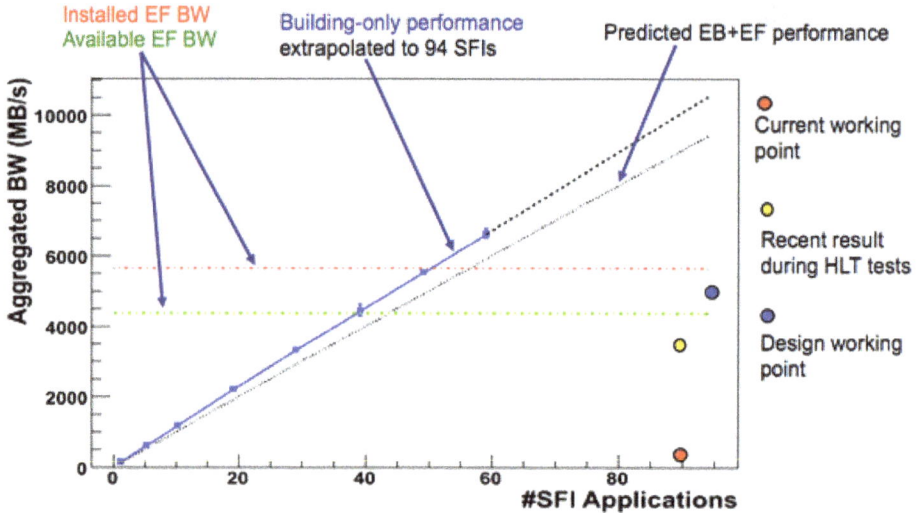

Fig. 8.   Measurement of the scalability of the event builder system.

taking of the experiment. It is designed following a layered component model. At the bottom are common base libraries, which include a custom-developed database — the object kernel support (OKS)[15] — and the libraries for the interprocess communication (IPC) based on the common object broker architecture (CORBA).[16] The intermediate layer consists of services, including the process manager and the message reporting service. The application layer contains the run control, the diagnostics and verification system, and the expert system. At the top there is a set of graphical user interfaces.

In order to deal with a large heterogeneous system whose components have to operate in a coordinated fashion, special care has been taken in the design of a control system capable of guaranteeing a coherent behavior and maintaining the overall TDAQ in a coherent state at every phase of operation. The organization of the ATLAS detector into detectors and subdetectors leads to a hierarchical organization of the control system, based on a large number of controllers distributed in a hierarchical tree following the functional composition of the TDAQ system. Figure 9 shows the overall architecture of the TDAQ control and its relationship with the detector control system (DCS; in ATLAS the DCS is operationally independent of the TDAQ, and it is not covered in this chapter).

In the full readout configuration, the number of controllers is ~370, controlling 15,000 applications (reaching ~50,000 in the final HLT configuration).

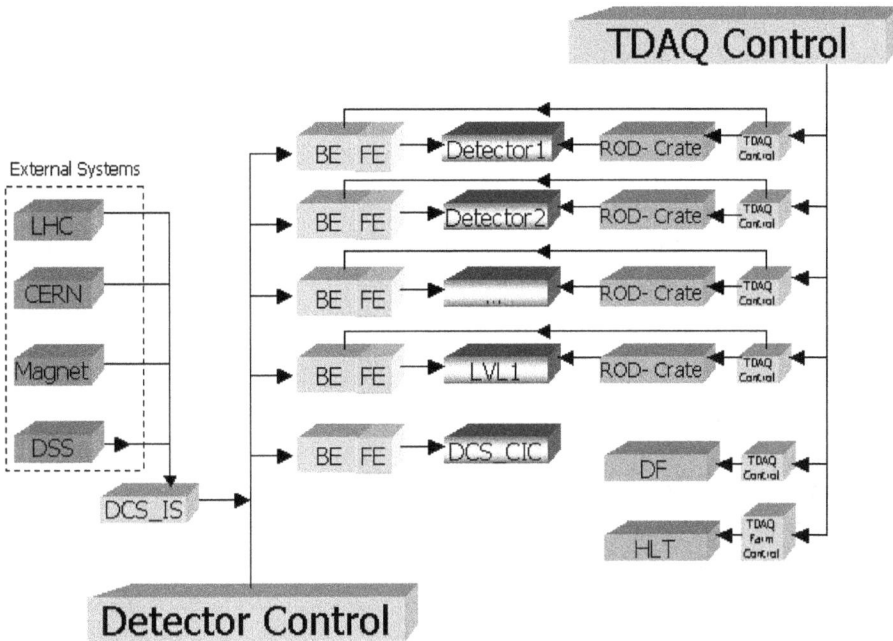

Fig. 9. The ATLAS control hierarchical architecture (ATLAS HLT-DAQ TDR[1]).

The basic element for the control and supervision is a controller. It is composed of two well-separated aspects: the skeleton, defining the interfaces of the controller with the entities it controls, unique throughout ATLAS; and the core of the controller, defining the actions to be taken at the occurrence of an event, customizable and configurable for each system.

This structure is very generic and allows any kind of command to be distributed to a system. In order to provide an abstraction of the control activity to controllers higher up in the control tree and to the operator, a common state machine model has been introduced in the core. Each state has a defined set of authorized transitions that bring the system into a new state. The main strength of this approach is that the system is always in a well-defined state so that the state machine gives an overview of the state of the whole system and keeps the overall coherence.

7.1. *Evolution of the initial system*

From the initial design described in the Technical Design Report (TDR) in 2003, the system had to undergo a substantial evolution. The initial system based on a complete design was validated following the ATLAS DAQ development strategy, on real detector readout at test beams. Scalability to large-scale systems was assessed on a large computing farm with simulated data. Functionality and performance met requirements, with the exception of fault tolerance. Other features were overlooked at the moment of the TDR, such as computing security, or underwent a big requirement change, such as scalability.

Although an analysis of the extended requirements revealed that an overall redesign of the full system was not required, it was decided nevertheless to proceed to an upgrade of the design and a partial reimplementation, mainly in the three areas below:

- *Computing security* was never really considered an issue in previous high-energy-physics experiments. However, in a collaboration of more than 2000 people unintentional mistakes on such a complex system can soon became a serious security threat, let alone the potential catastrophic consequences of malicious intrusions. Security requirements were gradually clarified only very late in the development process and have been addressed via a number of new features, such as traceability, identification of actions, a secure mechanism for remote monitoring and role-based access management.
- *Scalability.* At the time of the TDR the system was expected to be able to control and configure some 5000 software processes. The introduction of multicore technologies in the high-level trigger farms progressively scaled up the number of processes by one order of magnitude. Despite the fact that the evolution could be addressed without a change in the overall architecture, which would have provoked a major disruption to the experiment's commissioning, a redesign of some low-level components and the introduction of remote database servers had to be done to cope with the increased load.
- The *error recovery* of the TDR design was limited to the capability of ignoring or restarting some nonessential failing processes. A complete redesign and reimplementation of run control and expert system components was deemed necessary in order to provide means of recovering from complex errors. A major improvement of fault tolerance and error recovery came with the introduction of a "*stopless recovery*" mechanism. While by design the system can dynamically mask one or more faulty HLT or event builder nodes, the failure of a single element in the readout system, in contrast, would set up a busy signal that would block the entire data acquisition. A mechanism was therefore introduced to continue the run in the case of failures that do not involve a major reconfiguration of

the experiment, consisting in automatically removing from the readout chain items that block the busy system. It is then left to the operator to assess the level of degradation of the data quality and whether to let the run continue or stop it for repair.

8. Online Monitoring

The ATLAS online monitoring[17] is organized as a distributed modular hierarchical system, which includes several applications, ranging from low-level information sharing services to high-level analysis frameworks and graphical interfaces. This organization offers high flexibility in terms of accommodating various types of monitoring information as well as in configuring analysis algorithms and managing their outcome.

The monitoring system uses the Gbit Ethernet control network. Its overall structure is shown in Fig. 10. The heart of the online monitoring is the *information service* (IS), which provides a configurable number of shared information channels between any application in the ATLAS environment abstracting the underlying complexity of the distributed environment and adopting network and CPU load minimization algorithms.

A number of services interface to the IS:

- *Event monitoring* (Emon), to provide statistical sampling at different levels of the data flow chain; *event analysis frameworks*, to produce histograms and other results; *data quality monitoring framework* (DQMF),

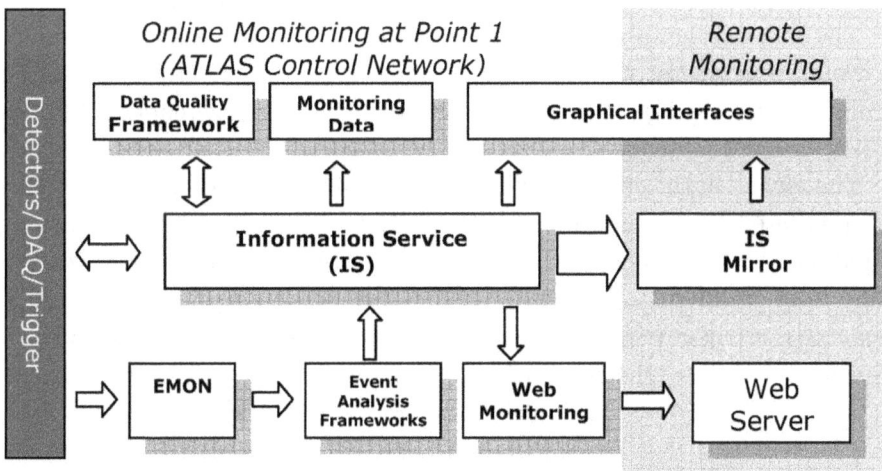

Fig. 10.   The ATLAS online monitoring architecture.

to apply quality-checking algorithms to results; *monitoring data archiving*
(MDA), to store, manage and retrieve monitoring data.

- *Remote monitoring.* Particular emphasis has been given in ATLAS to providing means of monitoring the experiment from anywhere in the world, in
  a complete and efficient, but also safe manner. Two types of remote access
  are provided. The *web monitoring interface* (WMI) converts a subset of
  monitoring information to static HTML pages at regular time intervals
  accessible via a standard web server running on the CERN Global Public
  Network (GPN).

  For advanced monitoring a mirror copy of the IS is maintained in real-
  time in the CERN GPN (*IS mirror*), so that a limited number of external
  users can monitor the status of the data taking in real time without logging
  onto the ATCN. This approach provides the necessary security level and
  also decouples the online system from any extra load, while enabling the
  use of the same *graphical interfaces* as in the main ATLAS control room.

9. Conclusions

In this chapter we have given an overview of the major challenges experi-
enced by today's HEP collider experiments at the leading edge of the field.
We have focused on the source of the challenges, *why* the ATLAS TDAQ
system followed certain principles, and some of the solutions adopted, *how*
the ATLAS TDAQ system "looks like," to make the TDAQ system, and the
experiment in the end, float against the maelstrom of the challenges.

The source is simply represented by the three main constraints: the rate
of events, the size of the events, the limited (in the end by financial constraints) mass storage and computer power that can be made available to
the experiment.

The *how* — in the end the definite success of the ATLAS TDAQ system —
is represented by the application, throughout the system, of the paradigm
of parallelism and the performance evolution of IT hardware and software.
They were blended, to optimally match the system requirements and the
available funding envelope, through many years of R&D, design, prototyping,
implementation and use in real life applications.

The result is a very large distributed system, which creates, so to speak,
its own challenges, such as the management of the description of the system,
the coordination of all elements (so that they operate for the common result),
the reliability (of hardware and software) and the system tolerance to faults,
and many others.

Acknowledgments

Design principles, technology and realization are part of the success story. The other component of success, the one that much too often remains in the shadow of technical papers, is the human factor. Each of the challenges highlighted in this chapter, and many others, were met only because of the talent, competence and dedication of the worldwide team that has made the list of requirements a successful reality.

We, the authors of this chapter, acknowledge and admire the quality of work of the ATLAS trigger and DAQ community. We thank them and hope that, with what we have written, we have done justice to their achievements.

References

1. ATLAS Collaboration, High-Level Trigger, Data Acquisition and Controls Technical Design Report. CERN/LHCC/2003-022 (2003).
2. ATLAS Collaboration, ATLAS Level-1 Trigger Technical Design Report. CERN/LHCC/98-014 (1998).
3. F. Butin (ed.), ATL-TECH-PUB-2008-002.
4. C. Bee *et al.* A scalable data taking system at a test beam for LHC. CERN-LHCC-94-47; DRDC-RD13-Rev.
5. ATLAS Collaboration (G. Aad *et al.*), The ATLAS experiment at the CERN Large Hadron Collider, *JINST* **3** (2008) S08003.
6. S. Ask *et al.* The ATLAS central level-1 trigger logic and TTC system, *JINST* **3** (2008) P08002.
7. R. Achenbach *et al.* The ATLAS level-1 calorimeter trigger, *JINST* **3** (2008) P03001.
8. F. Anulli *et al.* The level-1 trigger barrel system of the ATLAS experiment at CERN, *J. Instrum. Methods* **4** (2009) P04010; Y. Okumura *et al.*, The commissioning status and results of ATLAS Level 1 Endcap muon trigger system, *Proceedings of the Workshop on Electronics for Particle Physics* (TWEPP 2008), pp. 556–560.
9. E. van der Bij *et al.* Recommendations of the Detector Interface Group–ROD Working Group. EDMS Note, ATL-D-ES-0005 (2001).
10. ATLAS Collaboration, ATLAS DAQ, EF, LVL2 and DCS Technical Progress Report. CERN/LHCC/98-016 (1998).
11. E. Pasqualucci *et al.* Muon detector calibration in the ATLAS experiment: data extraction and distribution, in *15th Int. Conf. Computing High Energy and Nuclear Physics* (Mumbai, India, 13–17 Feb. 2006), pp. 134–137.
12. S. Gameiro *et al.* The ROD crate DAQ of the ATLAS data acquisition system, in *Real Time Conference 2005*. 14th IEEE–NPSS volume, issue 10, 10 June 2005, p. 5.
13. R. Cranfield *et al.* The ATLAS ROBIN, *JINST* **3** (2008) T01002.

14. G. Lehmann Miotto *et al.* Configuration and control of the ATLAS trigger and data acquisition. ATL-COM-DAQ-2009-065.
15. R. Jones *et al.* The OKS persistent in-memory object manager, *IEEE Trans. Nucl. Sci.* **45**(4) (1998) 1958–1964.
16. CORBA, http://www.omg.org/corba.
17. W. Vandelli *et al.* Strategies and tools for ATLAS online monitoring, *IEEE Trans. Nucl. Sci.* **54** (2007) 609–615.

Chapter 16

REMOVING THE HAYSTACK — THE CMS TRIGGER AND DATA ACQUISITION SYSTEMS

Vivian O'Dell

Fermilab, P.O. Box 500, Batavia, IL 60510, USA
odell@fnal.gov

The CMS Trigger and Data Acquisition Systems have been installed and commissioned and are awaiting data at the Large Hadron Collider. In this article, we describe what factors drove the design and architecture of the systems.

1. Introduction

As described earlier in this book, the CMS detector is comprised of many constituent detectors designed to examine different details of each proton–proton collision. The complementary information from all the sub-detectors assembled for each collision of interest allows us to reconstruct LHC collisions on an event by event basis. The role of the CMS Data Acquisition system is to collect, assemble and record for further analysis the information from all the constituent detectors of CMS for each proton–proton collision of interest. The LHC will deliver an average of 20 proton–proton collisions every 25 ns. Below we will describe how we selectively record the few drops of potentially interesting interactions from this sea of collisions. We will describe the general requirements and the design choices. The reader is referred to Refs. 1–3 for a more detailed list of requirements and implementation.

1.1. *Interaction rates at the LHC*

By using theoretical models to extrapolate from previous collider measurements to the LHC center of mass energy, we can estimate how many hard collisions per second we expect at the LHC. Figure 1 shows the total pp cross section as a function of center of mass energy. At the LHC center of mass energy of 14 TeV, the total inelastic pp cross section is predicted to be about 70 mb (= 70×10^{-27} cm^2), which means that for the LHC design luminosity of

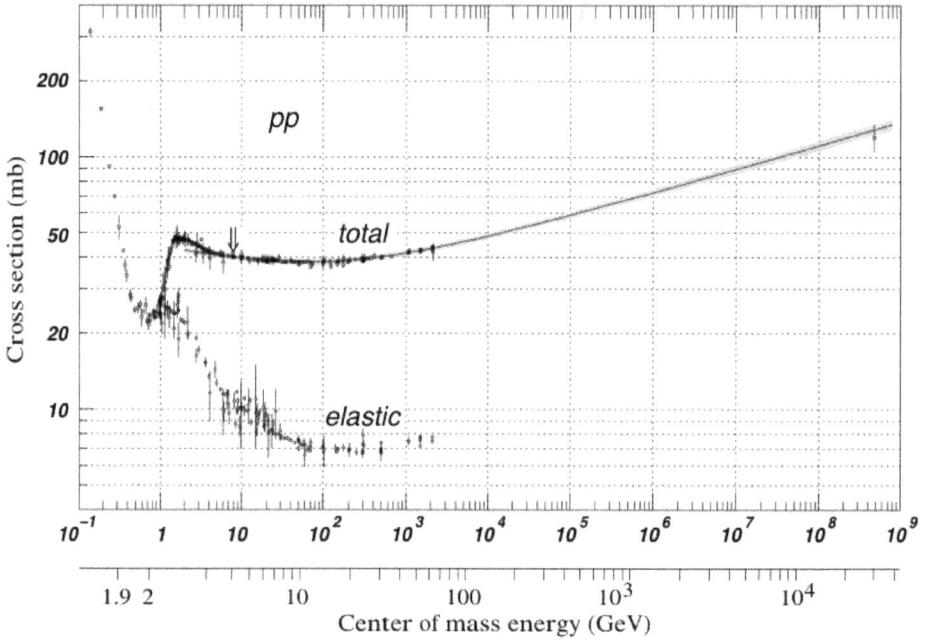

Fig. 1. The total pp cross section in millibarns as a function of center of mass energy.

10^{34} cm^{-2}s^{-1} the expected raw interaction rate is 7×10^8 Hz. The LHC beam is made up of RF bunches of protons: each bunch is separated by 25 ns which means two bunches cross in the CMS detector every 25×10^{-9} s. This also implies that each bunch crossing contains $(7 \times 10^8 \text{ Hz}) \times (25 \times 10^{-9} \text{ s}) = 17$ pp interactions on average. Not all bunches will be filled at the LHC and one can use this formula to calculate the approximate average in-time pileup per bunch crossing. Clearly, at the same luminosity, as the number of filled bunches goes down, the number of protons/bunch increases as well as the number of interactions/beam crossing.

Figure 2 shows the expected cross sections and rates of different physics processes of interest at the LHC. What is apparent from this plot is that interesting new physics such as SUSY or Higgs boson production is only a tiny fraction of the total inelastic pp cross section. For example, theory predicts the ratio of cross sections of the standard model $H \rightarrow \gamma\gamma$ with $M_H <\sim 200$ GeV/c, to the total inelastic cross section to be of the order of 10^{-12}. This means not only do we have to find the beam crossing with the $H \rightarrow \gamma\gamma$ production and decay, but that this beam crossing will have, on average, 16 additional pp interactions overlayed.

From this discussion, the challenge for the trigger and data acquisition system is formidable: somehow we must selectively record beam crossings

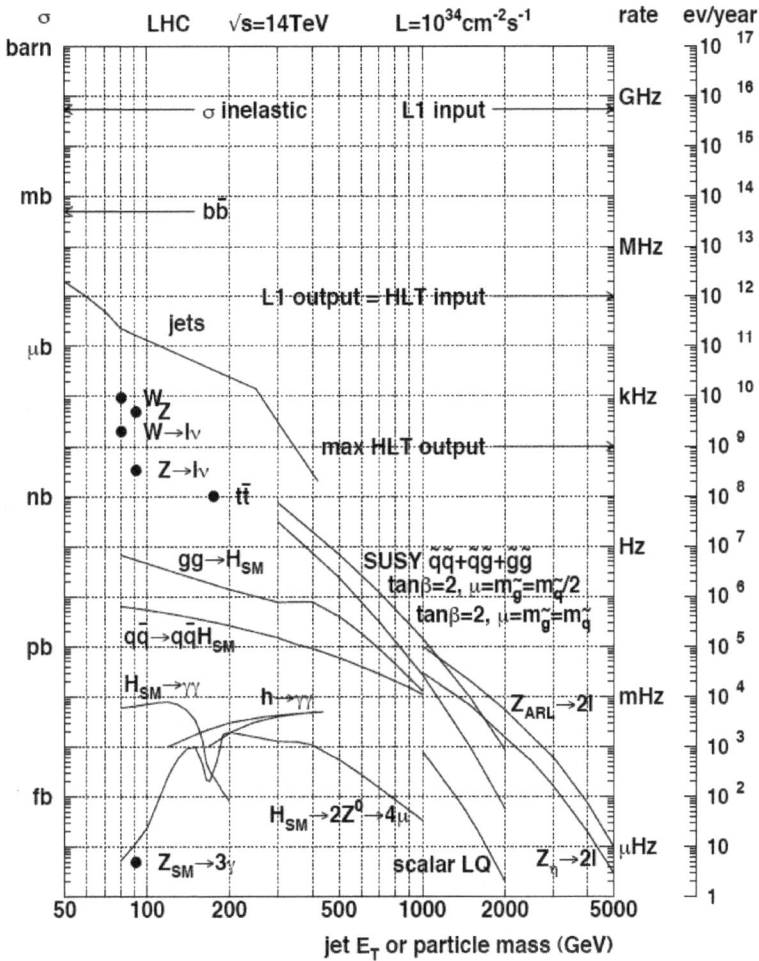

Fig. 2. Expected rate of physics processes at the LHC. The predicted cross section is on the y axis to the left and the resulting rate in Hz is shown on the y axis on the right. The predicted event rate for the total inelastic cross section is on the order of a GHz, whereas CMS is designed to search for rare events that occur at a rate $\ll 1$ Hz. The challenge of the trigger/DAQ system is to record all of the rare processes while reducing the overall event writing rate to 100 Hz.

with the highest probability of an interesting event and this must be done, on average, every 25 ns.

2. The CMS Trigger/DAQ System Requirements

2.1. *Introduction and overview of the system*

The CMS Trigger/DAQ system consists of a detector front end readout system, a triggering system and a global readout system that assembles all the detector data from one event. The detector readout system (or front ends) are synchronized with a timing system from the LHC accelerator which signals the 25 ns proton–proton bunch crossing rate. For each bunch crossing, detectors store their data in a front end buffer (pipeline) and in parallel send information to the L1 trigger system. The L1 trigger collates the trigger information from all the CMS subdetectors and makes a global decision. Once a detector gets a positive trigger from the L1 trigger, the data are read from the pipeline into the readout buffers. The L1 decision must be formed while the data are still in the detector front end pipelines, so the depth of the pipelines were designed to accomodate the trigger latency (or the time it takes for the trigger to form a decision and propagate it back to the detector front ends). While the L1 trigger logic is very flexible, any changes to it must be careful not to increase the trigger latency.

After a positive L1 trigger decision, the data are read into the event building network of the DAQ where the data from all the subdetectors for each event is assembled and sent to the Filter Farm for further evaluation. Software running on an array of computing processors reconstructs the full event and runs more complex algorithms to decide if the event is to be saved for further (offline) processing. At this stage the event can either be discarded or written to an online storage system, and eventually to an offline archive. Table 1 summarizes the overall requirements of the CMS Trigger and Data Acquisition System.

2.2. *The CMS L1 trigger system*

2.2.1. *Introduction and general strategy*

The first step in separating the events of interest from the 40 MHz event rate is at the Level 1 trigger. By making a list of interesting processes to search for and their likely decay products, a strategy has been developed for triggering on objects of interest. During the trigger design phase, tables were compiled of Higgs decay modes, SUSY signal modes, exotic particles and heavy quark decays (top and bottom) as well as events needed for standard model and "soft" physics. It became clear that a diverse physics program requires a flexible trigger in order to maximize the efficiency and acceptance for triggering on signal events.

The Level 1 trigger was designed to be efficient at triggering on physics objects while having the flexibility to increase or decrease thresholds in E_T and p_T in order to keep the over all event trigger rate at less than 100 kHz. The objects that can be defined at the L1 trigger level are jets, taus, leptons (muons and electrons), photons and missing E_T.

Table 1.   Overview of requirements for the CMS trigger and data acquisition system.

Summary of CMS Trigger and Data Acquisition System Requirements		
Bunch crossing rate	40 MHz	Defined by LHC parameters
L1 Trigger Latency	3.2 μs	Constrained by pipeline depth (tracker)
Deadtime	< a few %	Kept as low as possible
L1 Trigger Rate	100 kHz	Constrained by event building network
Trigger objects	Jets, muons, taus, e/γ, missing E_T, H_T	Defined by physics
Number of hits	500 k	Defined by CMS detector
Number of front ends	~ 600	Defined by event sie and switch architecture
Event size	1 MB	Defined by detector. Must be < 2 kB/FED (Front End Driver) (constrained by FED builder system)
Event storage rate	100 Hz	Defined by media/technology (online and offline event writing and/or processing rates)

2.2.2. *The CMS level 1 trigger*

An overall schematic of the Level 1 trigger is shown in Figure 3. Data flows from the bottom of the figure to the top. As can be seen in the figure, the trigger is divided into two distinct parts: the calorimeter trigger and the muon trigger. The Global Level 1 trigger receives input from these two subsystems, whose vertical architecture is very similar.

2.2.3. *An aside: Technologies used in the CMS level 1 trigger*

Each subdetector has designed its own trigger and readout systems, with the only caveat being that the data must be stored (pipelined) for at least 128 bunch crossings and that the total trigger latency must be less than 3.2 μs (128 bunch crossings). These requirements constrain the maximum allowed time each subdetector has to generate its trigger primitives, and that, for each positive L1 trigger decision at most 2 kB of data from that

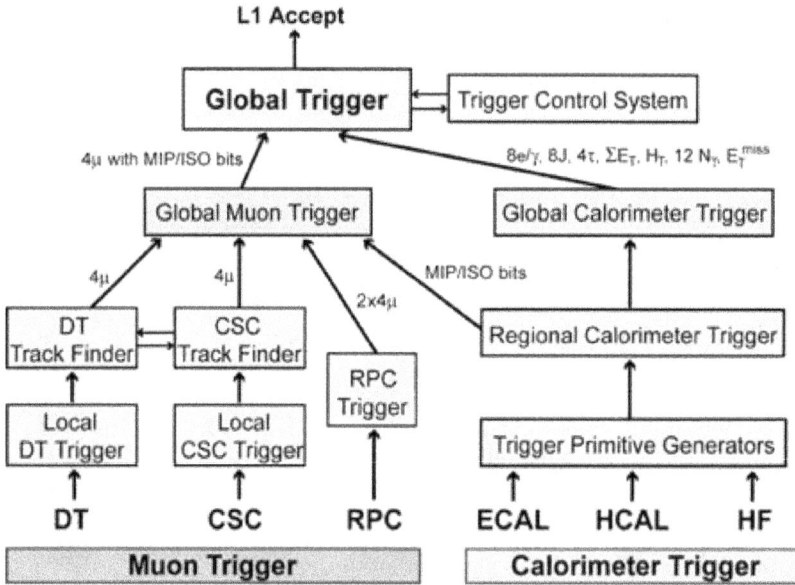

Fig. 3. The CMS Level 1 Trigger Architecture. The trigger pipelines operate at the beam crossing frequency (40 MHz), and the latency (decision time + time for the decision to propagate to the front ends) of the trigger is $<3.2\,\mu s$ (128 bunch crossings).

bunch crossing must be sent to the FED builder in the Data Acquisition System. Because each subdetector was then free to design their own trigger and readout system, this led to different choices of technology depending on the requirements and resources of each subdetector group. Three basic technologies are used in all systems: ASICs, FPGAs and Communication Technologies. Below is a brief description of each.

2.2.3.1. L1 trigger technologies: ASICs vs. FPGAs

ASICS (Application Specific Intgetrated Circuits) are the best performance option: they have high radiation tolerance and low power consumption. However they are costly to design and manufacture and lack flexibility. FPGAs (Field Programmable Gate Arrays) are commercial, widely used chips, that can be programmed for a variety of applications. Their size and popularity has increased at least as quickly as commercial CPUs: early FPGAs in 1987 held only 9,000 logical operations (or gates), whereas current FPGAs host millions of gates. Compared to ASICs they are slower and less energy efficient, however their cost and flexibility is a huge advantage and R&D in the FPGA field continues to close the performance gap between the two.

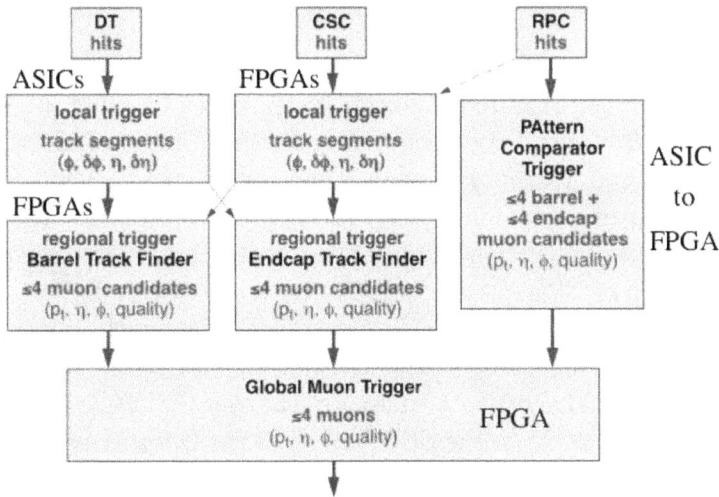

Fig. 4.   The CMS muon trigger system showing the hardware choices at each stage.

Although as mentioned above, different detectors have chosen different technologies, in general as the data handling tasks move from readout to forming complex trigger objects, the technology moves from ASICs to FPGAs. A typical example from the muon system is shown in Fig. 4.

2.2.3.2. L1 trigger technologies: communications

The second major area of technology challenge for the L1 trigger is communication. In order to process and send a decision for each event, the L1 trigger must be able to send a global accept or reject every 25 ns, the bunch crossing time. The trigger communications are broken into two groups: communicating between cards in different crates and communication between cards in the same crate. Intercrate communication is done with high speed copper or optical fiber serial links. Several different technologies are used in the trigger system: LVDS, which stands for Low Voltage Differential Signaling uses the difference in voltage between two wires to encode information and can run at 400 Mb/s on cables shorter than 10 meters.[4] Other technologies employed in the CMS trigger are the HP G-link[5] and the Vitesse transceiver.[6] All of these support high speed data transfer between crates. Communication between cards in the same crate can be done over the crate backplane, which has an advantage in that there are a large number of connections and backplanes can operate at ∼160–200 Mb/s.

2.2.4. *The trigger logic and data flow*

Both the calorimeter and muon trigger systems take data from the detector and produce trigger primitives. For all systems participating in the trigger, digitized data from the subdetectors are sent both to the Data Acquisition System and to the Trigger Primitive Generator systems. In both cases the data are stored in a pipeline until a L1 trigger decision is formed. Again, depending on the subsystem the trigger primitive generators may be based on either ASICS or FPGAs or a combination of the two. A digital filter at the detector signal input extracts the peak signal and the bunch time information in order to assign the correct bunch crossing to each signal. Once a positive Level 1 Accept signal arrives, the data are sent over CERN designed S-LINKs (Simple Link Interface)[7] to the FED builder of the Data Acquisition System. The S-LINKs will be described in more detail in the DAQ section.

2.2.4.1. The muon trigger system

The CMS raw single muon rate at nomimal ($L = 10^{34}$ cm^{-2} s^{-1}) LHC luminosity ranges from about a few 10^6 Hz at low p_T (1 GeV/c) to a few Hz at high p_T (100 GeV/c). The challenge of the CMS muon trigger is to reduce the rate of measurable muons in the 4–50 GeV/c range to about 10 kHz (i.e. a reduction factor of about 10^3) without introducing any deadtime, that is making a decision for every 25 ns beam crossing. There are several CMS muon detectors that all participate in the global muon trigger: the Resistive Plate Chambers (RPCs), the Drift Tubes (DTs) and the Cathode Strip Detectors (CSCs). While all three detectors can in general assign the correct bunch crossing number at nominal luminosity, because the RPC time resolution is on the order of only a few ns, it is an important trigger timing ingredient.

Figure 4 shows the data flow of the muon trigger. The DT trigger finds track candidates in each of four stations by extrapolating hits forming inner station segments in order to correlate them with segments in the outer stations. This step is done mainly in ASICs that are located on the detector itself. The next step, forming trigger level tracks using the segments reported from each station, is done in the regional trigger using look up tables. In this step the track p_T, position and charge are assigned.

The CSC trigger works in a similar way, using the 6 planes of detector in each station to form track segments which are then forwarded to the regional trigger to reconstruct tracks and assign p_T, ϕ and η.

The four highest quality muon candidates from both DTs and the CSCs are forwarded to the Global Muon Trigger, along with quality bits for each muon candidate based on the number of muon stations used to reconstruct the muon candidate. While the CSCs and the DTs use local and regional trigger components to find muon candidates, the RPCs directly compare the pattern of hits in the detector with predefined patterns corresponding to various p_T bins and send the four highest p_T muon candidates along with the candidate quality code to the Global Muon Trigger.

The Global Muon Trigger selects up to four muon candidates for each bunch crossing using the susbsystem quality and p_T information. In addition it receives isolation and minimum ionizing bits for each 0.35×0.35 in an η, ϕ calorimeter region from the Global Calorimeter Trigger (see Figure 5 for a definition of these bits). All of this information is merged, when possible, to form new kinematic parameters for each candidate. The muon candidates are then ranked according to p_T, η, and quality using (programmable) look up tables and the four best candidates are forwarded to the Global Level 1 Trigger.

2.2.4.2. The calorimeter trigger system

Figure 5 shows the logic for finding calorimeter objects at Level 1. On the left of the figure is a block diagram showing the data flow of the calorimeter

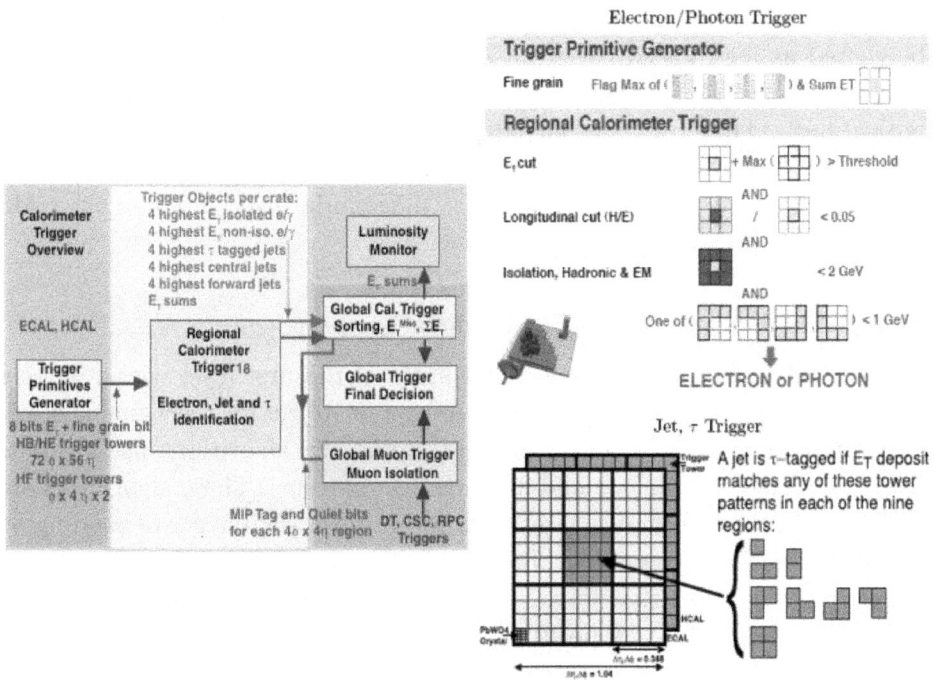

Fig. 5.   An overview of the CMS calorimeter trigger system.

trigger. Trigger primitives, containing trigger tower energy sums from the Electromagnetic Calorimeter (ECAL) and the Hadron Calorimeter (HCAL), are generated in ASICS in the detector front end readout boards for the ECAL and in FPGAs for the HCAL. Additionally, the ECAL also calculates an additional bit indicating the transverse size of the energy deposit which is used to distinguish electromagnetic objects (e.g. electrons and photons) from hadronic (jet) showers.

The trigger primitives are then forwarded to the Regional Calorimeter Trigger (RCT) whose job it is to sum the energies from the trigger primitives across trigger towers to identify electrons, photons, jets and taus. The RCT is made up of 18 crates, each crate containing 7 receiver cards, 7 electron identification cards and 1 jet/summary card. Each crate processes a $0.7\phi \times 5.0\eta$ region. The logic of the RCT is almost exclusively in ASICs but also uses programmable memory lookup tables for mapping the trigger primitive information to calorimeter E_T. The output of the RCT is a sorted (by E_T) list of trigger objects per crate. Because of memory space and time constraints only the best four of each object is sent.

The Global Calorimeter Trigger (GCT) receives information from the RCT and forms jet and electron objects using a sliding window algorithm to search for patterns of energy deposits. The GCT is technically challenging, as it has to perform a flexible triggering logic with low latency and data must be shared at high speeds between cards. The 23 large FPGAs which form the heart of the GCT communicate with each other over high speed (2 Gb/s) optical links. There are two output paths for the GCT trigger information: one goes to the Global Trigger to participate in the Level 1 trigger decision, and the other goes out on an S-LINK,[7] the standardized input to the CMS DAQ system which will be discussed below.

2.2.4.3. Putting it all together: The global level 1 trigger

The Global Level 1 Trigger is the last stage of the first level trigger system and is responsible for issuing a Level 1 Accept or Reject for every bunch crossing. The Global Level 1 Trigger applies up to 128 algorithms using inputs from the global calorimeter and muon trigger systems. These algorithms are flexible with programmable thresholds to optimize the trigger depending on luminosity and environmental conditions. The final L1 decision is sent to a trigger control board that forwards the decision to the front end electronics of all the subdetectors. A positive decision causes the front end electronics to read the data from the detector of the bunch crossing that generated the trigger as well as one bunch crossing before and one after in order to study out of time pileup for the event. In order to save readout time and memory, not all data from all channels are read out. In some cases the data are "zero suppressed", that is only signals above a programmable threshold are read out. In addition, the ECAL has implemented a system that allows only channels in a region around a significantly large energy deposit are read out (selective readout).

Figure 6 shows a block diagram of the Level 1 Trigger system, and Table 2 shows an example of a possible triggering scenario.

2.2.5. *CMS DAQ and higher level triggers*

2.2.5.1. Introduction, requirements and a little history

The job of the data acquisition system is to assemble all the subdetector fragments from each L1 triggered event, filter the data to reduce the overall rate to a manageable size and write the events onto local and central storage for further analysis.

Figure 7 shows the evolution of data acquisitions systems in High Energy Physics as a function of input trigger rate and event size. The LHC experiments are the most demanding both in terms of Level 1 rate and event size. For CMS, with an estimated maximum average Level 1 trigger rate of 100 kHz and an event size of 1 MB, we have to process nearly 1 Tb/sec of data.

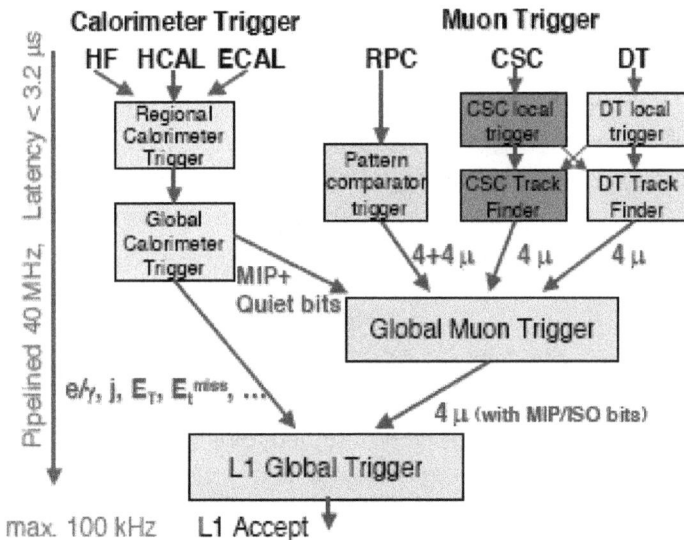

Fig. 6.   Block diagram of the CMS level 1 trigger system.

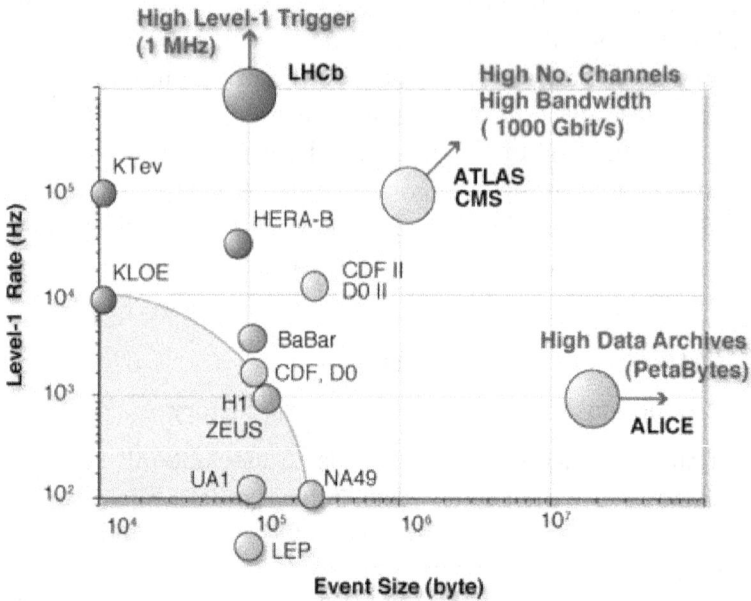

Fig. 7. Evolution of data acquisition Systems as a function of L1 accept rate and event size. Note that while the four LHC experiments have differing event sizes and data rates, their overall throughput requirements are roughly the same.

2.2.5.2. Considerations on how to build a DAQ

For building, filtering and storing data we need a system with a large (1 Tb/s) throughput capability, large amount of CPU to perform the event filtering and a high throughput to media storage in order to save the data for offline analysis. We have seen from the Level 1 trigger implementation that making a flexible, low latency trigger well matched to the CMS detector requires custom designed boards using rather expensive technology. In addition, between the design, production and commissioning phase of the trigger there is a long time lag which means that by the time the trigger is actually commissioned and being used in the experiment, the electronics are already obsolete. In order to ensure flawless operation of these custom boards, it is important to take care to over-order parts in danger of becoming obsolete and build enough spares for a 5–10 year lifecycle.

When the CMS DAQ was being designed, one of the considerations was to use commodity hardware whenever possible in order to reduce the cost of the system and also to increase its reliability and upgradeability. The backbone of the DAQ is the networking used to build the events and the CPU used to process them. When surveying the market to find suitable commodity technologies, it was clear that the networking market was being driven by the telecommunications field and the CPU market was being driven by the personal market. Luckily the analysis of large numbers of independent events is well suited to parallel processing, and the field of "farming" PCs is a well developed one, so it was natural to think of collecting all the data that passed the Level 1 trigger, building it, and sending it to a farm of PCs running scientific linux. The size of the PC farm is determined by the event rate into the Higher Level Trigger farm and the average CPU time needed to process each event. The media storage market is being driven by industry to host large databases of e.g. customer information, thus large, robust disk storage arrays with good bandwidth connectivity are easily found in the commercial market.

2.2.5.3. The CMS DAQ: custom vs. commercial

The DAQ was designed with the commercial market in mind, but as with the trigger, as we get closer to the demanding LHC clock, custom solutions had to be made. There are only two custom boards used by the DAQ: the CERN designed and built S-LINK that is the input to the DAQ and the FRL (Front End Readout Links). The S-LINK layer is the common interface of the DAQ to the subdetectors.

2.2.5.3.1. Detector front end. The subdetectors send their data over the custom S-LINK card,[7] which is directly plugged into the sub-detector FED. The S-LINK card has a 1.6 kB buffer for incoming data and an LVDS converter to sent the data to the FRL, the first step in the DAQ chain. The S-LINK also supports a bidirectional link for generating backpressure to the FEDs when DAQ buffers fill up.

The FRLs were designed and built by CMS. Their purpose is to receive the data from one, or (optionally) merge the data from two S-LINKs, and form the interface to the (first stage of the) event building network. It is implemented as a compact PCI card with an internal PCI bus to interface with a Network Interface Card (NIC) and an interface to the PCI backplane in order to control and monitor it. The FRL logic is implemented in FPGAs, the basic function being to check transmission errors over the S-LINK, to move data to the NIC interface and to merge data from 2 S-LINKs, if necessary.

2.2.5.3.2. Event building and the CMS event building switch. During initial stages of the event building design, it was unclear if there would be a switching fabric that supported the bandwidth needed to assemble event fragments from about 500 different sources and route them serially to 500 different syncs at a maximum average throughput of 1 Tb/s. The idea of only assembling parts of the event was developed. In this model, one would filter the events in stages: for example the first stage might be assembling the calorimeter data and running a more rigorous jet/photon object selection filter. For events that passed this first filter requirement, the muon data could be read out and finally for events that passed all the selections the full readout would be initiated. This methlod of "staging" the data readout would lower the total necessary event building throughput as most events would fail the filtering selection before the full event building stage. Thus the data selection running on the PC farm became known as the "Higher Level Triggers", since it incorporated a software version of a traditional Level 2 trigger.

By watching the market and keeping the event building design flexible, the final design of the event building system was constructed around technology already available and testable in 2005 when the first serious DAQ event building prototype was assembled. The design is summarized in Fig. 8. As can be seen in the figure, event building is a two step process: the first step, the so called FED builder, comprises 64 (8 × 8) switches, allowing event fragments from 8 input FRLs to be assembled and sent over 8 output links.

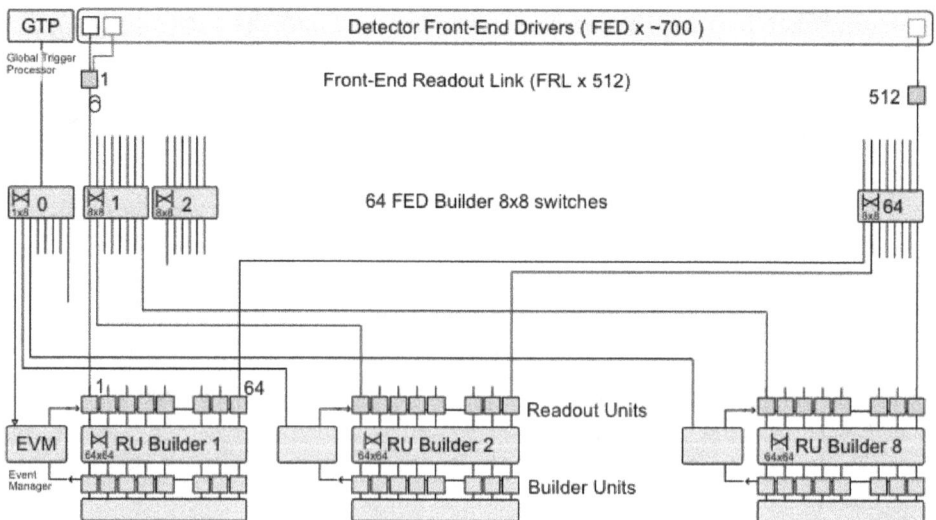

Fig. 8.   The CMS DAQ architecture. The DAQ is comprised of eight identical slices each supporting full event building and filtering at 12.5 kHz.

In this design, the event builder is made up of eight identical slices, each slice able to build the full 1 MB events at an average maximum rate of 12.5 kHz. Thus the 8 slice system supports the design requirements of event building at 100 kHz. The immediate advantage of this system was that 64×64 switches supporting at least 12.5 GB/s were readily available. However, another advantage to the 8 slice system is that the DAQ could be installed gradually according to the rate needs as the accelerator goes from low to high luminosity. In addition slices can be upgraded relatively independently as the technology ages.

2.2.5.3.2.1. Another aside: brief overview of networking issues. The rate of CMS event building depends on the rate that the data fragments from the input NICs can be sent through the event building switch to the output NICs. In order to understand the real rate a switch can support, it is important to understand how the switch works.

Buffering and routing. Switches may buffer at the input (Input Queuing or IQ), the output (Output Queuing or OQ) or both (Combined Input Output Queuing or CIOQ). Because there is no buffering on the output, IQ switches suffer from "head-of-line" blocking effect which occurs when packets in the input fifos must wait for an available output destination. This reduces the available throughput of the switch when, for example, the second packet in an input fifo has to wait for the first packet to be sent or when two packets on different inputs have the same output destination. OQ switches, since they buffer packets on the output of the switch, do not suffer from this problem, however to fully utilize the throughput of the switch, the requirements on the bandwidth of the output memory becomes large. For a completely non-blocking switch, the memory bandwidth for writing must be at least the speed of each input port times the number of input ports. For example for a switch configured for 500 1 Gb/s inputs would require a memory bandwidth for each output port of 500 Gb/s or 1ns access time for a 512 bit wide memory. By contrast, an IQ switch only has to buffer the inputs at 1 Gb/s. The CIOQ design, which utilizes memory on both input and output, can reduce the memory bandwidth requirements on the output significantly depending on the switch design.

Additionally, if the NIC itself is programmable, one can implement rudimentary packet "traffic shaping" to increase the switch efficiency. By understanding the typical application traffic patterns, some *a priori* ideas of how to decrease the incidence of blocking can be made. CMS employs barrel shifting traffic shaping for the time critical data collection from the FEDs.

Technologies chosen for event building. In designing the DAQ switching fabric, the requirements for stage one and stage two event building were considered separately. For stage one (the FED builder) the main considerations were bandwidth, flow control and reliability. The bandwidth requirement per input or output port is determined by the Level 1 trigger rate (100 kHz) and the average data fragment size in a single FED is 2 kB meaning each port must support a sustained rate of 200 MB/s. At the time the CMS DAQ was designed Gigabet Ethernet was the fastest widely available Ethernet option. Therefore at least three links/input would be needed to support the input bandwidth. In addition Ethernet NICs are not programmable, so any FED builder protocol or traffic shaping would have to be implemented upstream of the switch. Furthermore Ethernet does not provide a lossless packet transport; that is if an Ethernet switch becomes congested, it will drop packets unless there is an additional software layer (i.e. TCP/IP) which prevents it, adding complexity and additional latency. An alternative option was sought in the commercial market. The Myricom[8] company manufactured a switching fabric with programmable NICs, reliable transport with backpressure at the hardware level to prevent buffer overruns, a low latency message passing protocol and a bandwidth/port of 2 Gb/s. This technology was chosen for the first stage of event building (the FED builder).

In the second stage of event building, the RU builder, event "super" fragments arrive from each of the 64 FED builders. These fragments are on average 128 kB in size and must be combined over the 64 inputs and routed to one of the 64 outputs at an average maximum rate of 100 kHz (L1 trigger rate)/8 (DAQ slices) = 12.5 kHz giving a total of 160 MB/s. This is easily accommodated by either one Myricom link or two gigabit Ethernet ports. Gigabit Ethernet was chosen for this stage of event building, as the buffer depths in the RU machines were large enough that the latency overhead of TCP/IP to guarantee lossless event transmission was not an issue, and the wide availability of hardware and software in the commercial market was very attractive. As discussed above, each DAQ slice is easily upgraded with faster Ethernet technologies as needed.

For more details on switch throughput simulations and measurements see the CMS DAQ TDR.[3]

2.2.5.3.3. Event filtering. The final stage of online event handling before storing the events for offline analysis is done in the online farm of commercial PCs running final software selection software. Handing off events from the event builder to the filter farm PCs is done in the Builder Units, also commercial PCs with two Ethernet interfaces each. Each builder unit services up to four filter farm PCs which run the full CMS event reconstruction software.

2.2.5.3.3.1. *CMS higher level trigger strategy.* The CMS HLT is a purely software trigger run on the online computing farm (the Filter Farm). The strategy of the HLT implementation is to use offline software as much as possible in order to keep the software robust and maintainable. The main requirement of the software trigger is to satisfy a diverse physics program with high efficiency. This means that any event selection must be inclusive (to discover the unpredicted as well as predicted) and must not require precise knowledge of calibration/run conditions since precision detector calibrations lag behind data collection. In addition the event selection efficiency must be measurable from data alone. Finally, clearly all algorithms and event processors must be monitored closely as events failing the online selection will be lost forever.

The online selection code runs in a single processor and analyzes one event at a time. Its job is to lower the L1 100 kHz rate to an output selection rate of 100 Hz, that is, it can accept only 0.1% of the events. Unlike the L1 trigger, the HLT has access to the full event data and thus can make more stringent demands on the event. The main limitations of the online software trigger are that of available CPU time and the lack of precision of the calibration and alignment constants.

2.2.5.3.4. Saving the data. Once the online software selects an event it is marked and saved. Data are first saved to a large disk array located at the experiment, and then transferred to computing sites at CERN and worldwide.

2.2.5.3.4.1. *Data storage.* Each slice of the CMS data acquisition system has a dedicated online storage system. The storage system is made up of two logger nodes (commercial PCs) connected to the filter units by up to four Ethernet links and to a disk array by fiber channel. Four fiber channel switches are connected to each of the eight disk arrays and each logger node is connected to two of the switches via 2 4 Gb/s Fiber Channel connections, allowing failover in case one of the data loggers crashes. Each disk array contains 42 1 TB disks organized as 4 groups of 10 disks configured as RAID-6 (allowing the array to be robust against any two drives in a group failing), each group holding 8 TB of data. Thus there are 32 TB/disk arrays for a total online storage of over 250 TB. Data files are closed and transferred to the CERN offline storage area (the CERN Tier 0) every 93s, the definition of a CMS "luminosity section". Once the Tier 0 determines a data file is copied correctly, it is marked for deletion from the online storage. With 250 TB of online storage, CMS can run comfortably for many days even if the connection to Tier 0 fails.

The data are transferred to the CERN Tier 0 computing site over 2 10 Gb/s Ethernet links. Once at the Tier 0, the raw data are archived and, in parallel reconstructed then pushed out to seven Tier 1 sites located world-wide for secondary storage and reconstruction. The goal of the offline computing is to have a first full reconstruction of the data within 24 to 48 hours it is collected online. A small subset of the data, the "express stream" for events of particularly high interest, is prioritized and reconstructed within an hour.

CMS detector control and monitoring. An additional important system for CMS is the Distributed Control System (DCS), whose purpose is to ensure the detector is properly functioning while taking data and is in a safe mode when not taking data. The DCS controls and monitors the detector, the on and off detector electronics and the overall environment. While it does not have to react on a 25 ns time scale, it must communicate with the online DAQ and detector systems, with the LHC accelerator and with infrastructure experts in a timely fashion. Failure of the DCS could be disastrous for the CMS detector.

It was decided early on during the development of LHC DCS systems to encourage the use of a single commercial product, if possible, in order to leverage price, expertise and to have a uniform and well tested system. To this end, CERN formed the Joint Controls Project (JCOP)[9] as a collaboration between the LHC experiments and the relevant CERN support groups.

While choosing a commercial product as the backbone of the DCS is desirable, it is important to note the differences in DCS use in industry (which drives the commercial market) and HEP (which does not). In industry, the typical system to be controlled by a DCS (or SCADA: Supervisory Control and Data Acquisition) has fewer channels and thus generates much less data, is more homogeneous with little or no custom hardware or

interfaces, changes little after the initial installation, is developed and maintained by experts, does not need to know what running state the experiment
is in and is typically monitored and operated by experts. In addition in the
HEP world partitioning the system in order to let subdetectors run independently is crucial. Thus in looking for a commercial system it was important
to find a flexible solution that could be customized to the special CMS needs.

PVSS, an idustrial SCADA product from the Austrian company ETM[1]
was chosen for its flexibility in architecture and scalability. PVSS provides
a run time database to store and archive device data, alarm handling, a
graphical editor so that users can implement their own interfaces, a scripting
language to access the PVSS database and a graphical parameterization tool
which enables the user to define the database structure, the data being stored
and limits for generating alarms.

At the LHC, the concept of a piece of equipment or detector element
being in a particular stable or semi-stable state is important. Examples of
such states are "Operational", "Error", "Off" or "Standby". These states
are reached by well defined transitions. PVSS does not itself support the
state/transition concept, so the JCOP collaboration developed a framework
for DCS integrating it with a custom Finite State Machine (SMI++) [ref].
In addition the framework contains templates, standard elements and functions for controlling and monitoring hardware, ensuring a homogenous control layer and hiding the underlying tools as much as possible. Much of the
hardware used at the LHC is common across all experiments (such as CAEN
High Voltage supplies, rack controls, etc.) and therefore the framework templates help to reduce the duplication of effort.

3. Conclusions

In this article, we have attempted to describe what led CMS to the hardware
and software choices it made in the trigger and data acquisition systems. In
general commercial solutions were sought in order to obtain well-tested and
robust solutions. However, there are special demands at the LHC in terms
of scale and timing that made custom hardware necessary. In general, the
closer the electronics had to follow the LHC clock, the more custom the
solution: industry in general does not operate in a time pressure as exacting
as 25 ns. Whenever possible, and financially feasible, large buffering was put
between data handling stages in order to lengthen the allowed time latency
and thus move towards a commercial decision.

At the time of this writing, before even recording any data from the
LHC, the discussion of how the trigger and DAQ systems must evolve in

order to handle the rates from the "Super LHC", the upgraded LHC that will deliver an order of magnitude more luminosity has begun. The general strategy of using commercial solutions where possible will be followed, but custom solutions in both the trigger and the DAQ will likely be necessary to handle the much larger hit occupancies and more complex event structure coming from higher luminosity. Following and understanding market trends and who drives them will again be a critical part of the CMS DAQ strategy.

References

1. The CMS experiment at the CERN LHC, JINST 3 S08004, 2008.
2. The Trigger and Data Acquisition Project, Vol. I, The Level-1 Trigger, CERN/LHCC 2000-038, CMS TDR 6.1, 15, December 2000.
3. The TriDAS Project, TDR, Vol. II, Data Acquisition and High Level Trigger, CERN/LHCC 02-26, CMS TDR 6, 15 December 2002.
4. For a detailed description, see www.national.com/appinfo/lvds/files/ownersmanual.pdf
5. See, http://findarticles.com/p/articles/mi_m0HPJ/is_n5_v43/ai_12676958/
6. See, http://www.vitesse.com
7. See, http://hsi.web.cern.ch/HSI/s-link/
8. See, http://www.myri.com/
9. See, http://itco.web.cern.ch/itco/Projects-Services/JCOP/welcome.html
10. See the ETM website, http://www.etm.at/index_e.asp

www.ingramcontent.com/pod-product-compliance
Lightning Source LLC
Chambersburg PA
CBHW072009230326
41598CB00082B/6896